Fuels and Petrochemicals Division

Core Programming Topic at the 2012 AIChE Spring Meeting and 8th Global Congress on Process Safety

Houston, Texas, USA
1-5 April 2012

ISBN: 978-1-62276-558-4

Printed from e-media with permission by:

Curran Associates, Inc.
57 Morehouse Lane
Red Hook, NY 12571

Some format issues inherent in the e-media version may also appear in this print version.

Printed by Curran Associates, Inc. (2012)

For permission requests, please contact AIChE
at the address below.

AIChE
3 Park Avenue
New York, NY 10016-5991

Phone: (203) 702-7660
Fax: (203) 775-5177

www.aiche.org

Additional copies of this publication are available from:

Curran Associates, Inc.
57 Morehouse Lane
Red Hook, NY 12571 USA
Phone: 845-758-0400
Fax: 845-758-2634
Email: curran@proceedings.com
Web: www.proceedings.com

TABLE OF CONTENTS

() Future Fuels and Chemicals: Grand Challenges and Opportunities

Joseph B. Powell, Shell Projects & Technology, Houston, TX

To insure a Sustainable Future, CO_2emissions must be reduced despite strong future growth in demand. "Easy oil" from conventional production will be less abundant, though coal and natural gas, as well as unconventional hydrocarbon reserves will remain in strong supply for decades to come. Carbon footprint, cost, and convenience will be critical drivers for future energy choices. Basis recent advances in horizontal drilling and hydraulic fracturing, enhanced production of tight gas, shale gas, and coal bed methane will enable natural gas to be both an affordable and abundant option for meeting short and long-term sustainable energy goals. Sustainable production of biofuels, with consideration of well-to-wheels carbon intensity and other resource impacts, will play a major role in reducing carbon intensity in the transportation sector. Carbon capture and sequestration along with renewable energy sources and enhanced energy efficiency will play increasing roles in reducing future greenhouse gas emissions across the diversity of resources that will be deployed to meet local and global stakeholder needs for energy and security. Advantaged routes to chemicals will be impacted by new feedstocks and performance requirements for fuels, providing opportunities for innovation in gas to chemicals, biobased chemicals, and selective conversions associated with hydrocarbon portfolio balancing. Significant time and investment will be required to bring forth new alternative energy systems to materiality, driving the development of new paradigms for efficient de-risking and deployment of technology.

This presentation will address opportunities for the chemical engineering profession to differentiate and excel in sustainably meeting future needs in fuels and chemicals.

SK KBR TECHNOLOGIES

Advanced Trans-Alkylation (ATA)

As a Means of Capturing Value:

Productivity Improvement in Xylene Production

April 2, 2012

Nicholas J. Kim

SK KBR TECHNOLOGIES

Contents

- ✓ SK-KBR Introductions
- ✓ Market Situation
- ✓ Ways to Capture Value
- ✓ ATA: Process Concept
- ✓ Technical Advantages of ATA
- ✓ Commercial Experience
- ✓ Summary

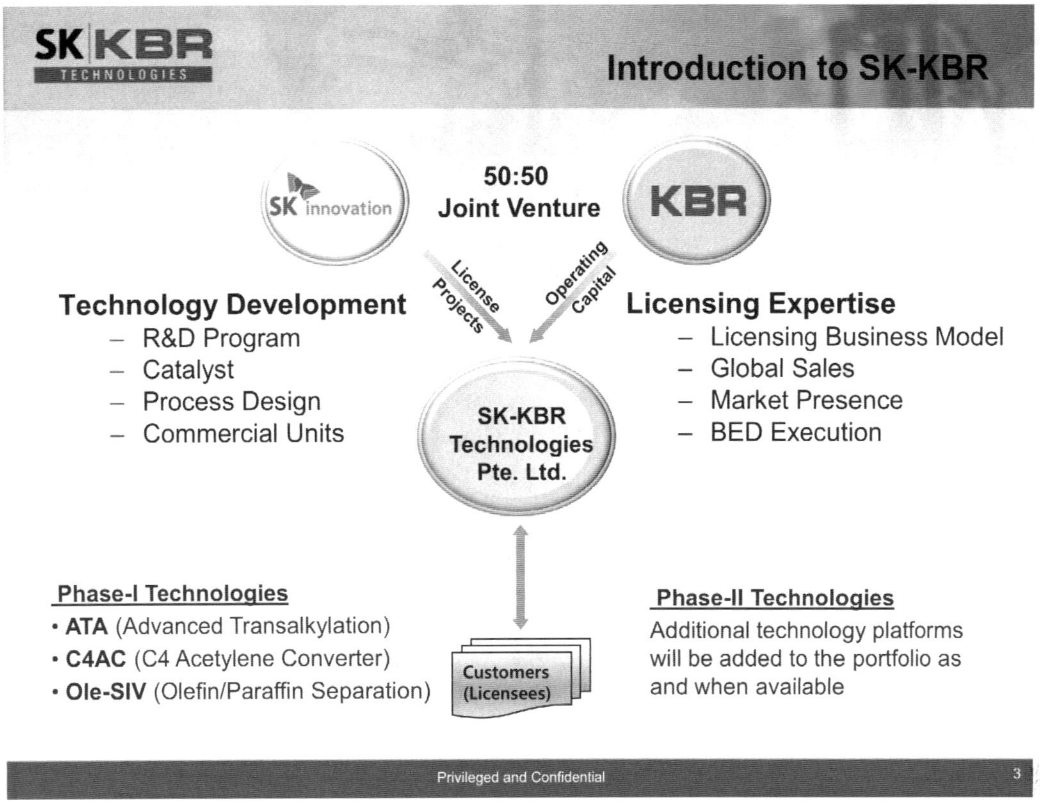
SK KBR
TECHNOLOGIES
Introduction to SK-KBR
SK innovation
50:50
Joint Venture
KBR
License Projects
Operating Capital
Technology Development
– R&D Program
– Catalyst
– Process Design
– Commercial Units
SK-KBR Technologies Pte. Ltd.
Licensing Expertise
– Licensing Business Model
– Global Sales
– Market Presence
– BED Execution
Phase-I Technologies
• ATA (Advanced Transalkylation)
• C4AC (C4 Acetylene Converter)
• Ole-SIV (Olefin/Paraffin Separation)
Customers (Licensees)
Phase-II Technologies
Additional technology platforms will be added to the portfolio as and when available
Privileged and Confidential
3

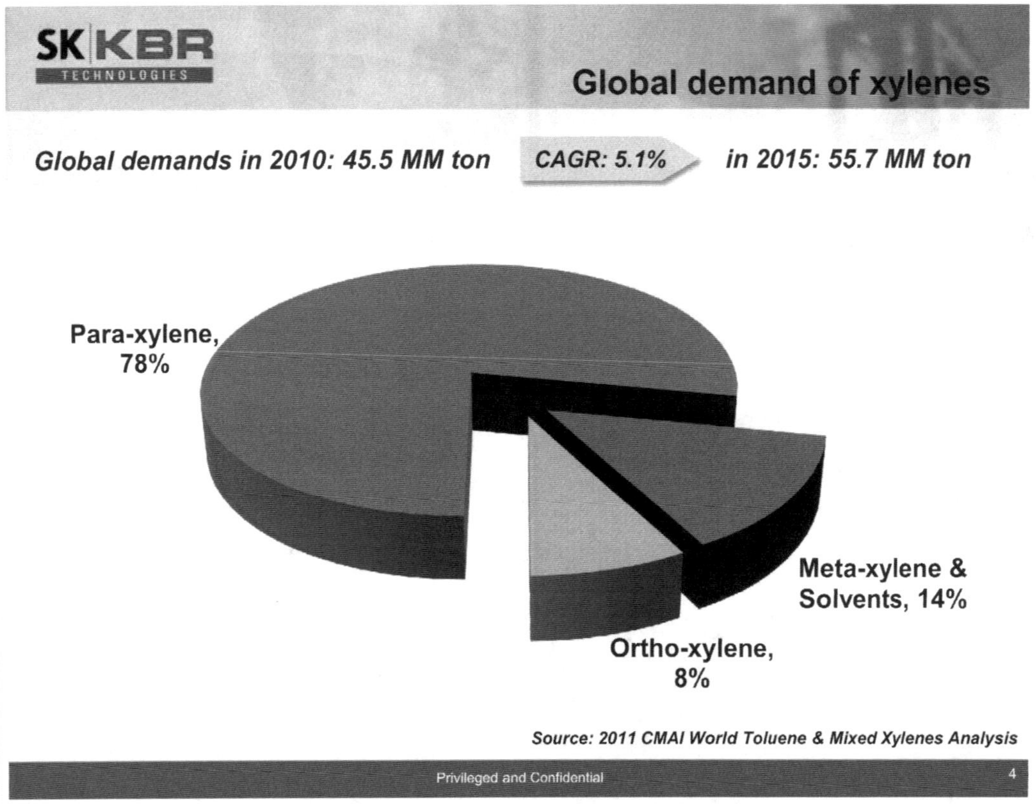
SK KBR
TECHNOLOGIES
Global demand of xylenes
Global demands in 2010: 45.5 MM ton
CAGR: 5.1%
in 2015: 55.7 MM ton
Para-xylene, 78%
Meta-xylene & Solvents, 14%
Ortho-xylene, 8%
Source: 2011 CMAI World Toluene & Mixed Xylenes Analysis
Privileged and Confidential
4

Global capacities of xylenes

Global capacities in 2011: 63.3 MM ton **CAGR: 4.8%** *in 2016: 79.9 MM ton*

STDP/TA capacity in 2011: 14.1 MM ton **CAGR: 8.9%** *in 2016 21.6 MM ton*

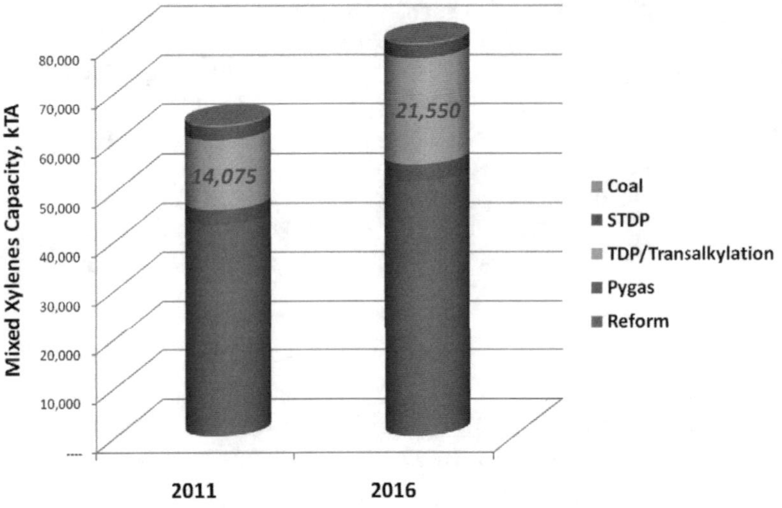

Global para-xylene consumption

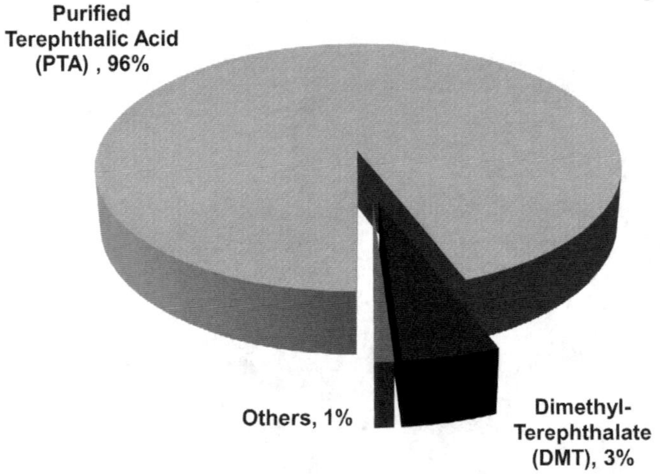

Source: Nexant ChemSystems , 20/7/10, p721

SK KBR TECHNOLOGIES

Global outlook of para-xylene

✓38% growth in demand '10~'15
✓11~14 million ton capacity increment expected '10~'15

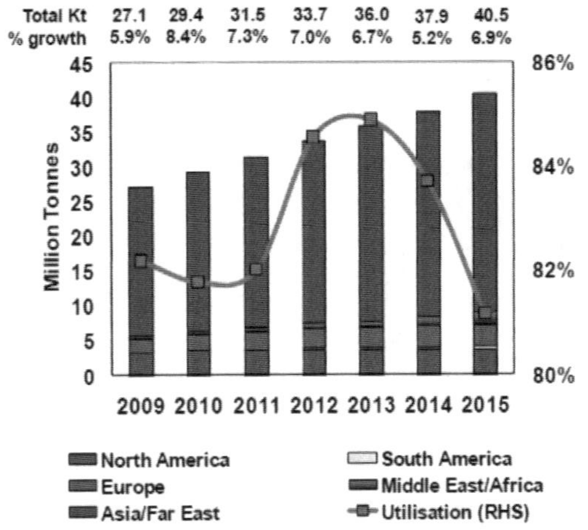

✓Shows capacity additions to '13
✓Probable slow down after '13 due to delayed projects

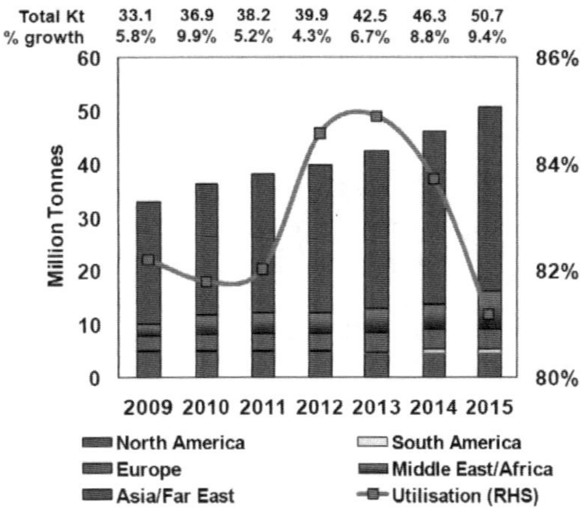

Source: PCI Xylenes & Polyesters, Indian Petchem 2010

SK KBR TECHNOLOGIES

Trading balance in Asia Pacific

Para-xylene in thousand metric tons

	2004	2005	2006	2007	2008	2009	2010
China	(1,015)	(1,565)	(1,742)	(2,651)	(2,956)	(3,365)	(3212)
Japan	2,046	2,282	2,292	2,330	2,238	2,488	2306
South Korea	377	300	622	538	95	6	(208)
Taiwan	(1,908)	(1,848)	(1,740)	(1,334)	(732)	(1,066)	(1288)
Indonesia	(869)	(806)	(698)	(379)	(545)	(496)	(171)
Malaysia	(24)	50	56	59	125	108	43
Philippines	0	0	0	0	0	0	0
Singapore	665	755	760	785	680	653	644
Thailand	142	125	(275)	(444)	(356)	259	146
India	124	23	339	557	602	383	(374)
Other Asia	(311)	(326)	(328)	(316)	(308)	(348)	(250)
Australia	0	0	0	0	0	0	0
New Zealand	0	0	0	0	0	0	0
Total Asia Pacific	(772)	(1,010)	(714)	(853)	(1,159)	(1,378)	(2364)

Source: ChemSystems Training, Nexant

Para-xylene outlook in China

- ✓ PTA shortage likely prolongs in China
- ✓ Entire Asia will be affected

- ✓ Para-xylene capacity cannot keep up with increasing demand from PTA

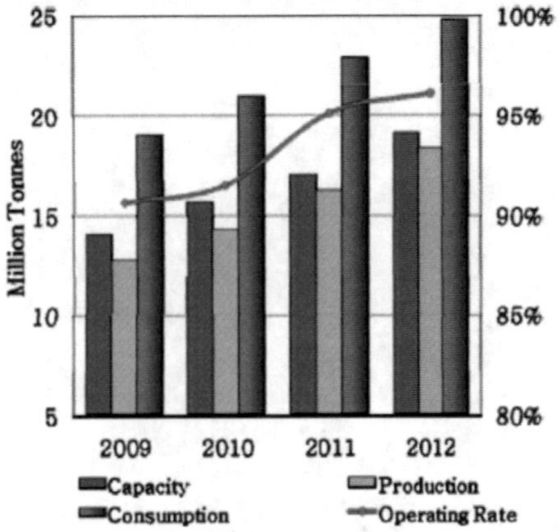

Source: PCI Xylenes & Polyesters, Indian Petchem 2010

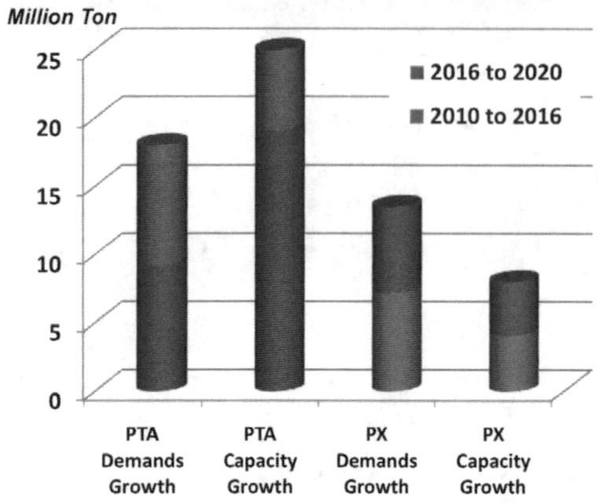

Source: CMAI Industry Workshops 2011

Ways to capture value

- ✓ **Operations excellence**
  - Relentless pursuit of safe and reliable operations
- ✓ **Margin capture**
  - Crude optimization – process more difficult crudes
  - Product optimization – premium quality rather than total volume
  - Technology improvements – catalyst selection, advanced distillation
- ✓ **Cost control**
  - Energy reduction – tighter furnace operation, increase heat integration
  - Non-energy savings – maintenance improvements, staff costs
- ✓ **Value chain optimization**
  - Larger cargo sizes, scheduling to minimize demurrage
- ✓ **Investment management**
  - Prudent investment in low cost / high value projects
  - Upgrade to add value rather than simply replace assets

SK KBR TECHNOLOGIES

Additional value to be captured

Margin capture opportunities can generate 50-75 c/bbl

Energy improvements can generate 20-25 c/bbl

Non-Energy cost improvements can generate 20-25 c/bbl

✓ *TOTAL VALUE OF USD 30-40 MM PER YEAR FOR 100,000 BBL/D*

SK KBR TECHNOLOGIES

Capturing margin by ATA

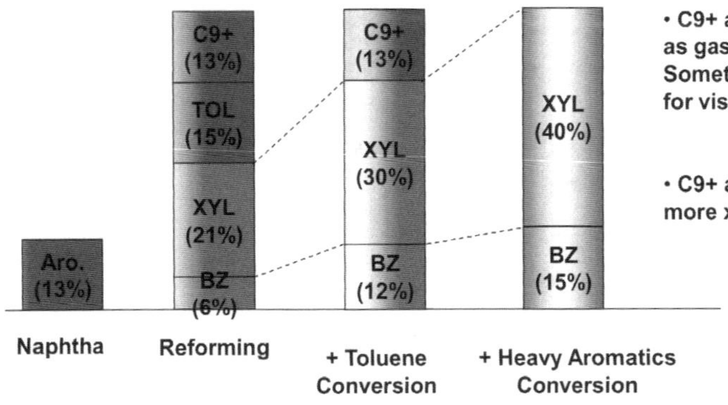

- C9+ aromatics are normally used as gasoline or diesel blending. Sometime they are sent to bunker-C for visbreaking (especially C10+).
- C9+ and C10 conversion will boost more xylene production.

Market Needs → ***To convert fuel-grade C9/C10 aromatics into chemical BTX***

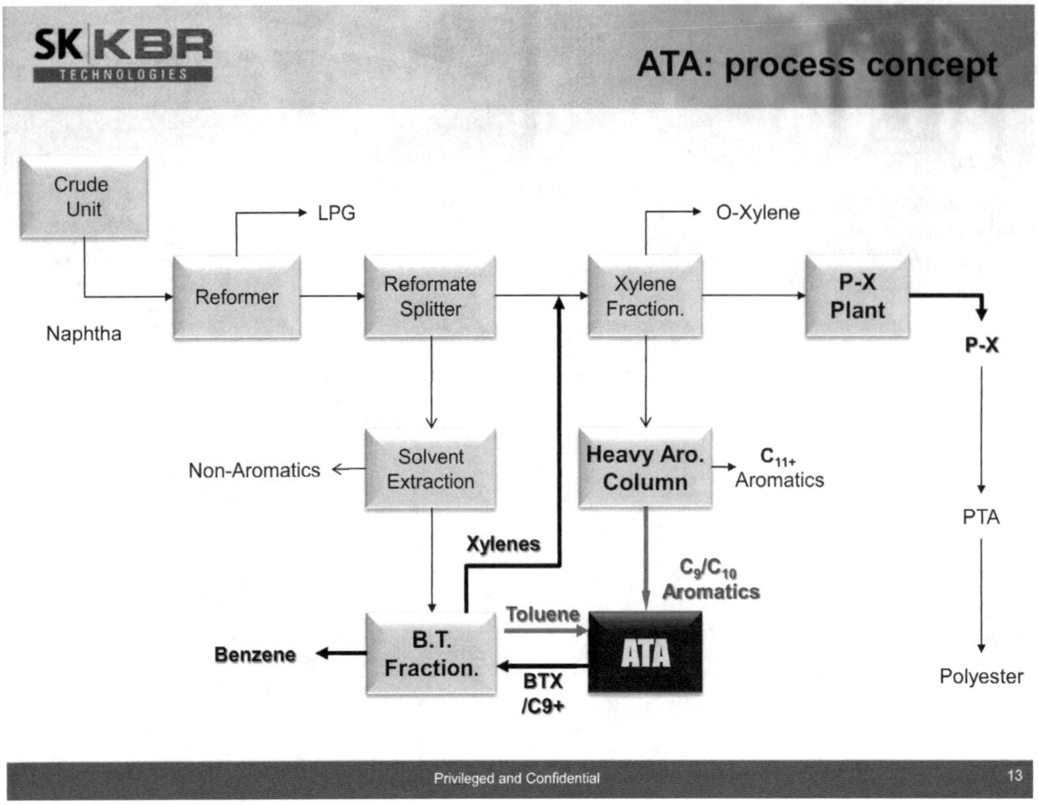
SK KBR TECHNOLOGIES
ATA: process concept
Crude Unit
Naphtha
Reformer
LPG
Reformate Splitter
Xylene Fraction.
O-Xylene
P-X Plant
P-X
PTA
Polyester
Solvent Extraction
Non-Aromatics
Heavy Aro. Column
C_{11+} Aromatics
C_9/C_{10} Aromatics
Xylenes
Toluene
B.T. Fraction.
Benzene
ATA
BTX /C9+
Privileged and Confidential
13

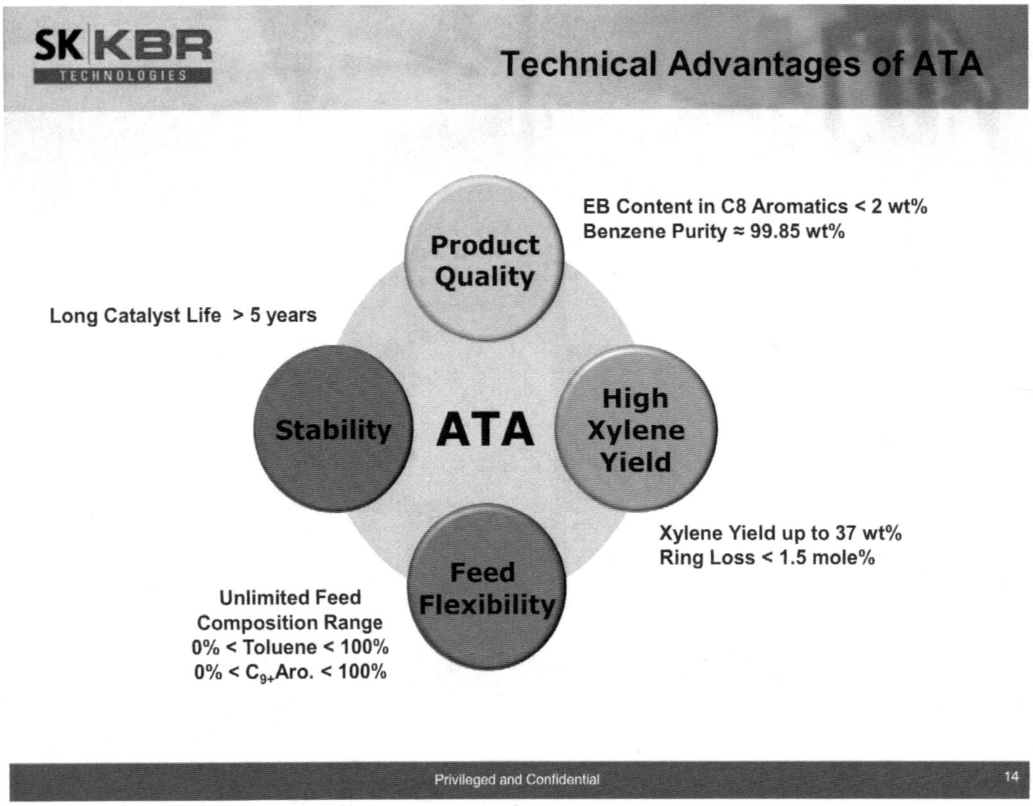
SK KBR TECHNOLOGIES
Technical Advantages of ATA
Product Quality
EB Content in C8 Aromatics < 2 wt%
Benzene Purity ≈ 99.85 wt%
Long Catalyst Life > 5 years
Stability
ATA
High Xylene Yield
Xylene Yield up to 37 wt%
Ring Loss < 1.5 mole%
Feed Flexibility
Unlimited Feed Composition Range
0% < Toluene < 100%
0% < C_{9+}Aro. < 100%
Privileged and Confidential
14

SK KBR TECHNOLOGIES

Technical advantages: feed flexibility

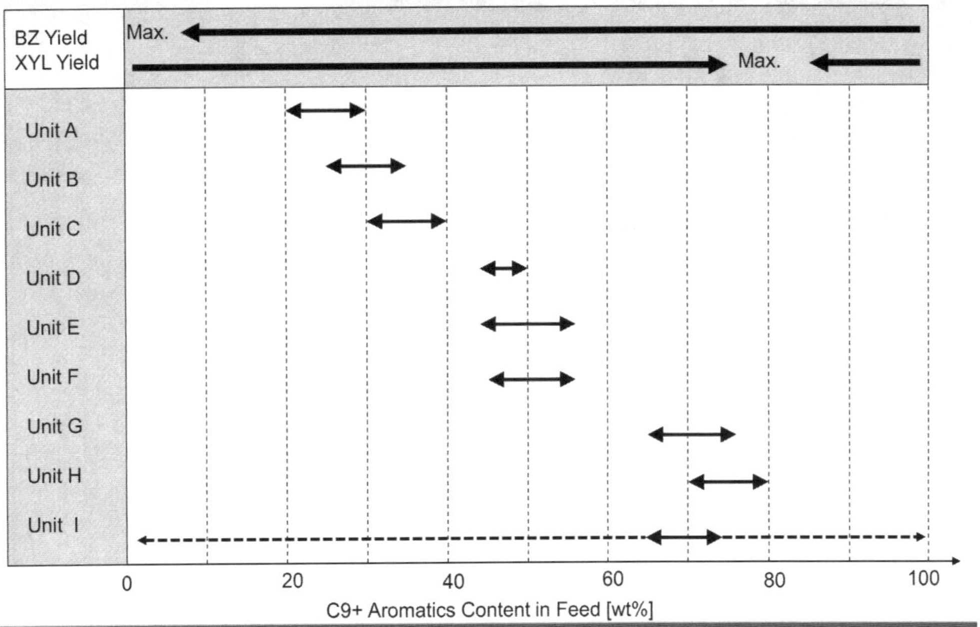

SK KBR TECHNOLOGIES

Technical advantages: feed flexibility

Feed compositions can be adjusted according to market situation.

Feedstock Mode	*C9+ Aro. Rich*	*Intermediate*	*Toluene Rich*
Feed			
Toluene, wt%	20	60	80
C_9 + C_{10} Aro, wt%	80	40	20
Yields			
Benzene, wt%	4	10	14
Xylene(C_8 Aro), wt%	37	33	29

SK KBR TECHNOLOGIES

Technical advantages: bi-functional catalyst

ATA-12 is a bi-functional catalyst that comprises highly-balanced metal on strong acid zeolite.

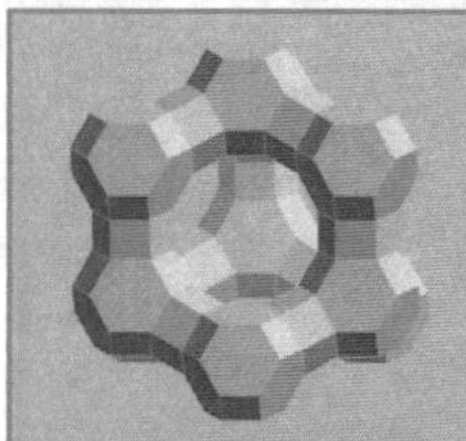

Zeolite

Acid Function

Chemical bonds between benzene ring and alkyl groups are weakened so that:

- **Methyl group is transferred** (Transalkylation / Disproportionation)
- **Ethyl, propyl and heavier alkyl groups are removed** (Dealkylation)
- **Long chain paraffins and naphthenes are cracked** (Hydrocracking)

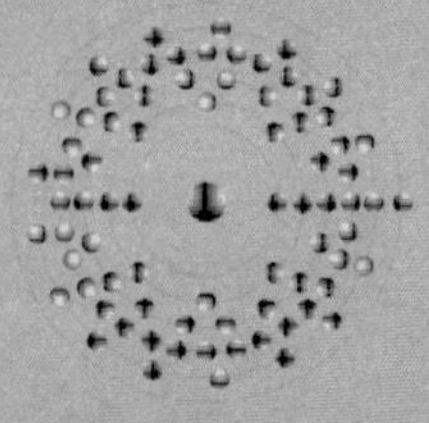

Platinum

Metal Function

Light olefins are immediately hydrogenated so that:

- **Re-alkylations are suppressed** (Dealkylation)
- **Oligomerizations are suppressed** (Hydrocracking)
- **Only olefins are selectively hydrogenated** (Selective Hydrogenation)
- **Aromatics are not saturated** (Minimize Ring Loss)

Privileged and Confidential 17

SK KBR TECHNOLOGIES

Technical Advantage: flowsheet optimization

✓ **Innovative flowsheet optimization that enable**

- **Capital savings up to 15%** capital savings for grassroots applications
- **Energy savings up to 10%** less energy than conventional column sequence
- **Lower Plot Size up to 15%** less space required
- **Simpler control and operation**

Privileged and Confidential 18

SK KBR TECHNOLOGIES

Commercial experience

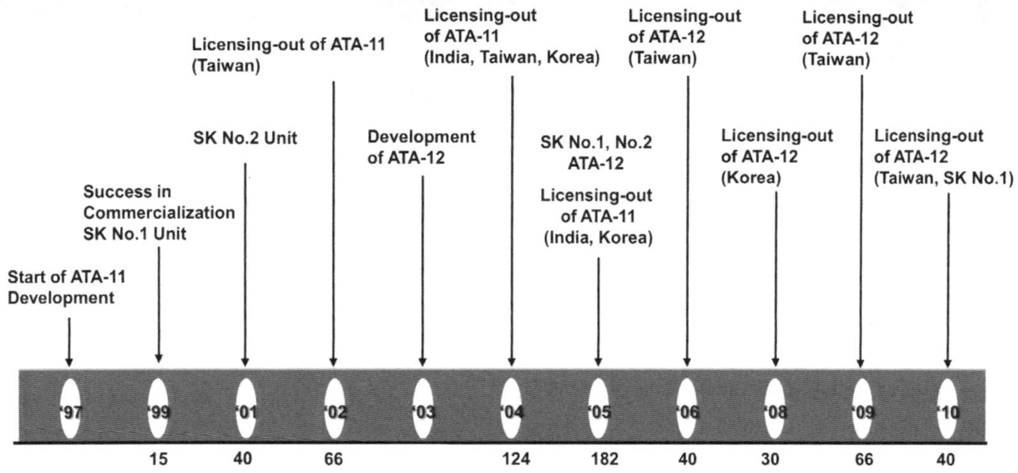

- 16 commercial applications in 10 different units with total of 601 tons
- Customers have been retained

SK KBR TECHNOLOGIES

Commercial experience

No.	Unit	Start-up	Toluene in Feed [wt%]	Remark
1	SK No.1	Jul, 1999	20~70	5 years of 1st cycle life (TA*)
2	SK No.2	Sep, 2001	50	4 years of 1st cycle life (TA)
3	Taiwan A	Oct, 2002	70	**7 years of 1st cycle life**
4	India A	Feb, 2004	30	**6 years of 1st cycle life**
5	Korea A	Mar, 2004	**0~100**	Very flexible operation
6	India B	May, 2004	50	C_8 Aro yield up to 70wt%
7	Taiwan B	Jun, 2004	70	**6 years of 1st cycle life**
8	SK No.2	May, 2005	40	Replaced by TA plan
9	SK No.1	May, 2005	20~70	Replaced by TA plan
10	Korea B	Jun, 2005	60~90	Flexible operation
11	India C	Nov, 2005	50	Very stable operation
12	Taiwan C	Jun, 2006	70~80	Flexible operation
13	SK Incheon	Aug, 2008	60~90	Flexible operation
14	Taiwan A	Nov, 2009	70	**Select ATA for 2nd Time**
15	SK No.1	Jan, 2010	20~70	Replaced by TA plan
16	Taiwan B	Apr, 2010	70	**Select ATA for 2nd Time**

* TA : Turn Around

SK|KBR TECHNOLOGIES

Summary

- ✓ **Additional 11MM tons of xylenes needed worldwide over next 5 years**
  - Driven by huge polyester demand especially in China
- ✓ **Greater focus on margin capture and energy reduction can deliver significant value to operators**
  - Added value of ***USD 30-40 MM per year for 100,000 bbl/d***
- ✓ **ATA is a proven technology for creating and sustaining added value for operators**
  - Product quality, high yields, feed flexibility and long life of catalyst
  - Lower plot ratio, reduce CAPEX and/or OPEX
  - Strong installed base

SK|KBR TECHNOLOGIES

PREPARATION OF CORE-SHELL MICROPARTICLES BY CRYOTROPIC GELATION OF CHITOSAN-BASED BIOPOLYMERS

Wiwut Tanthapanichakoon , Tokyo Institute of Technology, Tokyo, Japan
Kyuya Nakagawa, University of Hyogo, Himeji, Kobe, Japan
Nataporn Sowasod and Tawatchai Charinpanitkul, Chulalongkorn University, Bangkok, Thailand

Abstract

Made via sol-gel transition around the encapsulated ingredients, hydrogel is often employed in the formation of shell matrix composed of semipermeable membrane envelope. In the present study, the encapsulation of an oil-soluble food nutrient, curcumin, in hydrogels of natural polymer chitosan was carried out in the absence of any cross-linking agent. A curcumin-loaded oil-in-water emulsion was prepared and incorporated into a polymeric suspension consisting of chitosan and a suitable ratio of κ-carrageenan and carboxymethylcellulose sodium salt (NaCMC). The encapsulation of curcumin in chitosan was achieved by converting the resulting colloidal suspension into hydrogels via cryotropic gel formation. The frozen hydrogel samples were freeze-dried for preservation. The yield of curcumin encapsulation was found to depend on both the cooling rate and the ratio of κ-carrageenan to NaCMC. As the cooling rate increased, the encapsulation yield rose in tandem. Similarly, the release behavior of the encapsulated curcumin in an aqueous phase was shown to depend on the same two factors despite a difference in the oil-phase composition. Two types of release behavior were observed: a burst release and a moderate first-order release rate. In fact, when the ratio of κ-carrageenan to NaCMC was kept constant, the cooling rate was found to determine the type of release. Generally, the higher the number of cross-linkings in the polymer matrix and the stronger the bonding strength, the slower the release rate should become. A more balanced ratio of κ-carrageenan to NaCMC was shown to yield a pseudo-first-order rate release and by implication a stronger microstructure.

Introduction

In the last decades, nano-micro encapsulation technology has attracted keen interest from food and pharmaceutical industries in various applications such as preservation of efficacy and controlled release of active ingredients such as vitamin, polyphenol etc. (Lakkis, 2007). Hydrogel often plays an important role as shell matrix, in which membrane layers are formed on the surface of oil droplets via sol-gel transition. Two major disadvantages of the hydrogel method are that a change in the pH and/or temperature would significantly influence the gel formation and a cross-linking agent is needed to form good-quality gel. These drawbacks can be overcome by using cryotropic gel formation, as reported by Lozinsky *et al.* Cryotropic gelation is defined as a hydrogel formation process in which sol-gel transition is induced by localized concentration increase of the substrate due to ice formation during freezing (Lozinsky et al., 1986; Lozinsky & Damshkaln 2000, Giannouli, & Morris, 2003; Orrego, &, Valencia, 2009). In this study, the encapsulation of an oil-soluble food nutrient into a cryogel system was experimentally investigated. A curcumin-loaded oil-in-water emulsion was prepared from a chitosan colloidal suspension, and the emulsion was converted into hydrogel by cryotropic gel formation. The frozen samples obtained under various conditions were freeze-dried for preservation. Curcumin was subsequently extracted from the dried samples by immersing them in a phosphate-buffered solution to determine the curcumin encapsulation yield in the matrix. The transient release behavior of the encapsulated curcumin in an aqueous phase was investigated, and the influence of the freezing condition on the release characteristics was reported.

Materials & Methods

Materials

The chitosan granules used in this work were supplied from Tokyo Chemical Industry Co., Ltd., Japan. As measured by the authors using ^{1}H-NMR, the degree of deacetylation of this chitosan was about 82% (the evaluation method was obtained from a report of Hirai, Odani, & Nakajima (1991)). Curcumin powder was purchased from Thai-China Industry Co., Ltd. in Thailand. Commercial-grade carboxy methylcellulose sodium salt (NaCMC) and κ-carrageenan powder were supplied from Sigma Co., Ltd., Germany. All other chemicals used in this work, Tween 80, acetic acid, ethanol, triolein and sodium chloride, were of analytical grade. All chemicals were used as received

Preparation of frozen and freeze-dried cryogels

The preparation process of cryogel samples was schematized in **Figure 1.** First, a triolein solution consisting of 0.3 wt% of curcumin (oil phase) was mixed with an aqueous suspension of 3 wt% chitosan containing 2 wt% acetic acid and 5 wt% Tween80. Then the mixture was emulsified with a high-speed homogenizer (Cleamix, M-Technique Co. Ltd., Japan) at 20,000 rpm for 3 min. The ratio of the oil phase to the aqueous phase was 10:90 (v/v). Next 5 % (w/v) of sodium chloride was added to the prepared o/w emulsion before mixing it with an equal volume of another polymer suspension consisting of κ-carrageenan, NaCMC and 5% of sodium chloride. The resulting mixture was then homogenized at 20,000 rpm for 3 min to obtain a uniform colloidal suspension. As noted in Figure 1, two types of the polymer suspensions with different ratios of carrageenan to NaCMC (1:9 and 4:6 for sets A and B, respectively) were investigated.

Preparation of each cryogel specimen was carried out with a freezing system comprising a contact plate heat exchanger. In this setup, the temperature of the sample was regulated by circulating the coolant in the heat exchanger. A sample holder with a cylindrical hole (made from PTFE, diameter D=10mm, Height H=10mm) was set on the cooling plate, and the above-prepared colloidal suspension solution was poured into the sample holder. The solution was cooled at a selected cooling rate (-0.5, or -2.0 K/min) and kept frozen at -40°C. The completely frozen specimen was subsequently freeze-dried for preservation.

Characterization of specimens

Yield of gel fraction

To determine the extent of cryotropic gelation and find out the effect of incubation after freezing, colloidal suspensions without the oil phase in Fig. 1 were prepared and frozen at a cooling rate of interest with and without incubation at -5°C for 10 hours. Then 0.5 g of the frozen specimens were thawed and added with 2 mL of DI water. The resulting suspension was centrifuged to obtain a supernatant solution and the polymer concentration in it was analyzed with UV-Vis spectroscopy at 300-nm wavelength. The yield of gel fraction (%) is calculated from $((C_0-C_s)/C_0) \times 100$, where C_0 is the similarly obtained polymer concentration of the colloidal suspension before freezing and undergoing cryotropic gelation, and C_s is the supernatant polymer concentration obtained after freezing and thawing.

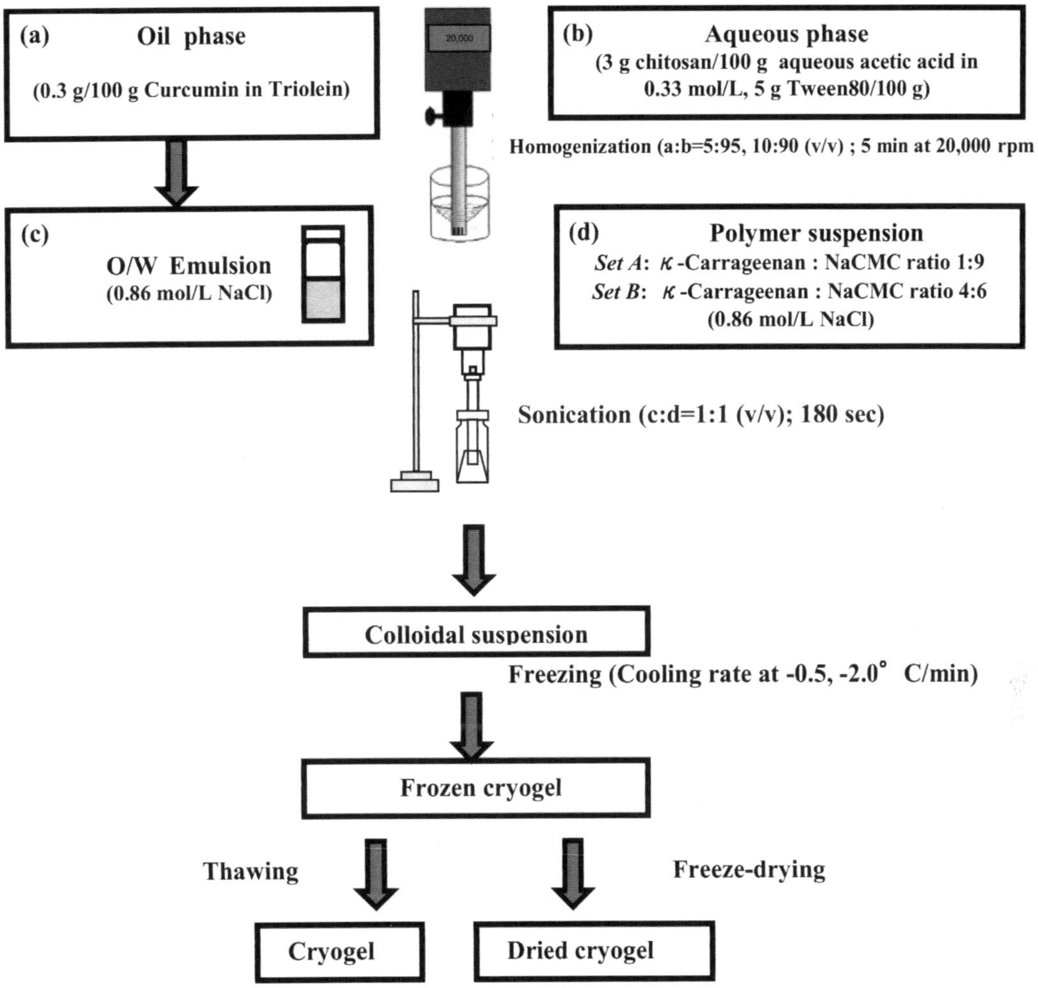

Figure 1. Preparation scheme

Yield of encapsulation (Y) of curcumin

The yield of curcumin encapsulation was determined by measuring the amount of curcumin extracted from a freeze-dried sample soaked in 2-ml ethanol solution for 24 hours. The mass of extracted curcumin was divided by the total mass of curcumin initially loaded in the specimen to obtain the encapsulation yield value.

In-vitro release curves of curcumin

The release curves of curcumin were obtained in phosphate buffer solutions at pH 7.4 (1.3 mM phosphate). A freeze-dried cryogel specimen was placed in a test tube containing 5 mL of the buffer solution. Then the test tube was tightly capped and installed in a shaking water bath whose temperature was maintained at 37ºC. One mL of the solution was withdrawn periodically, and the same make-up volume of fresh buffer solution was added to the test tube. The curcumin concentration in each withdrawn sample was analyzed with HPLC (high performance liquid chromatography) and the obtained value was used to calculate and obtain a data point for the accumulated curcumin released on the curve. To facilitate the pattern identification, the released amount of curcumin at any time t was normalized by the total mass of curcumin introduced at the start of each preparation and shown as percentage. The measurements were made in triplicates and the standard deviation of the release tests shown here was around 2.2 %.

Results & Discussion

Hydrogels were formed by ionic interactions, which occurred spontaneously via electrostatic interactions between the positive charges and the negative charges of chitosan and carrageenan, respectively. In other words, the gelling of chitosan and carrageenan promptly formed dense gels by strong attraction between the positive charges of the amino groups of chitosan and the negative charges of the sulfonate groups of the polymer suspension consisting of κ-carrageenan and NaCMC. First, the present ternary mixtures (sets A and B) without the oil phase were frozen at a cooling rate of -0.5 or -2.0 °C/min with and without incubation at -5°C for 10 hours. Then the frozen specimens were thawed to determine the gel fraction yields as shown in **Figure 2.** The specimen ID is given in **Table 1**, where the first character, A or B, denotes the polymer set defined in Fig. 1, and the second character, F or S, denotes a fast or slow cooling rate. It was evident from Fig. 2 that the freezing protocol clearly influenced the sol-gel transition of the present polymer suspensions. In the absence of incubation, polymer set B always yielded specimens with a higher gel fraction than polymer set A regardless of the rate of cooling, apparently because of the more balanced ratio of κ-carrageenan and NaCMC. In case of polymer set A, the gel fraction was slightly enhanced by a faster cooling rate, whilst the reverse was true for polymer set B. The incubation process at subzero temperature allowed further gel formation. As a result, a slower cooling rate always yielded a higher gel fraction than a faster rate when the type of polymer set is the same. It should be noted that a higher gel fraction did not necessarily lead to a slow release of curcumin.

Table 1. Influence of the polymer set and cooling rate on the encapsulation yield of curcumin

Sample ID	Oil phase (%)	Mixed polymer Type A and B	Cooling rate (K/min)	Encap. yield (%)
AS-5	5%	A	-0.5	89.45
AF-5	5%	A	-2	99.89
BS-5	5%	B	-0.5	91.94
BF-5	5%	B	-2	99.64
AS-10	10%	A	-0.5	83.89
AF-10	10%	A	-2	99.93
BS-10	10%	B	-0.5	95.15
BF-10	10%	B	-2	99.37

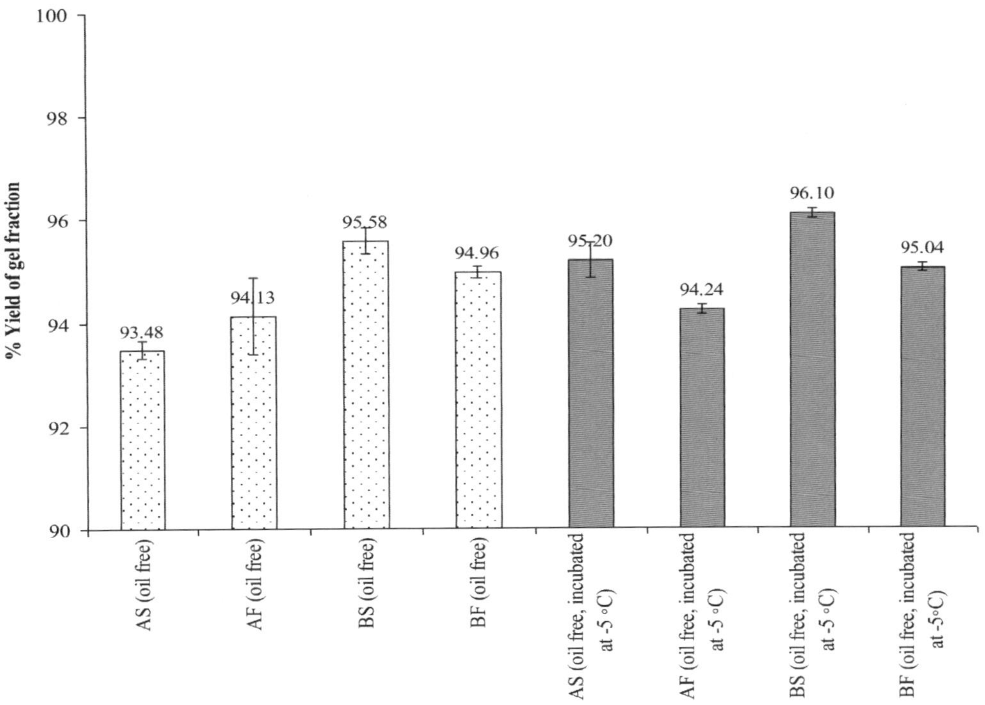

Figure 2. Influence of the polymer set, freezing rate and incubation on the yield of gel fractions

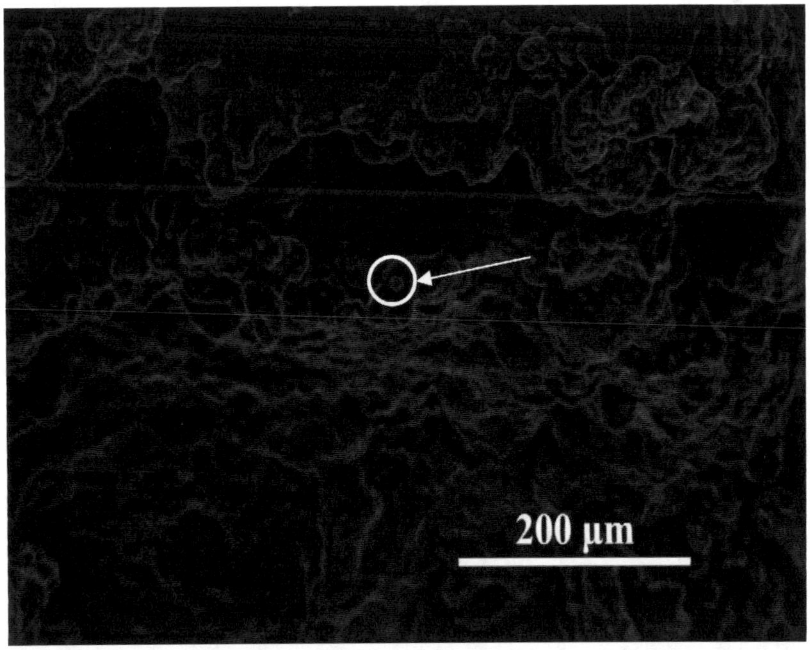

Figure 3. SEM micrograph of a cross-section of freeze-thawed sample showing the encapsulated oil phase

Table 1 shows the influence of the polymer set and cooling rate on the encapsulation yield of curcumin. Here the cryogel samples were not incubated. As seen from Table 1, the cooling rate had a significant effect on the yield. For the same polymer set and oil phase composition, as the cooling rate increased, the yield of encapsulation always rose in tandem apparently because the dispersed oil phase had less time to be squeezed out of the growing ice crystals. When the cooling rate was slow, polymer set B always showed higher yield than polymer set A at the same oil phase composition. This was consistent with the above finding that polymer set B always yielded specimens with a higher gel fraction than polymer set A because of the more balanced ratio of κ-carrageenan and NaCMC. However, the yields became

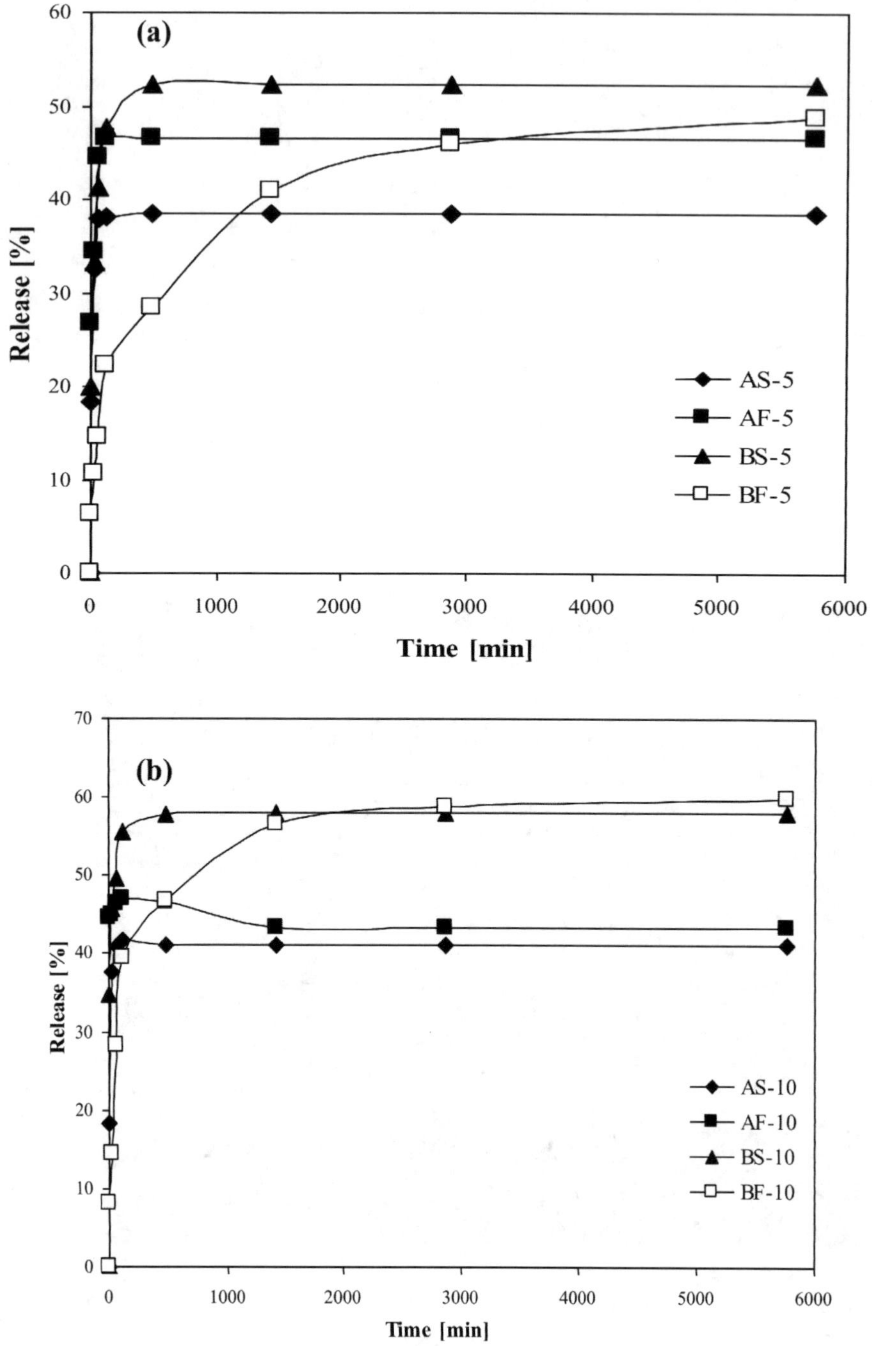

Figure 4. Time release curves of curcumin (measured in phosphate buffer pH 7.4) (a) 5% (b) 10% oil phase

essentially the same when the cooling rate was fast. This indicated that the cooling rate had more effect on the encapsulation yield than the type of polymer set. **Figure 3** is a SEM micrograph showing a typical example of the microstructure of a freeze-thawed cryogel sample containing an encapsulated oil phase.

Figure 4 shows the effect of polymer set type and cooling rate on the curcumin release pattern. Here the cryogel samples were not incubated. Two types of release patterns were observed: a burst release (typified by AS-5, -10) and a moderate first-order release rate (typified by BF-5, -10), essentially irrespective of the introduced oil phase composition. Obviously, the first-order release (controlled release) was achieved only at the faster cooling rate and only for polymer set B. In principle, the more numerous crosslinkings and higher bonding strength the polymer matrix has, the slower the rate of release should become. Since specimens BF-5 and -10 were made of polymer set B with a carrageenan:CMC ratio of 4:6, a more balanced ratio was again shown to be necessary for strong microstructure. Since the total release amounts for BS-5 vs. BF-5 as well as BS-10 vs. BF-10 were found to be asymptotically equal, it may be considered that the cooling rate could influence the microstructure formation of the present cryogels (encapsulants), and as a consequence, control the mass transfer resistance of the shell matrix.

Conclusion

A ternary system of chitosan, κ-carrageenan, and NaCMC was used for hydrogel encapsulation of curcumin through cryotropic gel formation. It found that the cooling rate had a significant positive effect on the encapsulation yield. When the cooling rate was slow, polymer set B always showed higher yield than polymer set A at the same oil phase composition. Two types of release patterns were observed: a burst release and a first-order release rate, essentially irrespective of the introduced oil phase composition. The first-order release (controlled release) was achieved only at the fast cooling rate and only for polymer set B. It may be considered that the cooling rate could influence the microstructure formation of the present cryogels (encapsulant) and control the mass transfer resistance of the shell matrix.

References

1. Giannouli, P. and E. R. Morris (2003), "Cryogelation of xanthan," *Food Hydrocolloids*, 17, 495-501.
2. Lakkis, J.M. (2007), *Encapsulation and Controlled Release Technologies in Food Systems*, Blackwell Publishing Ltd.
3. Lozinsky, V. I., E. S. Vainerman, S. A. Ivanova, E. F. Titova, M. I. Shtil'man, E. M. Belatseva and S. V. Rogozhin (1986), "Study of cryostructurization of polymer systems. VI. The influence of the process temperature on the dynamics of formation and structure of cross-linked polyacrylamide cryogels," *Acta Polymerica*, 37, 142-146.
4. Lozinsky, V. I. and L. G. Damshkaln (2000), "Study of cryostructurization of polymer systems. XVII. Poly(vinyl alcohol) cryogels: Dynamics of the cryotropic gel formation," *Journal of Applied Polymer Science*, 77, 2017-2023.
5. Orrego, C. E. and J. S. Valencia (2009), " Preparation and characterization of chitosan membranes by using a combined freeze gelation and mild crosslinking method," *Bioprocess and Biosystems Engineering*, 32, 197-206

(22c) Removal of Phenol in Water by Alpha-, Beta- and Gamma-Cyclodextrin Polymer Particles

Qingchuan Chen, Xiaolei Dong, **Xuhong Guo** and Li Li, School of Chemical Engineering, East China University of Science and Technology, Shanghai, China

In this work, well-defined spherical particles of alpha-, beta-, and gamma-cyclodextrin polymers were prepared by reverse suspension polymerization using epichlorohydrin as crosslinker. They can selectively absorb phenol by inclusion association in water and desorption in ethanol. Their adsorption capacity of phenol in water was found to be beta-CD polymer > alpha-CD polymer > gamma-CD polymer. Equilibrium isotherms for the adsorption of phenol were analyzed according to Langmuir and Freundlich equations. The results showed that the Freundlich adsorption isotherm was more suitable to fit the experimental data. The negative value of free energy change indicated the spontaneous nature of the adsorption of phenol by all three CD polymers.

Polyethersulfone (PES)/ metal oxides nanocomposite membranes: A review of performance improvement for water treatment

Fateme S. Shariatmadar, Department of Chemical Engineering, University of Kashan, Kashan, Iran
Mohsen Mohsen-Nia, Department of Chemical Engineering, University of Kashan, Kashan, Iran, Department of Chemical Engineering, California Institute of Technology, CA, USA

Abstract

The development of synthetic membranes had always been inspired by the fact that the selective transport through biological membranes is enabled by highly specialized novel polymer membranes. Due to good performance, good chemical resistance, wide pH tolerances, easy to be fabricated into membrane and broader range of pore sizes, polyethersulfone (PES) has been widely used for different industrial applications. But inside these advantages, PES membrane fouling is a considerable problem that instigates researchers to solve this via different manners. Recently, successful attempts have been made to add nanoparticles with different particles sizes in polymer structures in membrane synthesis. This paper reviews the role of metal oxide nanoparticles in PES membrane structures and improving their applications in water treatment.

Keywords: polymeric membrane; Polyethersulfone; metal oxide Nanoparticles; nanocomposite membrane; fouling reduction

Introduction

Membranes play critical roles in a number of life-sustaining processes [1]. A membrane can be defined as an interphase between two phases acting as a selective barrier. Membranes separation can be performed isothermally at lower temperatures with lower energy consumption as compared to other thermal separation processes [2]. Considering the large diversity of membranes suited for technical applications, they classify as material, cross-section, preparation methods and shape of membranes.

Table 1. Classify of membranes [2, 3].

classify	**explanation**
Membrane materials	Organic polymers, inorganic materials (oxides, ceramics, metals), mixed matrix or composite materials
Membrane cross-section	symmetric, asymmetric, bi- or multilayer, thin-layer or mixed matrix composite
Preparation method	Phase separation (phase inversion) of polymers, sol–gel process, interface reaction, stretching, extrusion, track-etching, and micro-fabrication
Membrane shape	Flat-sheet, hollow fiber, spiral and tubular

The most common problems of polymeric membranes are high hydrophobicity, exposure to befouling, low fluxes and low mechanical strength that reduction the performances of polymeric membranes. Researcher's attempts have focused on these disadvantages in order to remove them and improve the performance of membrane [2].

Due to better control of the pore forming mechanism, higher flexibility, and smaller spaces required for installation and lower costs compared to inorganic membranes, current researches focus on preparation of mixed matrix or composite membranes for the synergistic combination of different functions by different (polymeric) materials [4].

In this study, at first we will discuss about PES membrane as a polymeric membrane and in continuance, hydrophobicity and fouling of PES membrane as the most common disadvantage of this polymer and the methods for removing these problems will be express.

Polyethersulfone

PES is a commercially available polymer and a well-known transparent amorphous polymer for its dimensional stability, and resistance to steam, boiling water, and mineral acids. Other desirable properties include creep resistance, inherent flame resistance and approved for use with food, water and medical applications. All these properties make PES as an attractive and an important separation membrane material, but its hydrophobicity leads to a low membrane flux and poor anti-fouling property, which has great effect on its application and usage life. It is necessary to modify the PES membrane surface by physical or chemical methods in order to improve its hydrophilicity and fouling [5]. Achieving the optimum surface characteristics without diminishing the advantageous properties of PES membrane is an important goal for the membrane industry [6].

Polyethersulfone gets its name because they have ether groups and sulfone groups in their backbone chains. PES polymers have high glass transition temperatures (T_g) and flexible chain in order of softening of polymer at a reasonable temperature because of the stiff sulfone groups and flexible ether linkages, respectively. PES have different structures with a range of T_g about 190-230 $^{\circ}$C. Depending on membranes application and amount of mechanical and chemical strength that is needed, we can use any of PES structures as a polymeric membrane.

Reduction of PES membrane's fouling

The most common techniques used to remove fouling include back-flushing and chemical cleaning. But, these methods are not ideal because in order of membrane cleaning, filtration process must be shut down regularly and it is necessary to dispose the waste of this cleaning. A number of approaches are available to reduce the membrane fouling which an increase in membrane hydrophilicity appears to be a good appropriate.

Different methods were investigated to impart a higher hydrophilicity to PES membranes such as Surface modification [7], [8]; Plasma treatment [5], [7-12]; Grafting [5-8], [10-22]; Sulfonation [6], [10], [20], [23-27] and Blending with hydrophilic monomers or polymers and inorganic nanoparticles [5-9], [12], [15-17], [19], [20], [22], [26], [28-50].

Among these methods, blending with inorganic materials, especially nanoparticles, has attracted much interest because by the way of blending, the modified membrane can combine basic properties of organic and inorganic materials and offer specific advantages for the preparation of membranes with excellent separation performances, good thermal and chemical resistance and adaptability to the harsh wastewater environments.

Metal oxide nanoparticle/PES membranes

Variety types of inorganic materials are used in polymeric membrane structures. One of them is metal oxide nanoparticles [12]. The studies that focus on metal oxide nanoparticles/PES nanocomposite membranes are contain using TiO_2, Al_2O_3, ZrO_2, SiO_2 and Ag nanoparticles in PES structure that will investigate in continuance.

Table 2. Different studies about metal oxide nanoparticles/PES nanocomposite membranes

Metal oxide /PES nanocomposites	Reference
TiO_2/PES	Kim et al. [30, 43], Luo et al. [32-34, 43, 48], Tak et al. [31, 39, 43], Bae et al. [35, 39], Yang et al. [43, 51], Lee et al. [36, 43], Wu et al. [5, 43, 48], Rahimpour et al. [9, 43, 48], Li et al. [38, 43, 48], Li et al. [43, 48], Li et al. [43, 48], Razmjou et al. [48]
Al_2O_3/PES	Maximous et al. [2, 12, 41, 42, 45, 47, 50]
ZrO_2/PES	Maximous et al. [41, 45, 47, 50]
Ag/PES	Wang et al. [37, 46], Basri et al. [44, 46]
SiO_2/PES	Jiang et al. [52]

TiO_2/PES nanocomposites

The most common procedure for preparation of TiO_2/PES nanocomposites is self-assembly of TiO_2 nanoparticles on membrane surface. The self-assembly procedure for TiO_2 nanoparticles on to PES membranes is schematically shown in Figure 1. For Polyethersulfone, self-assembly can be due to a coordination of the sulfone group and the ether bond to Ti^{4+}, or by a hydrogen bond between the sulfone group and the ether bond and a surface hydroxyl group of TiO_2 [32, 43].

TiO_2 nanoparticles are photocatalytic particles which are capable to decompose organic chemicals on the membrane surface and kill bacteria and affect the surface morphology greatly. Analysis showed that by loading TiO_2 nanoparticles on membrane structure, the contact angle of the composite membrane decreased and the hydrophilicity of the membrane surface improved remarkably. The appropriate addition of TiO_2 improved the anti-fouling and fouling-resistant ability of the membrane. Different studies showed that the water flux through a PES-TiO_2 membrane was significantly enhanced, but this effect was dependent to the concentration of particles. Above a threshold concentration, the water flux decreased again. However, agglomeration of TiO_2 nanoparticles exists due to the poor interfacial interactions between the hydrophilic nanoparticles and hydrophobic polymer and high surface energy of particles which should be removed.

Al_2O_3/PES nanocomposites

Al_2O_3 nanoparticles were used in PES structure in order to improve the performance of PES membrane for wastewater filtration. Al_2O_3 entrapped membrane showed lower flux decline compared to neat polymeric membrane. PES membrane characteristics and performances were changed by the addition of Al_2O_3 nanoparticles to the casting solution and fouling decreased during the activated sludge filtration.

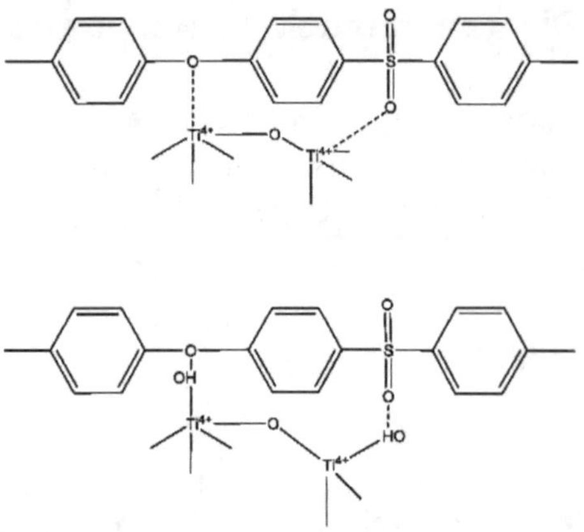

Figure 1. Self-assembly of Ti nanoparticles to polyethersulfone membranes by coordination of sulfone and ether bond (upper scheme) or by hydrogen bonding (lower scheme) [32, 43].

ZrO_2/PES nanocomposites

Small amounts of ZrO_2 particles were added into the PES casting solution in order to improve the performance of PES membrane for wastewater and activated sludge filtration. These composites are more stable than titania and alumina composite membranes and are suitable for liquid phase applications under harsh condition. ZrO_2/PES nanocomposites are potential membrane materials for high temperature catalytic reactions but the toxicity of ZrO_2 is high compare to TiO_2 and Al_2O_3 so this disadvantage diminish using ZrO_2 composites in water treatment.

Ag /PES nanocomposites

Polyethersulfone–silver composite membranes were fabricated by using silver nitrate ($AgNO_3$) as an antibacterial agent and polyvinylpyrollidone (PVP) of different molecular weights as dispersant in the dope formulation. Ag particles have high conductivity and antibacterial properties and Increased the hydrophilicity and flux of membranes. But the most important disadvantage of these particles was Leaching from membrane structure which Exhibited toxicity in drinking water application.

Because of leaching problems due to the low resistance to attrition which in turn caused silver particles to fall off from the membrane, addition of polyvinylpyrollidone (PVP) and 2, 4, 6-triaminopyrimidine (TAP) overcome silver leaching during fabrication as well as to achieve better silver distribution in PES–$AgNO_3$ membrane.

SiO_2 /PES nanocomposites

In PES/SiO_2 composite membranes, the cross-section structure of membrane was obviously affected by addition of SiO_2 nanoparticles. Also, the pure water fluxes and raw water fluxes were increased. The membrane hydrophilicity was enhanced because of hydrophilic SiO_2 nanoparticles and there was higher water flux recovery ratio of the SiO_2 modified PES membrane compared to pure PES membrane after the same cleaning method for surface water filtration. The appropriate addition of SiO_2 improved the anti-fouling ability of the membrane during raw water and the SiO_2 modified ultrafiltration membrane was found to be suitable as pretreatment process during surface water nanofiltration.

Conclusions

Polymeric materials are the ideal choice for membrane fabrication because of the lower cost and more comfortable performance against inorganic membranes. PES is a polymer with appropriate properties which has been used widely in membrane technology but PES is hydrophobic which lead to membrane fouling. Thus, it is worthwhile to investigate different ways of modification PES membranes for reduction fouling problem. Modification by adding different nanoparticles indicated to have more advantages over pure polymeric materials alone in most cases.

Ag as an antibacterial material is incorporated into the PES membrane matrix but leaching these particles out of the membrane, causes the anti-biofouling properties decreases. Also leaching of silver nanoparticles could pose the danger of water poisoning if the membranes are used in drinking water purification processes. Catalytic TiO_2 nanoparticles reduce the biofouling problem of membranes but agglomeration of these particles is high. Increase in ZrO_2 concentration leads to an increase of the permeate flux at the expense of retention which this property is similar to Al_2O_3 nanoparticles.

Among different nanoparticles, SiO_2 had received most attention because of their perceived characteristics. Moreover, silica nanoparticles are more suitable to be incorporated into the membranes which will be used in drinking water application because silica exhibits lower toxicity and is environmentally inert. Incorporating SiO_2 nanoparticles in PES membrane structure decreases the membrane fouling and improves PES membranes efficiency and performance.

References

1. M. Gholami, S. Nasseri, C.Y. Feng, T. Matsuura, K.C. Khulbe (2003), "the effect of heat-treatment on the ultrafiltration performance of polyethersulfone (PES) hollow-fiber membranes", *Desalination,* pp.293-301.
2. L. Y. Ng, A. W. Mohammad, C.P. Leo, N. Hilal (2010), "Polymeric membranes incorporated with metal/metal oxide nanoparticles: A comprehensive review", *Desalination,* in press, accepted manuscript.
3. A. Idris, N.M. Zain, M.Y. Noordin (2007), "Synthesis, characterization and performance of asymmetric polyethersulfone (PES) ultrafiltration membranes with polyethylene glycol of different molecular weights as additives", *Desalination*, pp.324–339.
4. M. Ulbricht (2006), "Advanced functional polymer membranes", *Polym.,* pp.2217–2262.
5. G. Wu, S. Gan, L. Cui, Y. Xu (2008), "Preparation and characterization of PES/TiO_2 composite membranes", *Appl. Surf. Sci.,* pp.7080–7086.
6. A. Rahimpour, S.S. Madaeni (2010), "Improvement of performance and surface properties of nano-porous Polyethersulfone (PES) membrane using hydrophilic monomers as additives in the casting solution", *J. Membr. Sci.,* pp.371–379.
7. A. Rahimpour, S.S. Madaeni, S. Mehdipour-Ataei (2008), "Synthesis of a novel poly(amide-imide) (PAI) and preparation and characterization of PAI blended polyethersulfone (PES) membranes", *J. Membr. Sci.,* pp.349–359.
8. A. Rahimpour (2011), "UV photo-grafting of hydrophilic monomers onto the surface of nano-porous PES membranes for improving surface properties", *Desalination*, pp.93–101.
9. A.Rahimpour, S.S.Madaeni, A.H.Taheri, Y.Mansourpanah (2008), "Coupling TiO_2 nanoparticles with UV irradiation for modification of polyethersulfone ultrafiltration membranes", *J. Membr. Sci.,* pp.158– 169.
10. A. Rahimpour, S.S. Madaeni, Sh. Ghorbani, A. Shockravi, Y. Mansourpanah (2010), "The

influence of sulfonated polyethersulfone (SPES) on surface nano-morphology and performance of polyethersulfone (PES) membrane", *Appl. Surf. Sci.*, pp.1825–1831.

11. W. Xi, W. Rong, L. Zhansheng, A. G. Fane (2006), "Development of a novel electrophoresis-UV grafting technique to modify PES UF membranes used for NOM removal", *J. Membr. Sci.*, pp.47–57.
12. N. Maximous, G. Nakhla, W. Wan, K. Wong (2009), "Preparation, characterization and performance of Al_2O_3/PES membrane for wastewater filtration", *J. Membr. Sci.*, pp.67–75.
13. M. Wang, L.G. Wu, J.X. Mo, C.J. Gao (2006), "The preparation and characterization of novel charged polyacrylonitrile/PES-C blend membranes used for ultrafiltration", *J. Membr. Sci.*, pp.200–208.
14. B. Kaesele, P. Kingshott, G. Jonsson (2002), "Influence of the surface structure on the filtration performance of UV-modified PES membranes", *Desalination*, pp.265-271.
15. T. Wang, Y. q. Wang, Y.l. Su, Zh.Y. Jiang (2005), "Improved protein-adsorption-resistant property of PES/SPC blend membrane by adjustment of coagulation bath composition", *Colloids Surf., B.*, pp.233–239.
16. D. He, H. Susanto, M. Ulbricht (2009), "Photo-irradiation for preparation, modification and stimulation of polymeric membranes", *Prog. Polym. Sci.*, pp.62–98.
17. B. Fang, Q. Ling,W. Zhao, Y. Ma, P. Bai, Q. Wei, H. Li, Ch. Zhao (2009), "Modification of polyethersulfone membrane by grafting bovine serum albumin on the surface of Polyethersulfone/poly(acrylonitrile-co-acrylic acid) blended membrane", *J. Membr. Sci.*, pp.46–55.
18. Z. Ma, Zh. Lan, T. Matsuura, S. Ramakrishna (2009), "Electrospun Polyethersulfone affinity membrane: Membrane preparation and performance evaluation", *J. Chromatogr. B.*, pp.3686–3694.
19. Zh. Yi, L.P. Zhu, Y.Y. Xu, Y.F. Zhao, X.T. Ma, B.K. Zhu (2010), "Polysulfone-based amphiphilic polymer for hydrophilicity and fouling-resistant modification of polyethersulfone membranes", *J. Membr. Sci.*, pp.25–33.
20. G. Wei, L. Xu, Ch. Huang, Y. Wang (2010), "SPE water electrolysis with SPEEK/PES blend membrane", *int. j. hydrogen energy,* pp.1-6.
21. Q. Shi, Y. Su, X. Ning, W. Chen, J. Peng, Zh. Jiang (2010), "Trypsin-enabled construction of anti-fouling and self-cleaning Polyethersulfone membrane", *Bioresource Technol.,* in press, accepted manuscript.
22. W. Zhao, J. Huang, B. Fang, Sh. Nie, N. Yi, B. Su, H. Li, Ch. Zhao (2010), "Modification of polyethersulfone membrane by blending semi-interpenetrating network polymeric nanoparticles", *J. Membr. Sci.,* in press, accepted manuscript.
23. N.N. Krishnan, H.J. Kim, M. Prasanna, E. Cho, E.M. Shin, S.Y. Lee, I.H. Oh, S.A. Hong, T.H. Lim (2006), "Synthesis and characterization of sulfonated poly(ethersulfone) copolymer membranes for fuel cell applications", *J. Power Sources,* pp.1246–1250.
24. Y.S. Oh, H.J. Lee, M. Yoo, H.J. Kim, J. Han, T.H. Kim (2008), "Synthesis of novel crosslinked sulfonated poly(ether sulfone)s using bisazide and their properties for fuel cell application", *J. Membr. Sci.,* pp.309–315.
25. D.W. Seo, Y.D. Lim, S.H. Lee, Y.G. Jeong, T.W. Hong, W.G. Kim (2010), "Preparation and characterization of sulfonated amine-poly(ethersulfone)s for proton exchange membrane fuel cell", *int. j. hydrogen energy,* pp.13088-13095.
26. X. Cao, M. Tang, F. Liu, Y. Nie, Ch. Zhao (2010), "Immobilization of silver nanoparticles onto sulfonated polyethersulfone membranes as antibacterial materials", *Colloids Surf. B.,* pp.555–562.
27. Ch. Klaysom, B.P. Ladewig, G.Q.M. Lu, L. Wang (2011), "Preparation and characterization of

Sulfonated Polyethersulfone for cation-exchange membranes", *J. Membr. Sci.*, pp.48–53.
28. A. Rahimpour, S.S. Madaeni, Y. Mansourpanah (2010), "Fabrication of polyethersulfone (PES) membranes with nano-porous surface using potassium perchlorate ($KClO_4$) as an additive in the casting solution", *Desalination*, pp.79–86.
29. A. Rahimpour, S.S. Madaeni (2007), "Polyethersulfone (PES)/cellulose acetate phthalate (CAP) blend ultrafiltration membranes: Preparation, morphology, performance and antifouling properties", *J. Membr. Sci.*, pp.299–312.
30. S.H. Kim, S.Y. Kwak, B.H. Sohn, T.H. Park (2003), "Design of TiO_2 nanoparticle self-assembled aromatic polyamide thin-film-composite (TFC) membrane as an approach to solve biofouling problem", *J. Membr. Sci.*, pp.157-165.
31. T.H. Bae, T.M. Tak (2005), "Effect of TiO_2 nanoparticles on fouling mitigation of ultrafiltration membranes for activated sludge filtration", *J. Membr. Sci.*, pp.1-8.
32. M.L. Luo, J.Q. Zhao, W. Tang, Ch.Sh. Pu (2005), "Hydrophilic modification of poly(ether sulfone) ultrafiltration membrane surface by self-assembly of TiO_2 nanoparticles", *Appl. Surf. Sci.*, pp.76–84.
33. J.B. Li, J.W. Zhu and M.Sh. Zheng (2006), "Morphologies and Properties of Poly(phthalazinone ether sulfone ketone) Matrix Ultrafiltration Membranes with Entrapped TiO_2 Nanoparticles", *DOI:10.1002/app. 2006.25428. wiley interscience*, in press, accepted manuscript.
34. M.l. Luo, W. Tang, J.q. Zhao, C.s. Pu (2006), "Hydrophilic modification of poly(ethersulfone) used TiO_2 nanoparticles by a sol–gel process", *J. Mater. Process. Technol.*, pp.431–436.
35. T.H. Bae, I.Ch. Kim, T.M. Tak (2006), "Preparation and characterization of fouling-resistant TiO_2 self-assembled nanocomposite membranes", *J. Membr. Sci.*, pp.1–5.
36. H.S. Lee, S.J. Im, J.H. Kim, H.J. Kim, J.P. Kim, B.R. Min (2008), "Polyamide thin-film nanofiltration membranes containing TiO_2 nanoparticles", *Desalination*, pp. 48-56.
37. H.Wang, T.Yu, C.Zhao, Q.Du (2009), "Improvement of hydrophilicity and blood compatibility on polyethersulfone membrane by adding polyvinylpyrollidone", *Fibers Polym.*, pp.1–5.
38. J.H. Li, Y.Y. Xu, L.P. Zhu, J.H. Wang, C.H. Du (2009), "Fabrication and characterization of a novel TiO_2 nanoparticle self-assembly membrane with improved fouling resistance", *J. Membr. Sci.*, pp. 659-666.
39. V. V. Tarabara (2009), "Multifunctional Nanomaterial-Enabled Membranes for Water Treatment", *Nanotechnol. Appl. Clean Water.*, pp.59–75.
40. M. M. Hasani-Sadrabadi, E. Dashtimoghadam, S.R. Ghaffarian, M.H. Hasani Sadrabadi, M. Heidari, H. Moaddel (2010), "Novel high-performance nanocomposite proton exchange membranes based on poly (ether sulfone)", *Renewable Energy*, pp.226–231.
41. N.N. Maximous, G.F. Nakhla, W.K. Wan (2010), "Removal of Heavy Metals from Wastewater by Adsorption and Membrane Processes: a Comparative Study", *World Academy Sci. Eng. Technol.*, pp.594-599.
42. N. Maximous, G. Nakhla, K. Wong,W. Wan (2010), "Optimization of Al_2O_3/PES membranes for wastewater filtration", *Sep. Purif. Technol.*, pp.294–301.
43. J. Kim, B.V. Bruggen (2010), "The use of nanoparticles in polymeric and ceramic membrane structures: Review of manufacturing procedures and performance improvement for water treatment", *Environ. Pollut.*, pp.2335-2349.
44. H. Basri, A.F. Ismail, M. Aziz, K. Nagai, T. Matsuura, M.S. Abdullah, B.C. Ng (2010), "Silver-filled polyethersulfone membranes for antibacterial applications— Effect of PVP and TAP addition on silver dispersion", *Desalination*, pp.264–271.
45. N. Maximous, G. Nakhla, W. Wan, K. Wong (2010), "Performance of a novel ZrO_2/PES membrane for wastewater filtration", *J. Membr. Sci.*, pp.222–230.
46. H. Basri, A.F. Ismail, M. Aziz (2010), "Polyethersulfone (PES)–silver composite UF membrane:

Effect of silver loading and PVP molecular weight on membrane morphology and antibacterial activity", *Desalination*, in press, accepted manuscript.

47. N. Maximous, G.Nakhla, W.Wan, K.Wong (2010), "Effect of the metal oxide particle distributions on modified PES membranes characteristics and performance", *J. Membr. Sci.*, pp.213–222.
48. A. Razmjou, J. Mansouri, V. Chen (2010), "The effects of mechanical and chemical modification of TiO_2 nanoparticles on the surface chemistry, structure and fouling performance of PES ultrafiltration membranes", *J. Membr. Sci.*, in press, accepted manuscript.
49. F.J. Wang, W. Li, M.S. Xue, J.P. Yao, J.S. Lu (2011), "$BaTiO_3$–polyethersulfone nanocomposites with high dielectric constant and excellent thermal stability", *Compos., B.*, pp.87–91.
50. W. Zhao, J. Huang, B. Fang, Sh. Nie, N. Yi, B. Su, H. Li, Ch. Zhao (2011), "Modification of polyethersulfone membrane by blending semi-interpenetrating network polymeric nanoparticles", *J. Membr. Sci.,* pp.258–266.
51. Y. Yang, H. Zhang, P. Wang, Q. Zheng, J. Li (2007), "The influence of nano-sized TiO_2 fillers on the morphologies and properties of PSF UF membrane", *J. Membr. Sci.,* pp.231-238.
52. J. Shen, H. Ruan, L. Wu, C. Gao (2011), "Preparation and characterization of PES–SiO2 organic–inorganic composite ultrafiltration membrane for raw water pretreatment", *Chem. Eng. J.,* pp.1272–1278.

(22e) A Comparative Study on Double Alkylation of Naphthalene with Long Chain Olefins Using Different Catalysts

Ming Zhang, Zhiqiang He and Yun Fang, School of Chemical & Material Engineering, Jiangnan University, Wuxi, China

A comparative study on double alkylation of naphthalene with long chain olefins using differentcatalysts

Ming Zhang, Zhiqiang He,Yun Fang*

School of Chemical and Material Engineering, Jiangnan University, Wuxi, Jiangsu 214122, China

Abstract

A series of double long-chain alkylnaphthalenearesynthesized by naphthalene and long-chain alkenes in this dissertation.A comparative study has been made over these catalysts including proton acid, Lewis acid, HY, HMS and ionic liqulds in the Friedel-Crafts reaction of naphthalene. Traditional catalysts have several disadvantages£-corrosion,low selectivity, separation of the final product from the catalyst, difficulty in catalyst recovery, environmental pollution. High acid amount zeolites gave poor catalytic performance due to their diffusional constraints. As one of novel media,ionic liqulds could be used as the catalyst in the process of alkylation of naphthalene with high activity and high selectivity; on the other hand the products and unreacted reactants do not dissolve in the ionic liquids, and therefore they could be isolated easily by decantation. In this paper, ionic liquid systems prepared with $AlCl_3$-Et_3NHCl, $SnCl_4$-Et_3NHCl, $FeCl_3$-Et_3NHCl were used as reaction media for the alkylation of naphthalene and the influences of various parameters, such as type and dosage of catalyst, molar ratio of naphthalene to alkenes, reaction temperature, and reaction time were studied. By adjusting the acidity of ionic liquid catalyst, excellent conversion and selectivity have been achieved and the catalyst could be recycled for several times without any obvious loss of its activity.

Keywords: Catalyst; Alkylation; long-chain alkylnaphthalene; green synthesis

(24a) CO2 Capture Via Potassium Carbonate/MEA Solution

Yee Soong, National Energy Technology Laboratory, U.S. Department of Energy, Pittsburgh, PA, Robert Dilmore, U.S. Department of Energy, Pittsburgh, PA, Sheila Hedges, U.S. DOE/NETL, Pittsburgh, PA and D. E. Allen, Salem State College, Salem, MA

A great deal of concern has been expressed with regard to global climate change and its link to growing atmospheric concentrations of carbon dioxide. To decrease the impact of anthropogenic CO_2on global climate, several strategies are under development that will potentially remove $CO_2$from the atmosphere or decrease CO_2emission. One such strategy involves the capture of $CO_2$from large point sources (such as pulverized coal (PC) plants) and the long-term storage of CO_2underground. The**post-combustion**capture of $CO_2$is a challenging application because of (1) the low pressure and dilute concentration (10 to 15 %) of the carbon dioxide dictate a high volume of gas to be treated and (2) trace impurities in the flue gas tend to reduce the effectiveness of the CO_2absorbing processes.

Much work has been studied in this area and has included development of amine-based liquid sorbents, solid amine-based sorbents, metal carbonate sorbents, ionic solution, chemical looping, membrane system and other novel technologies. The aqueous amine system is one of the commercial technologies for CO_2capture for PC power plants.

The commonly used amine in scrubbing solutions contains monoethanolamine (2-aminoethanol) ($NH_2CH_2CH_2OH$) (MEA). The disadvantages of using MEA solutions include energy-intensive, amine degradation and corrosive. The typical reaction is illustrated as follows:

2 MEA + H^++ HCO_3^-¬> MEA& Middot;H^++ MEA& Middot;HCO_3^-

Work also has been conducted in the area of using carbonate sorbents to remove carbon dioxide from flue gas. A carbon dioxide capture process using regenerable sorbents such as an alkali metal oxide or an alkali metal carbonate has been reported. Potassium carbonate (K_2CO_3) can be supported on a high surface area of activated alumina (Al_2O_3). The product formed from absorption of carbon dioxide is taken aside and heated to regenerate the sorbent. No significant chemical degradation is observed, but there is physical attrition. Loading occurs between from about 60^0C to 100^0C according to the following reaction equation.

K_2CO_3+ H_2O + CO_2¬> 2 $KHCO_3$

Thermal desorption of potassium bicarbonate ($KHCO_3$) and the alumina support takes place at temperatures around 145^0C to 150^0C.

In this paper, a CO_2capture process (US patent 7,842,126) is presented where $CO_2$from combustion flue gas dissolves into an aqueous solution of amine and soluble potassium carbonate. Potassium carbonate rapidly reacts with amine-associated carbamate and water to form less-soluble potassium bicarbonate that precipitates from solution. This reaction serves both to chemically regenerate the amine solution by removing the carbamate and to separate CO_2in the form of solid bicarbonate precipitate. Precipitated potassium bicarbonate is further physically separated from bulk solution and heated to temperatures greater than approximately 150 ºC to thermally regenerate back to solid potassium carbonate and release concentrated CO_2gas. Because of the absence of bulk aqueous solution, regeneration enthalpy requirements for separated bicarbonate solid is predicted to be low as compared to those reported for regeneration in traditional MEA-based absorption processes. Following thermal regeneration of the separated solid, potassium carbonate is re-dissolved in potassium carbonate-lean aqueous amine solution, and the reactive solution is re-exposed to flue gas to close the process loop. This process concept has the potential to achieve high CO_2capture efficiency with relatively low regeneration energy requirements.

In addition to details description of capture methodology, preliminary laboratory results will be presented.

(24b) Where Are the Opportunities for CO_2 Conversion?

Lubo Zhou, UOP LLC, a Honeywell Company, Des Plaines, IL

Emission of green house gas CO_2 has caught the attention of people in most recent years. Conversion of CO_2 to other products has been proposed by many scientists to reduce the CO_2 emission to atmosphere. However, there is a big challenge for the CO_2 conversion to any product: large amount of energy or hydrogen will be used during CO_2 conversion. In this paper, thermodynamics and energy efficiency will be analyzed for CO_2 conversion reactions. Life cycle CO_2 emission will also be analyzed. The objective of the paper is to identify the opportunities of CO_2 conversion that can actually reduce CO_2 emission in a life cycle of the whole process, which includes energy and hydrogen production.

CO_2 Reforming for Variable Chemical Process

Osamu Hirohata, Tomoyuki Mikuriya and Mitsunori Shimura
Chiyoda Corporation, Yokohama, Japan

1. Introduction

Production of synthesis gas from natural gas and conversion of synthesis gas to fuels and chemicals would become increasingly attractive from the viewpoint of energy resource diversification and increasing price of crude oil. The technology of synthesis gas production by reforming of natural gas with CO_2 is one of the effective solutions for the issues of energy resource and environmental preservation.

Chiyoda CO_2 reforming catalyst can generate synthesis gases by CO_2 and H_2O reforming under lower H_2O/Carbon and CO_2/Carbon ratios in feed gas compared with the conventional steam reforming catalyst. Therefore, this catalyst produces synthesis gases with valuable H_2/CO ratios more efficiently. This is because this catalyst has higher resistance against carbon deposition under the severe condition. The commercial size CO_2 reforming catalyst has been demonstrated its activity and stability in the GTL demonstration plant tests in Japan.

In this paper, a feasibility study on applying Chiyoda CO_2 reforming to methanol plant and CO production plant has been conducted. CO_2 reforming can increase production of methanol and produce CO as by product. These increased methanol and CO can be utilized for the production of acetic acid and MMA.

2. Characteristics and performance of Chiyoda CO_2 reforming catalyst

(1) Favorable conditions of Chiyoda CO_2 reforming

Fig.1 shows a carbon limit diagram which is usually utilized for the operation of the steam reformer. The bold line in the center of this figure represents the carbon limit line at 600 ℃. On the left side of this line, the process gas has high thermodynamic potential for carbon formation at 600 ℃. Operation near this line cannot be carried out in conventional reforming process. Usually, conventional catalyst (Ni based catalyst) is operated under the condition of steam carbon ratio (S/C) in excess of around 2.0, in the absence of pre-reformer.

On the other hand, Chiyoda's CO_2 reforming catalyst (Noble metal catalyst) has higher resistance for carbon formation. So this catalyst can be used in low S/C conditions in addition to co-feeding of CO_2 on the left side of carbon limit line. High energy efficiency operation can be performed for the synthesis gas production by Chiyoda's CO_2 reforming process.

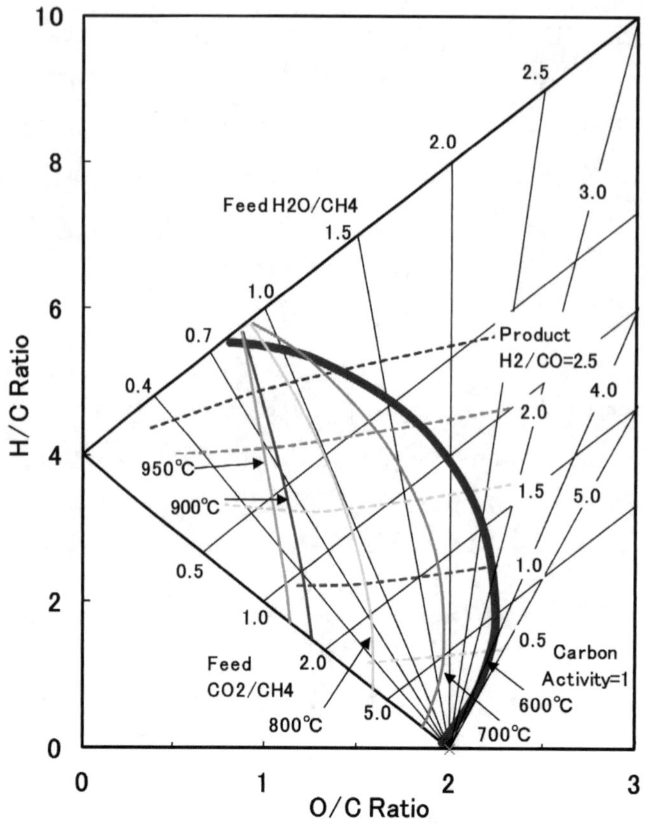

Fig.1 Carbon Limit Diagram

(2) Reaction mechanism

It is considered that the simplified reaction sequence for CO_2 reforming of methane may involve following two irreversible steps, namely, activation of methane followed by the surface reaction with adsorbed oxygen atoms[1]:

$$CH_4 + * = CH_3* = CH_2* = CH* = C* \quad \cdots\cdots \quad (1)$$

$$CHx* + O* = CO + X/2\ H_2 + 2* \quad \cdots\cdots \quad (2)$$

The adsorbed oxygen atom is formed by the dissociation of CO_2 on the catalyst support as shown in Eq(3).

$$CO_2 + * = O* + CO \quad \cdots\cdots \quad (3)$$

These reaction mechanisms are schematically represented in Fig.2

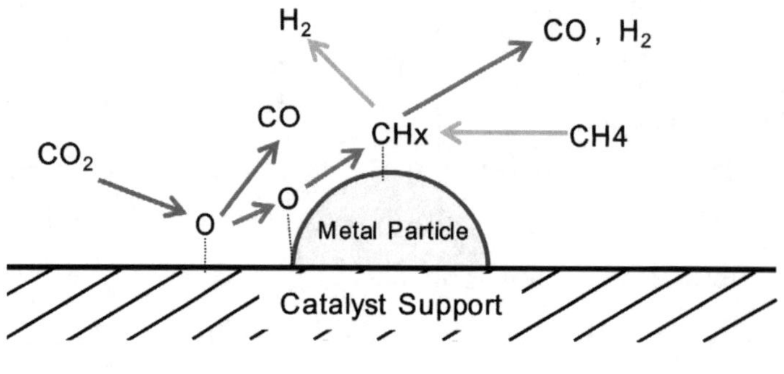

Fig.2 Reaction mechanism on CO_2 reforming catalyst

From this reaction mechanism, to prevent the carbon deposition on the catalyst surface that causes catalyst deactivation, it is important that the x-value of CHx can be maintained high and the rapid oxidation of CHx should be performed by O＊ species.

It was reported that the x-value of CHx species is related to the acidity of the catalyst support as shown in Table 1.[2] The basic support that is used for this catalyst can keep the x-value high.

Table.1 CHx species observed on Ni and Co during carbon dioxide reforming of methane

Catalyst	x-value
Ni/MgO	2.7
Ni/ZnO	2.5
Ni/Al_2O_3	2.4
Ni/TiO_2	1.9
Ni/SiO_2	1.0
Co/Al_2O_3	0.75

The rate of oxidation of CHx species is considered to be crucially dependent on the particle size of active metal.[3] If metal particles are small and interact with catalyst support, oxygen atom that is generated by the dissociation of CO_2 moves quickly to the interface of metal and support and is activated. The oxygen atom (O＊) is rapidly supplied to hydrocarbon species (CH_X) on the metal surface as shown in the Fig.2.

Based on the consideration from these investigation results, our CO_2 reforming catalyst was designed, in which highly dispersed noble metals are supported on the basic catalyst support .

(3) Performance

The demonstration plant operation of Chiyoda CO_2 reforming catalyst was conducted. Average synthesis gas production rate is around 20,000Nm3/hr with hydrogen to carbon monoxide ratio of 2.0. Feed natural gas contains around 90% of methane and the rest are almost higher hydrocarbon than methane. In the normal operation condition, the temperature of catalyst bed outlet was set around 870-880 ℃ as inlet temperature of waste heat boiler. Operation pressure was 1.94-2.09MPaG. And ratios of carbon dioxide to carbon and steam to carbon of the feed stream were controlled to 0.43-0.50 and 1.1-1.32 respectively. As shown in Fig.1, these conditions cause carbon formation with conventional catalyst. As shown in Fig.3, the stable operation for over 10,000 hours has been confirmed under the target reaction condition.

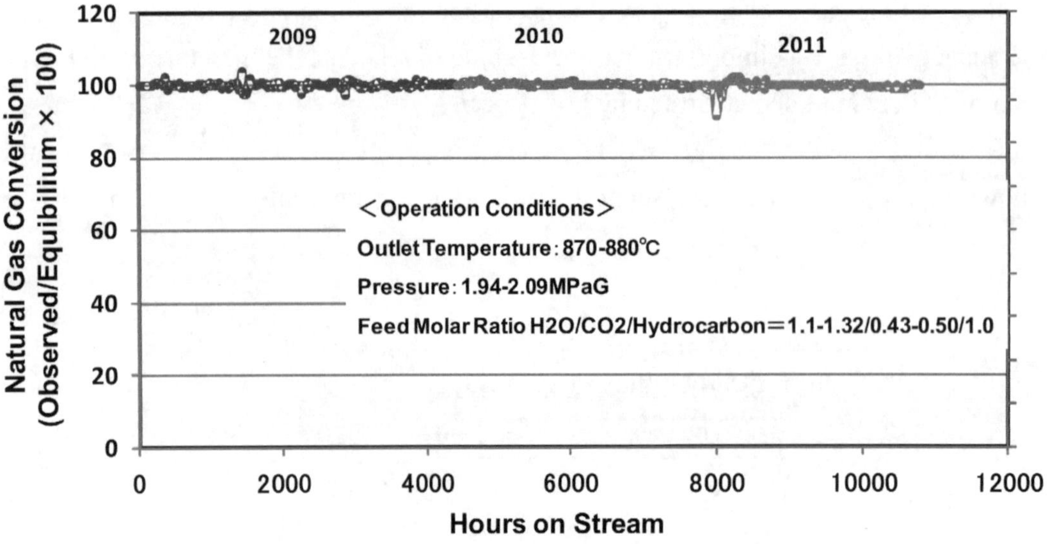

Fig.3 Demonstration Plant Result of Chiyoda CO_2 Reforming Catalyst

3. Study Results

(1) Case-1; CO_2 reforming for methanol production (Revamping case)

(a) Study basis

In Fig.4, the conventional methanol plant flow scheme is shown based on a plant with approx. 3000 t/d methanol production.

Compared with conventional steam reforming, CO_2 reforming case (Fig.5) has two additional systems; ie , import CO_2 line and PSA system. In revamping case, CO_2 is utilized as additional feed for reforming. PSA system can reduce methane concentration and increase H_2 concentration in the recycle stream of the methanol reactor to promote methanol synthesis reaction. As a result of CO_2 import and hydrogen recycle, energy for the reformer to produce excess hydrogen can be reduced and synthesis gas composition can be adjusted to the ideal composition for methanol synthesis.

Process study is conducted on same methanol capacity plant with these two flows.

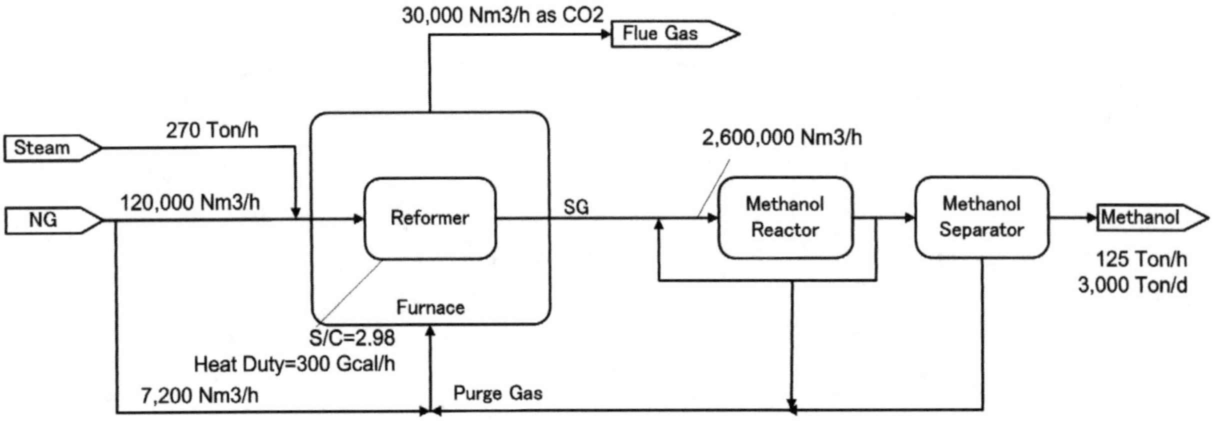

Fig.4 Conventional Case (Steam Reforming)

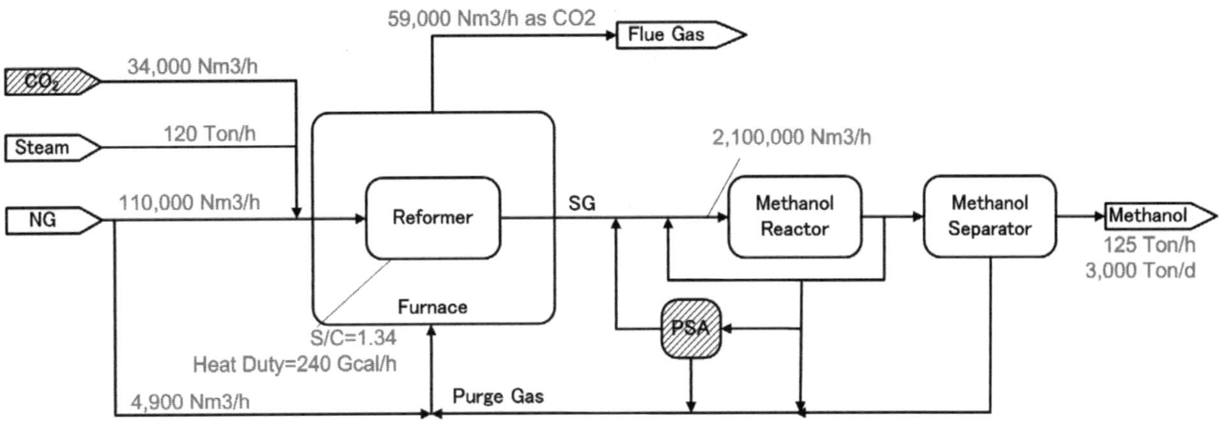

Fig.5 Revamping Case (CO_2 Reforming with PSA)

(b) CO_2 utilization and NG consumption

Methanol plant exhausts the CO_2 from reformer furnace. However, in the revamped plant with CO_2 import, CO2 is effectively utilized NG reforming to produce CO. Therefore, the net exhaust CO_2 can be reduced by approx.18% under the same methanol production capacity (Fig.6). NG consumption in revamped case is 4.4% lesser than conventional case by utilizing CO_2 as feedstock (Fig.7).

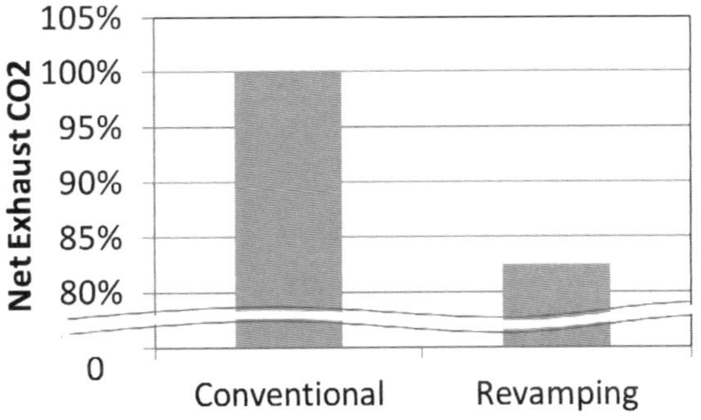

Fig.6 Net Exhaust CO_2

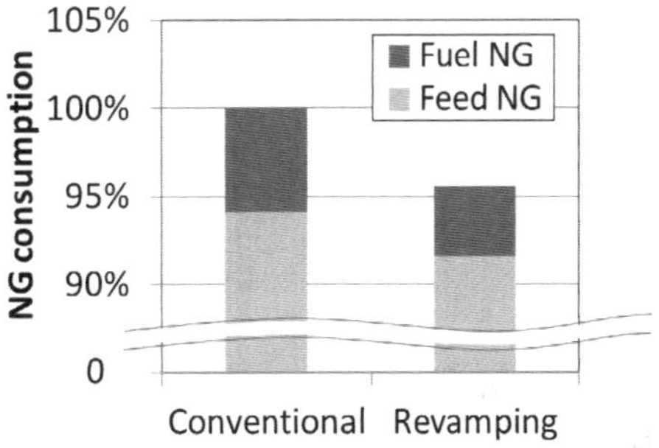

Fig.7 NG consumption

(c)Possibility of increased methanol production

As shown in Fig.8, in the revamped case reformer duty is 20% less than conventional case. Therefore, 25% increase in product can be expected in the revamped case at current reformer capacity (Fig.9).

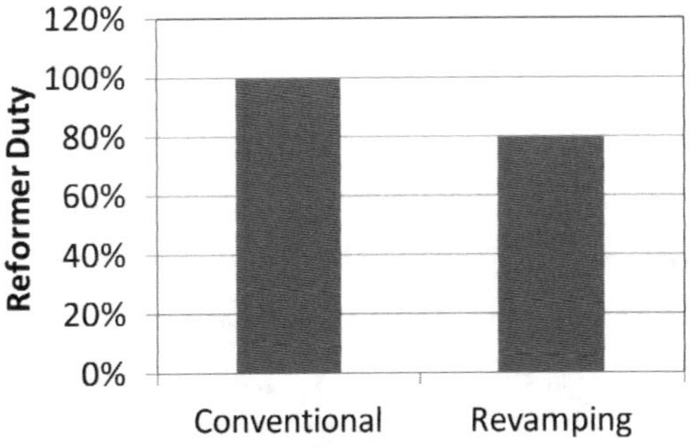

Fig.8 Reformer Duty

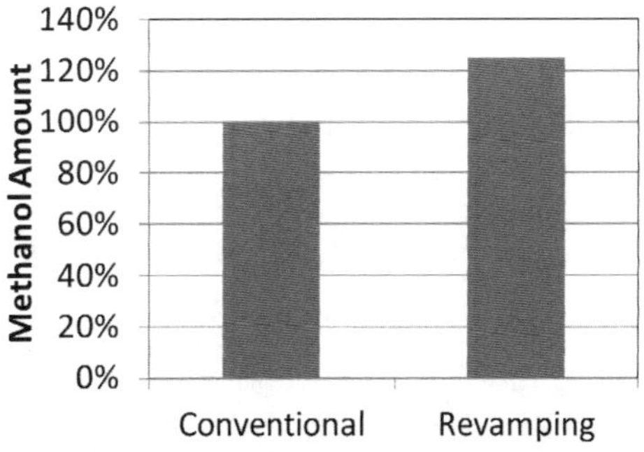

Fig.9 Product Methanol Amount (Same Reformer Duty)

(2) Case-2; CO_2 reforming for methanol and chemical production with one reformer (Revamping case)

(a) Study basis

From the study of case-1 (CO_2 reforming for methanol production plant), methanol product can be increased with CO_2 reforming. In case-2, this increase in production is used for another chemical synthesis.

The revamped flow is shown in Fig.10. After CO_2 reforming, some of the synthesis gas is fed to CO separation unit to produce CO for chemical synthesis. Separated CO_2 and H_2 are utilized in the reformer and methanol synthesis. Even though some synthesis gas is separated to produce CO, there still would be increase in methanol production. This increased methanol and separated CO are used as feedstock for chemical synthesis such as acetic acid or MMA.

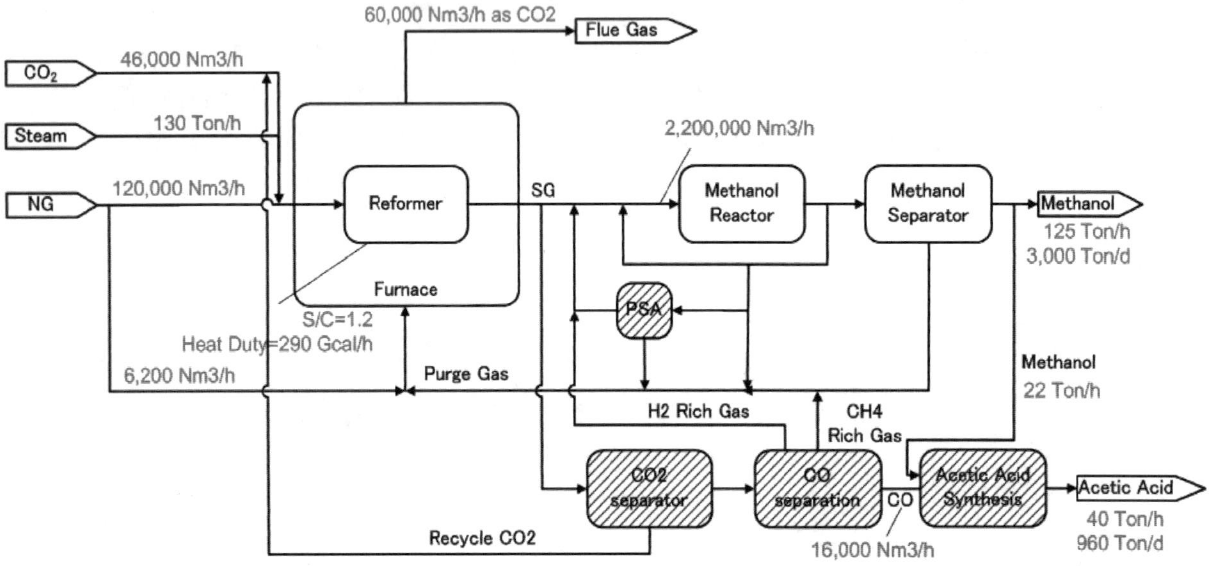

Fig.10 Revamping Case(Methanol and CO production with one reformer)

(b) Possibility of increased methanol production

From Fig.10, revamped case can produce 960 Ton/d acetic acid, with maintaining 3,000 Ton/d methanol production. Therefore, catalyst replacement and new construction of CO separation unit and chemical synthesis unit can produce new chemical products while maintaining current methanol production.

(3) Case-3; CO_2 reforming for CO production (Grass roots case)

(a) Study basis

Partial oxidation of NG is assumed as conventional case for CO production (Fig.11). With partial oxidation (POX), H_2/CO of synthesis gas is adjusted to 2.0. NG and O_2 which is separated from air are fed into POX reactor. After CO_2 removal, synthesis gas is fed into CO separation unit such as cryogenic unit

The block flow of CO_2 reforming case is shown in Fig.12. With CO_2 reforming, H_2/CO of synthesis gas is adjusted to 0.5, which means it can reduce energy required to produce excess hydrogen. After CO_2 is removed from synthesis gas, synthesis gas is fed into CO separation, such as cryogenic unit. Separated CO_2 is recycled to reformer, and off gas from CO separation unit is used as reformer fuel.

In this balance, the plant capacity is assumed as 16,000 Nm3/h CO which is the same amount as CO in Case-2.

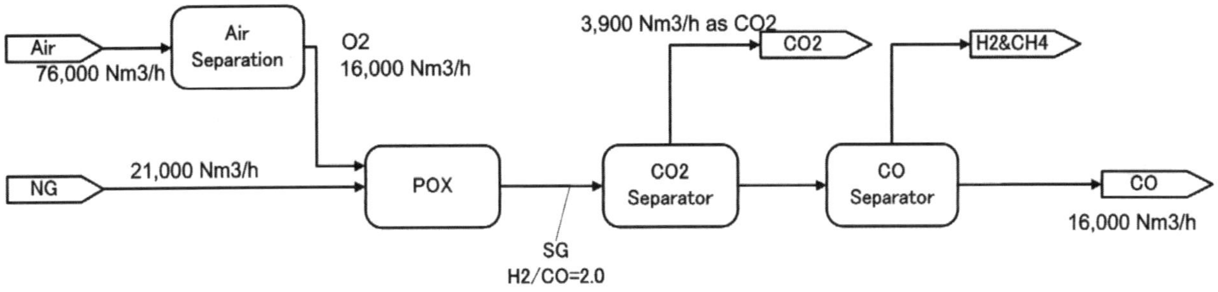

Fig.11 Conventional case (Partial oxidation of NG)

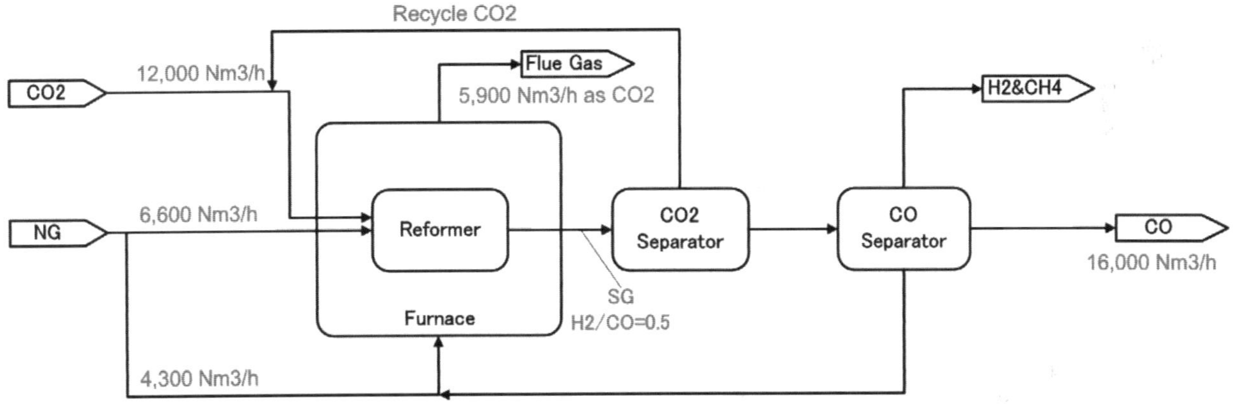

Fig.12 CO_2 reforming case

(b) CO_2 utilization and NG consumption

While CO2 is exhausted from CO_2 removal unit in conventional case, CO_2 is exhausted from reformer furnace in CO_2 reforming case. However, in CO_2 reforming case, because much CO_2 is utilized as feedstock for CO production, amount of import CO_2 is higher than exhaust CO_2 (This means zero emission because net exhaust CO_2 is a negative value (Fig.13)).

NG consumption in CO_2 reforming case is 48% lesser than conventional case by utilizing CO_2 as feedstock (Fig.14). The effect is much more than case-1, (CO_2 reforming for methanol production case). In general, lesser the H_2/CO of the synthesis gas, lesser the NG consumption in CO_2 reforming.

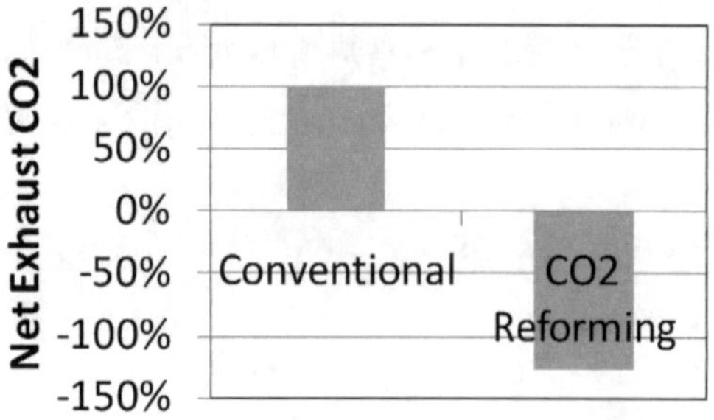

Fig.-13 Net Exhaust CO_2

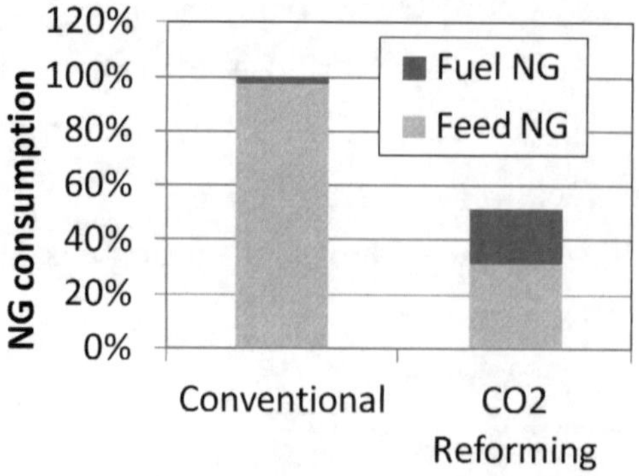

Fig.-14 NG consumption

4. Conclusion

Methanol product amount can be increased by 25% by revamping of the existing methanol plant with CO_2 reforming catalyst. Furthermore, with this increase and CO separation from synthesis gas, new chemical product can be produced while maintaining current methanol production.

In CO production case, CO_2 is mostly utilized as feedstock and 48% of NG can be saved compared to CO production by Partial Oxidation of NG.

5. References

1) J.R.Rostrup-Nielsen ,J-H. Ban Hansen, J.Catal. 144,38-49(1993)

2) T.Osaki, H.Masuda, T.Horiuchi, T.Mori, Catal.Lett., 34,59(1995)

3) K.Tomishige, K.Fujimoto, Catal. Surveys from Japan 2,3-15(1988)

(24d) Spectroscopic Studies of Gas-Liquid Reactions Using Microfluidics for the Study of CO_2 Conversion to High-Value Products

Jesse Greener[1], Rachelle Choueiri[1], Ethan Tumarkin[1], Milad Abolhasani[2], Axel Guenther[3]and Eugenia Kumacheva[1], (1)Department of Chemistry, University of Toronto, Toronto, ON, Canada, (2)Department of Mechanical and Industrial Engineering, University of Toronto, Toronto, ON, Canada, (3)Dept. of Mechanical and Ind. Engineering, Inst. of Biomaterials and Biomedical Engineering, University of Toronto, Toronto, ON, Canada

We present a new concept for spectroscopic studies of the kinetics of fast gas-liquid reactions involving CO_2. The strategy relies on the microfluidic generation of highly monodisperse gas bubbles in the liquid reaction medium. Characterization of the reaction system was achieved by monitoring time-dependent changes in (i) vibrational spectra of the liquid phase and (ii) changes to bubble dimensions. These measurements were achieved using*in situ*attenuated total reflection Fourier transform infrared (ATR-FTIR) spectroscopy and from the analysis of optical micrographs collected by a high speed camera, respectively. The method can be exploited to study important CO_2(g) reactions including: reactions with frustrated Lewis pairs, secondary amines, Grignard reactions and enzyme catalysis. This talk will demonstrate that the method enables rapid determination of reaction rate constants and the observation of transient species. The proposed strategy offers new possibilities in studies of rapid multiphase reactions in microfluidic reactors, and can also be combined with high-throughput optimization of reaction conditions for synthesis of high-value products from waste CO_2gas.

(24e) A Technoeconomic Analysis of Biodegradable Polymers Production From CO_2 and Waste Plastics

Michael Hartmann, Chemistry and Chemical Engineering Department, Fuels and Energy Development Section, Southwest Research Institute, San Antonio, TX, Maoqi Feng, Div. of Chemistry and Chemical Engineering, Southwest Research Institute, San Antonio, TX and Eloy Flores III, Chemical Engineering Department, Southwest Research Institute, San Antonio, TX

This paper evaluates the economic viability of manufacturing biodegradable polymers (mostly polycarbonates) from CO_2 and waste plastics. Waste plastics recycling are beneficial to environment and economy as well. Thermal processes have been used for feedstock recycling from waste plastics, such as polyethylene, polypropylene and polystyrene. Olefins are very useful feedstock in industry; however, it's expensive to produce olefins by using conventional hydrocarbons cracking technology. In our simulation, olefins are generated from pyrolysis of waste plastics with thermal energy supplied from a solar concentrator, thus no net CO_2 is produced in the pyrolysis. Biodegradable polymers are then produced from the polymerization of CO_2 and olefins directly in a catalytic reactor. This method provides an alternative method for CO_2 sequestration and utilization. A detailed economic analysis using VMGSim software is presented.

25a

Mixed Fluid Cascade, Experience and Outlook

Heinz C. Bauer, Linde Engineering, Pullach, Germany

Background

Starting in 1996 Linde AG and Statoil ASA developed a new process for natural gas liquefaction. The Mixed Fluid Cascade (MFC®) process was successfully commercialized in the Snøhvit plant with 4.3 mtpa LNG capacity, which is located in Hammerfest/Norway. After some teething problems, which will be discussed in some depth, the plant is running since early 2009 beyond its design capacity and has passed the performance test in all aspects successfully.

The main features of the Hammerfest LNG concept include

- MFC® process with three cascaded mixed refrigerant cycles
- Direct sea water cooling for all compressor coolers and the main refrigerant condenser
- All electric concept with power generation in highly efficient aero-derivative gas turbines
- Full sequestration of CO_2 contained in the plant feed gas
- High purity nitrogen (99.99 vol%) rejection from end-flash gas
- Waste heat recovery from gas turbines

In total the actual energy consumption of the liquefaction plant is as low as 243 kWh/t_{LNG}, which is unsurpassed by any other base load LNG plant worldwide. Together with the low carbon foot print this marks an impressive achievement with respect to overall plant efficiency.

Plant concept

Since the early 1980ies large natural gas reservoirs had been identified in the Tromsø patch, which is located in the Barents Sea north of Norway (see Fig. 1). The recoverable resources include 190 BCM (billion cubic meters) of natural gas and 34 million cubic meters of condensate. A discovery of oil would have been more welcome at that time as a production concept for natural gas in arctic regions was not readily available. It took almost 20 years[1] until the offshore

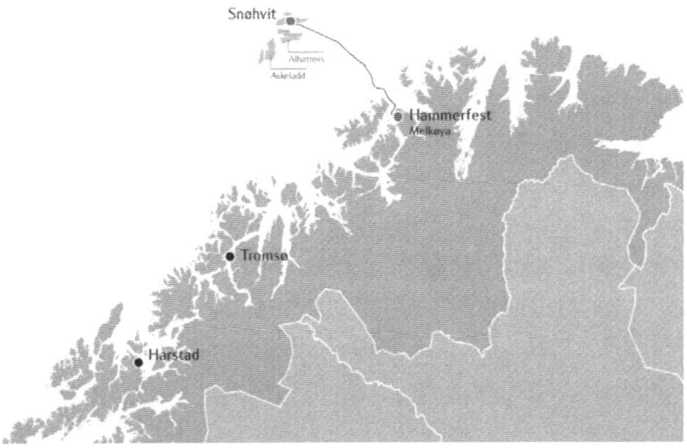

Fig. 1 Location of the Snøhvit field north of Norway
Courtesy Statoil

production technology matured sufficiently to face all challenges successfully.

As the Barents Sea is an ecologically very sensitive area, Snøhvit is the first major development on the Norwegian continental shelf without a fixed or floating unit. Instead, a subsea production system on the seabed feeds a land-based plant on the north-west coast of Melkøya, at the entrance to the shipping channel into Hammerfest via a 26" ID, 143 km multiphase pipeline. In addition two chemical lines, an umbilical and a separate pipeline for transporting carbon dioxide were laid in 2005. Both the subsea production system located on the field and pipeline transport are monitored and controlled from a control room at Melkøya via signals transmitted along fiber-optic cables and high-voltage electrical and hydraulic power lines.

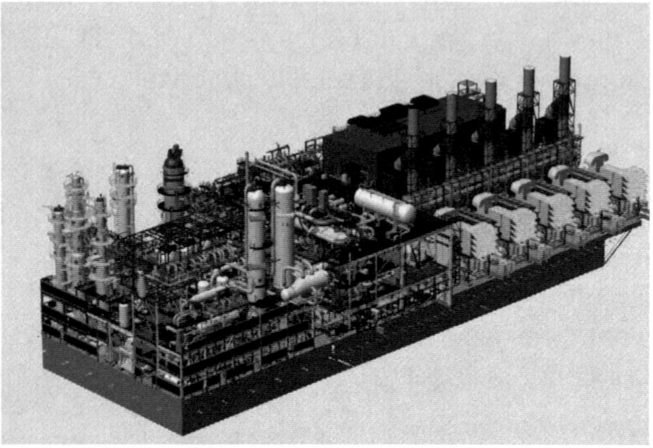

Fig. 2 Process barge

A stick-built concept for the process plant was not applicable as the available plot on Melkøya Island was very limited and as extensive construction work at site would have been very inefficient. Therefore, a modular design with the maximum possible prefabrication was selected. A process barge (see Fig. 2) with a 154x54 m footprint and a shipping weight of 35,000 tons accommodates pre-treatment, all large compressors and power generation with 5 LM6000 aero-derivative gas turbines.

When Statoil started to select the liquefaction technology, all relevant licensors had been invited to a detailed design competition. Eventually, the Mixed Fluid Cascade (MFC®) process[2] (see Fig. 3), a co-development by Statoil and Linde within a technology alliance, was selected for several reasons. (1) The use of three independent mixed refrigerant cycles provided the best conditions for excellent heat integration with a fair power distribution to the cycle compressors. In arctic conditions the temperature range of traditional propane pre-cooling would have been very much restricted by a sea water temperature below 10 degree C all over the year. (2) At that time the desired nameplate capacity of 4.3 mtpa LNG could not be reached by proven concepts, even if modern Mega Trains reach almost twice the capacity today.

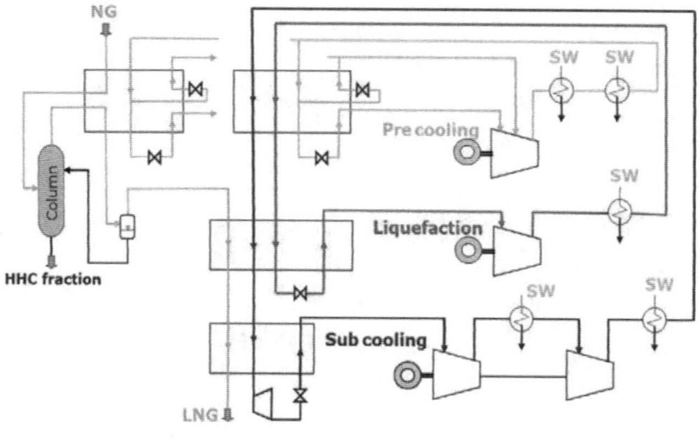

Fig. 3 Mixed Fluid Cascade (MFC®) process Courtesy Statoil

The Norwegian government was very clear with its request to authorize only the best available technique at a location like Melkøya. Hence, further actions had to be taken to minimize the environmental footprint. This includes

- direct sea water cooling for all compressor inter- and aftercoolers and the main refrigerant condenser with titanium tubes and tube sheets,
- all electric concept with power generation in highly efficient LM6000 aero-derivative gas turbines, connection to the national grid as back-up, waste heat recovery from gas turbines
- reinjection and sequestration of 700,000 tons per year of CO_2 from feed gas,
- high purity nitrogen (99.99 vol%) rejection from end-flash gas to the atmosphere,
- minimized flaring.

The result of this effort convinced the Norwegian Ministry of Petroleum and Energy and the Plan for Development and Operation (PDO) was sanctioned in early 2002. It took until September 2007 to complete the construction of the entire plant and to be ready for start-up.

Maturing phase

Shortly after the initial start-up of the plant, a major challenge surfaced. All sea water cooled heat exchangers had been designed in a special way as helix type exchangers[3] in order to save plot and weight on the process barge. However, the actual shell side pressure drop of gas coolers had been significantly beyond the guaranteed figures provided by the licensor[4] and a predicted helical flow did not develop in the main condenser of the pre-cooling refrigerant. Rather, differential condensation[5] of the mixed refrigerant components had to be observed. These shortcomings could be compensated only by an increased discharge pressure of the refrigerant cycle compressors and a corresponding increase of the energy consumption. Even more concerning was the fact that several of the HELIXCHANGER®s failed after a few months due to tube leakages. The failure mode was wear of the tubes against baffle plates caused by flow induced vibration. The resulting sea water ingress into the refrigeration cycles caused several times an extended shut-down of the plant. After provisional plugging of the leaking tubes all gas coolers have been replaced meanwhile by heat exchangers with a different design. Only the replacement main condenser (see Fig. 4) is still a HELIXCHANGER® in a 'robustified' version.

Fig. 4 Replacement of the main condenser Courtesy Statoil

The pre-cooling cycle uses vacuum brazed aluminum plate-fin heat exchangers for heat transfer. Temperature probes on the individual cores of the parallel exchangers showed significant temperature gradients perpendicular to the flow direction, which should not occur. A detailed root-cause analysis revealed uneven blockages of passages caused by solid particles and ambiguous fluid distribution due to the complex hydraulics[6] of mixtures in a two-phase regime. In 2009, all plate-fin heat exchangers have been exchanged by clean ones with an optimized hydraulic design.

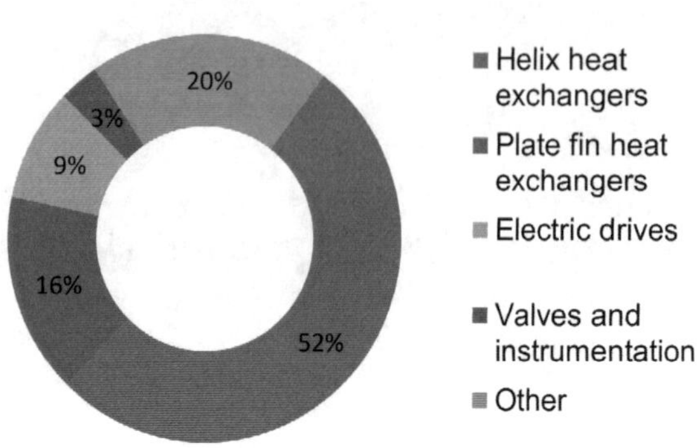

Fig. 5 Distribution of downtime causes since start-up
Courtesy Statoil

The all electric concept had been selected to improve the overall availability of the plant. In essence, this goal has been reached, but some lessons had to be learnt. For example, some details of the exciter machine for the motor of the end-flash gas compressor had to be optimized to tolerate the stress caused by high centrifugal forces and heavy thermal load. Further, the quantitative interaction between mechanical and electrical resonances (see Campbell diagrams[7]) in a compressor train with a variable speed drive (VSD) had to be evaluated. In the context of a short circuit damage part of the large (65 MW) drives had to be replaced by optimized components.

Further causes for downtime (see Fig. 5[8]) include the two coil wound heat exchangers for liquefaction and sub-cooling, which experienced a minor number of leaking tubes. In essence, all challenges of an arctic location with many pioneering concepts have been accepted and have been surmounted as per today.

In addition to equipment related improvements also the operation of the plant became more and more sophisticated. For example, the CO_2 emission from warm restart flaring could be reduced by about a factor 30, as cool-down is now based on the use of vaporized LNG instead of the use of feed gas.

Performance test

About four years after the initial start-up the plant was ready for the official performance test. The long interval between these dates was used to overcome all hurdles, which might have jeopardized the success of the project.

The main results from the plant performance test are

- The guaranteed plant capacity is exceeded by 4%. This translates into an absolute production quantity of 13,200 t/d of LNG (rising in tank).

- All guaranteed product qualities are fulfilled (including the CO_2 routed to offshore and the N_2 Tail Gas of the NRU released to atmosphere).
- The specific refrigeration power consumption is 5.4% lower than guaranteed value. The achieved level of 243 kWh/t_{LNG} is the lowest level for all LNG plants worldwide. Based on the thermal efficiency of the LM6000 gas turbines of 39% (measured at site) only about 5.5 wt% of the feed gas are used to generate the required power for the whole LNG plant.
- The measured total power consumption of the plant is 9.9% less than the guaranteed total power consumption.
- The calculated fuel gas consumption is less than the guaranteed volume.
- All CO_2 separated from the feed stock is routed to offshore. This CCS (carbon capture and sequestration) effort reduces the total CO_2 emission of the plant by about 45% to only 110 t/h CO_2.
- All N_2 separated from the feed stock is released to atmosphere as the remaining methane content is well below the specified 100 mol-ppm.
- The guaranteed levels of noise are deemed to be met.
- The guaranteed NO_x levels in the gas turbine exhaust gas are met.
- Furthermore, there wasn't any flaring and the operation of the plant was quite smooth and stable.

Outlook

Despite of the successful implementation of the MFC process in the arctic region other projects in a more moderate climate require different solutions. With this goal in mind a 2nd generation[9] of the MFC® concept with optional propane pre-cooling (MFC®3) has been developed, which will be described underneath (see Fig. 6). Traditionally, all heavier hydrocarbons (C_{5+}) and to some extent natural gas liquids (C_2 to C_4) are removed from the feed gas in a scrub column in order to avoid freezing of heavies, especially aromatic material, to meet residue specifications (C_4 content) of the LNG, and to provide make-up components (C_2 and C_3) for the refrigeration cycles.

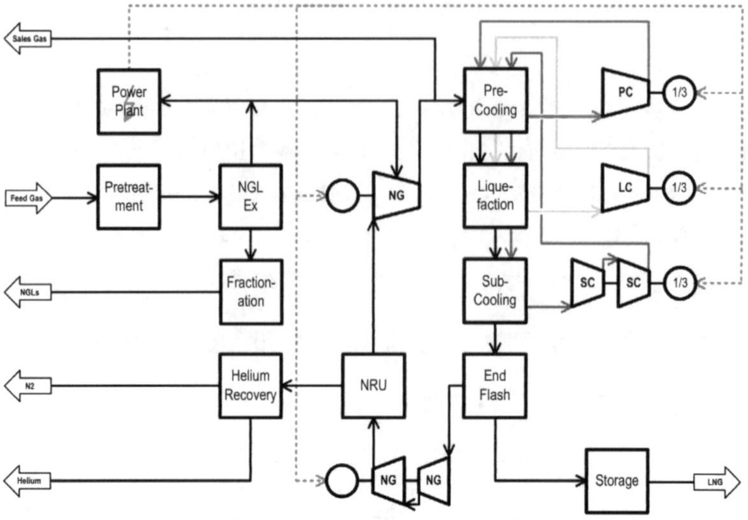

Fig. 6 Overall concept for a 2nd generation MFC® plant

If, however, deep LPG recovery or even C_{2+} recovery is economically recommendable, conventional scrub column configurations hit their limits. Close to the critical point high NGL recovery rates involve large quantities of co-recovered liquid methane in the scrub column bottoms due to the low relative volatility (k-value) between methane and ethane and/or propane. A more efficient and cost effective concept is a dedicated NGL recovery step between pre-treatment and liquefaction. It involves a cryogenic column (demethanizer or deethanizer), whose operating pressure is limited by the critical pressure of the bottoms to a range between 30 and 35 bar. Lean natural gas downstream NGL recovery is not only a perfect feedstock for an LNG plant, but can be sent also to a gas pipeline as sales gas. This requires recompression of the lean sales gas to typically about 100 bar. A common header for lean natural gas feeding both the pipeline grid and the LNG plant provides an attractive means to balance demand on either side.

Linde's base load LNG process is characterized by three independent refrigeration cycles. The intermediate (liquefaction) cycle and the coldest (sub-cooling) cycle use mixed refrigerants in any case, while the type of refrigerant in the warmest (pre-cooling) cycle depends mostly on the ambient temperature. The Mixed Fluid Cascade (MFC®) process is suffixed with a figure 3 in case propane is used for pre-cooling. Fig. 7 shows a simplified process sketch for the MFC®3 process with propane pre-cooling. As discussed above NGL extraction is not implemented via a scrub column,

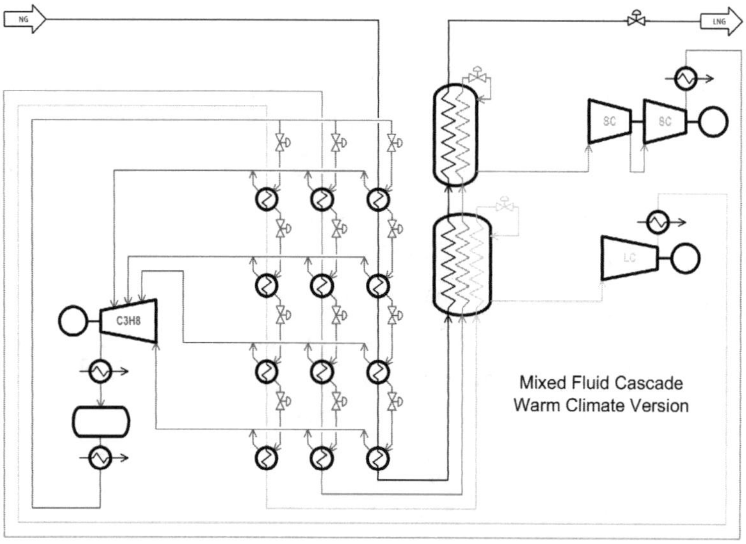

Fig. 7 Mixed Fluid Cascade (MFC®3) process for warm climate

but rather with an independent process step, which is not shown in Fig. 7. As a result of the upstream NGL extraction the lean and compressed natural gas does not cross a two-phase region. Now, there is no specific need to meet a certain process temperature by any of the three refrigeration cycles. Therefore, the load between the three refrigeration cycles can be balanced perfectly. That means the shaft power of all three compressor trains is identical. With such a configuration, which is unique amongst the present Base Load LNG technologies, the largest LNG capacities can be achieved with a given set of main compressor drives.

This load balanced concept is compatible with the use of propane pre-cooling only as long as the ambient temperature is not too low. In a warm Middle East climate a load balanced MFC®3 design results in a lowest operating temperature of the pre-cooling cycle with about -30 to -35 deg C. A lower limit for an economic propane pre-cooling temperature is -35 deg C. Below this limit propane compressor suction line sizes would grow significantly and the risk of air ingress into the closed cycle increases, when the propane system is operated too close to the atmospheric pressure. As soon as the relevant ambient temperature is consistently lower than above mentioned, either the propane pre-cooling compressor no longer can contribute with one third of the overall shaft power to the overall duty or the choice of the pre-cooling refrigerant needs to be reconsidered.

In a conventional plant design the end flash gas from the tail end of the LNG production frequently changes in quality and quantity, when operation requires process adjustments to external influences like plant load, ambient temperature, loading/holding mode, and so on. The result is modifications in the Wobbe index, which cannot be tolerated by highly efficient and clean burning (LoNox) gas turbines. Therefore, the end flash gas is not the preferred fuel source for modern gas turbines in any configuration. An alternative concept uses lean natural gas as gas turbine fuel and reprocesses the end flash gas in a nitrogen rejection unit (NRU). The NRU splits the end flash gas into clean nitrogen with less than 1 vol% methane, which can be sent to the atmosphere without further processing. The remaining hydrocarbon fraction will be re-liquefied either jointly with the feed stream to the LNG plant or separately within the NRU itself. The described method provides good quality fuel to the gas turbines at any required flow rate and maintains low nitrogen levels in the LNG even at high nitrogen concentrations in the natural gas. This enhanced operational flexibility does not impact the plant efficiency, but rather increases its availability.

As soon as the natural gas contains at least 500 vol ppm helium a commercial production can be considered. The best source for helium extraction in an LNG plant is the nitrogen fraction of the NRU. Technology for helium purification and liquefaction is widely available and has been implemented several times successfully.

Conclusions

Linde has successfully developed and implemented a new contender in the arena of world scale Base Load LNG processes. The new concept results in outstandingly low energy consumption, which is just 243 kWh/t_{LNG} in case of the Hammerfest LNG plant. This excellent efficiency can be achieved only with the implementation of three independent mixed refrigerant cycles.

The integration of carbon capture and sequestration (CCS) features reduces the overall CO_2 emission to about 110 t/h at 550 t/h LNG production, which is an exceptionally low value. Clever concepts for a warm restart with vaporized LNG further contribute to a very low carbon footprint. Nitrogen from the feed gas, which cannot be exported as LNG component, is cleaned up to less than 100 ppm hydrocarbons and can be vented without environmental concerns.

More recent concepts with propane pre-cooling (MFC®3) and perfectly balanced shaft power of all three refrigeration cycles have been applied to several projects and are in different phases of execution.

References

[1] Alf R. Jacobsen, Snøhvit, The history of oil and gas in the Barents Sea, Statoil publication, 2010

[2] e.g. US patent 6,253,574

[3] US Patent 6,827,138

[4] http://www.cbi.com/technologies-services/helixchanger-heat-exchanger/

[5] Jostein Pettersen, Experience with water cooled mixed refrigerant condensation, Proceedings of the 16th International Conference & Exhibition on Liquefied Natural Gas, Poster PO1-3, 2010, Oran, Algeria

[6] G. Skaugen, G. A. Gjøvåg, P. Nekså, P. E. Wahl, Use of sophisticated heat exchanger simulation models for investigation of possible design and operational pitfalls in LNG processes, Journal of Natural Gas Science and Engineering, Volume 2, Issue 5, November 2010, pages 235-243

[7] http://en.wikipedia.org/wiki/Campbell_diagram

[8] Jostein Pettersen, Øivind Nilsen, Snøhvit, A Successful Frontier LNG Project, paper presented at the 6th Annual LNG Tech Global Summit, 2011, Rotterdam, Netherlands

[9] Heinz Bauer, Modular Design of a Base Load LNG Plant, Proceedings of the International Gas Union Research Conference, Poster P4-11, 2011, Seoul, South Korea

AIChE 2012 Spring Conference
Control Panel - 12th Topical Conference on Gas Utilization (T6A), T6A00 LNG simulations and controls.

Fuel Gas System Design – Key to Reliable and Efficient LNG Plant Operation

Pei-Han (Jason) Yong, P.E.
Process Engineer
Bechtel Oil, Gas & Chemicals, Inc.

Christine Teh
Process Engineering Specialist
Bechtel Oil, Gas & Chemicals, Inc.

Bindupriya Bandari
Process Engineering Specialist
Bechtel Oil, Gas & Chemicals, Inc.

Robert A. Zinsmeister, P.E.
Control Systems Senior Engineer
Bechtel Oil, Gas & Chemicals, Inc.

Abstract

A well designed fuel gas system in an LNG liquefaction plant ensures a continuous, reliable, and properly specified fuel gas supply to all users while optimizing the thermal efficiency of the plant. Fuel gas system design development starts with identifying the fuel gas users, their needs, the applicable design standards or constraints, and Owner preferences. Fuel gas system design is also influenced by various plant operating modes such as plant start-up, normal operation, turn down, and upset conditions.

This paper will identify typical high pressure and low pressure fuel gas users in an LNG plant. Generally, the users with the most stringent fuel gas specifications are the gas turbines with their complex burner design. Therefore, the high pressure fuel gas is usually treated to meet the original equipment manufacturer (OEM) requirements. The fuel gas specification for several commercially available gas turbines used in LNG industry and its impact on fuel gas system design will be analyzed in this paper. The sources of fuel gas generally available within an LNG plant and their impact on overall thermal efficiency will be also be discussed. The process parameters used to tune the performance of the unit including supply pressure, temperature, composition, specific gravity, wobbe index and fuel gas heating value will be evaluated and described in detail.

A few case studies to demonstrate the robustness of the fuel gas system, advanced controls and analytical measurements required to facilitate various plant operating modes will also be addressed.

1.0 Introduction

An LNG liquefaction plant is designed for stable baseload production under different operating conditions. The continuous availability of fuel gas is very important for operating an LNG plant since it is used as fuel for the gas turbine mechanical drivers to drive the refrigeration compressors, gas turbine generators to generate electrical power, and fired heater to heat up the heating medium. The fuel gas system also supplies other miscellaneous users, such as flare pilots and purges.

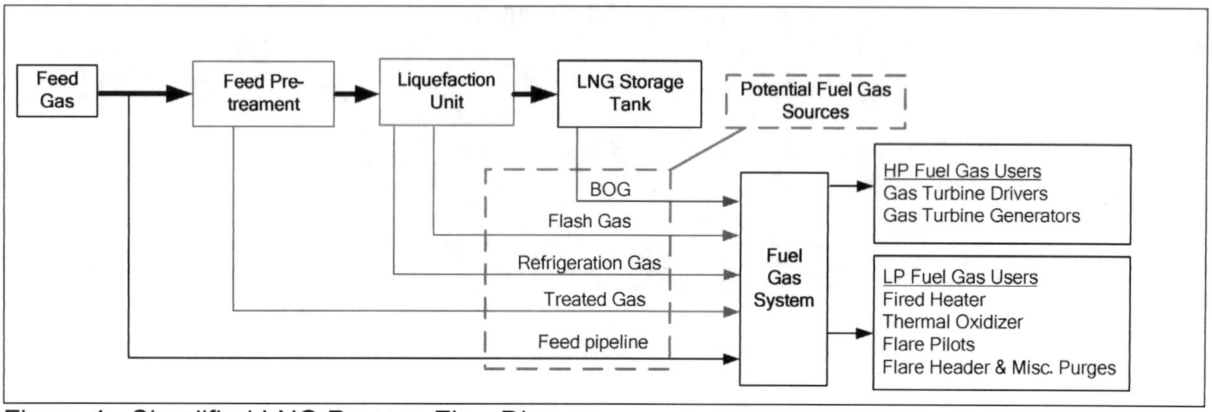

Figure 1. Simplified LNG Process Flow Diagram

Fuel gas for a stand-alone LNG liquefaction plant is supplied internally. There are several fuel gas sources within an LNG facility as shown in Figure 1. The typical fuel gas sources are from the incoming feed pipeline, feed pre-treatment, liquefaction unit, and BOG from LNG storage tank. The amount of heavy hydrocarbons sent to the fuel gas system are typically limited by the gas turbine combustor's specification, such that they need to be blended with the normal fuel gas supply to bring the heating value within the allowable range and allowable rate of change.

Fuel gas composition varies based on the various feed sources. The typical feed gas stream composition is approximately 90 mol% methane and the rest are inerts and heavy components.

2.0 Fuel Gas System Design Development

In general, the fuel gas system in an LNG plant is designed for at least two pressure levels: high pressure (HP) and low pressure (LP). The main users for HP fuel gas are the gas turbine mechanical drivers and gas turbine generators. Whereas, the main users for the LP fuel gas vary depending on the design of heating medium system and CO_2 / H_2S emission requirements. If the plant does not have a waste heat recovery (WHR) system to recover heat energy from turbine exhaust, the main user for the LP fuel gas is the fired heater providing heat to the heating medium. Another potential user for LP fuel gas is the flare system, which is governed by the emission requirements for CO_2 and H_2S. If the plant design includes a CO_2 sequestration unit, there will be no LP fuel gas demand from CO_2 and H_2S thermal oxidation.

The typical breakdown of HP and LP fuel gas users during normal operation are shown in Table 1. This table shows the typical breakdown for HP and LP fuel gas users in an LNG plant without a waste heat recovery unit. The Refrigeration Gas Turbines and Gas Turbine

Generators are the main users whereas only a small fraction of fuel gas is sent to the LP fuel gas system. During normal operation, the only LP users are the flare pilots and fired heater. LP fuel demand to the flare system can increase slightly during flaring events.

Table 1. Fuel Gas Breakdown in Absence / Presence of WHR

	Fuel Gas Breakdown	
	Without WHR	With WHR
LP Fuel	5%	2%
HP Fuel	95%	98%

Table 1 also shows the typical fuel gas breakdown in an LNG plant with a waste heat recovery unit. The LP fuel gas demand decreases slightly in terms of overall fuel gas demand due to the absence of a fired heater during normal operation. This translates to fuel gas savings and more LNG production. Technical and economical studies are performed for the selection of the waste heat recovery. The use of a fired heater is beneficial during startup when the gas turbines are unavailable to supply waste heat. Once the refrigeration gas turbine reaches the stable operating mode, the fired heater can be turned down gradually while ramping up the waste heat recovery system. The heating medium duty ultimately depends on the inlet gas components such as carbon dioxide and heavy hydrocarbons and also plant configuration. A higher carbon dioxide concentration will require a higher duty in the Amine Regeneration Unit to strip off the dissolved carbon dioxide from the solvent. The presence of heavier hydrocarbons will require a higher duty in the NGL Fractionation Units. These conditions may justify the use of the waste heat recovery system due to additional fuel cost savings and reduction in CO_2 emission.

Since the primary HP fuel gas users are the gas turbines for mechanical and power generation applications, there are specific compositional limits specified by the gas turbine manufacturers. Some of those limitations are supply pressure, temperature, MWI, and contaminant levels.

- The typical HP fuel gas supply pressure is between 25 barg to 45 barg.
- The HP fuel gas stream is superheated to at least 28°C (50°F) above the dew point temperature at the gas turbine fuel delivery pressure.
- A strict MWI range can limit the range of fuel gas composition and temperature to the combustors.
- Any contaminants (i.e. particulate, sulfur, liquid droplets) should be removed before the fuel gas is sent to the combustor.

These specifications, which generally vary from gas turbine types and manufacturers, will be discussed further in the next section.

Conversely, LP fuel gas compositional requirements are not as stringent as that for the HP fuel gas. Being one of the main LP fuel gas users, the flare unit is supplied with stable, dry and, continuous fuel gas to ensure smooth operation. The main source of LP fuel gas is from HP let down, when the pressure is let down to approximately 3.5 barg (50 psig).

Gas turbines with advanced combustor design have been developed to ensure that emissions meet the strict criteria for pollutants such as NO_x and CO. Specific emission requirements depend on the prevailing application and the local environmental permitting requirements and is checked on a project by project basis.

In order to lower NOx formation, DLN and DLE type combustors must operate at low flame temperature, shorter residence time in the flame zone, and lean fuel conditions as shown in Figure 2. The fuel is premixed with air to avoid forming hot spots in the flame during combustion [6].

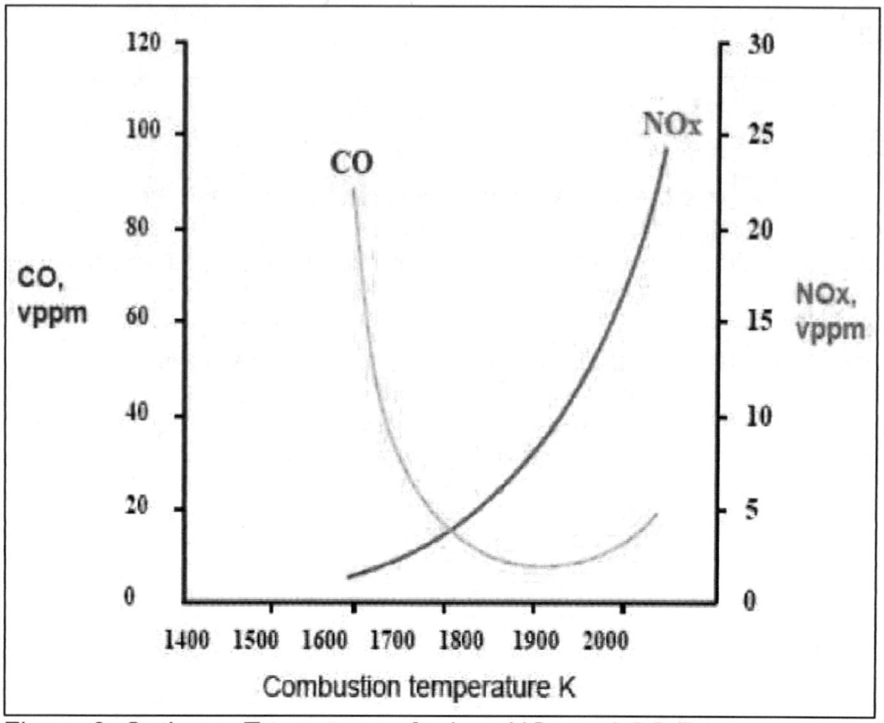

Figure 2. Optimum Temperature for Low NOx and CO Emission [6].

Since DLN and DLE combustor configurations are more sensitive to fuel gas composition variations, the following fuel gas specifications need to be met for uninterrupted operation:

2.1 <u>Temperature</u>

Fuel gas temperature at the fuel gas skid should be maintained at least 28°C (50°F) plus the higher of moisture or hydrocarbon dew point temperature at the gas turbine fuel delivery pressure to compensate for temperature drop when the pressure is reduced between the fuel gas filtration skid and the combustion chamber. Both hydrocarbon and moisture superheat requirements should be met to prevent any liquid droplet formation that can affect gas turbine performance and/or cause damage to combustor hardware. Superheat requirements should be analyzed on a case-by-case basis. For example: when water saturated regeneration fuel gas from the dehydration bed is used as the high pressure fuel gas source, moisture superheat will govern the superheat requirement for the fuel gas stream. Hydrocarbon superheat can become the governing case when there are heavy hydrocarbons present in the fuel gas stream, such as C6+. It is important to obtain a detailed fuel analysis and not just calculate the dew point based on "C4+" type analyzes. The dew point can be very sensitive to the heavier hydrocarbons if they are present.

2.2 Modified Wobbe Index (MWI)

MWI is a function of fuel LHV, temperature, and fuel gas molecular weight. MWI is defined as below:

$$MWI = \frac{LHV}{\sqrt{SG_{gas} x T_{gas}}}$$

(Equation 1)

where:

MWI	= Modified Wobbe Index (Btu/scf-°$R^{1/2}$)
LHV	= Lower Heating Value of fuel gas (Btu/scf)
SGgas	= specific gravity of fuel gas with respect to air
Tgas	= Fuel gas temperature (°R)

The supplied fuel gas MWI range must meet the gas turbines OEMs specifications. If two different fuel gases have the same MWI at the same pressure, the energy output is similar. The lower range of Modified Wobbe Index limits the amount of inerts in the fuel gas. If too many inerts are present in the fuel gas, this can affect the combustion and the energy output. The upper range of MWI limits the amount of heavy hydrocarbons (C2+) in the fuel gas. Fuel gas with heavier hydrocarbons produces more NOx during combustion. In addition, the presence of heavy hydrocarbons can lower the dew point of the fuel gas stream, which is undesirable considering liquid droplets are not allowed in the gas turbines.

The required MWI range for DLN combustors is between 40 and 54 Btu/scf-°$R^{1/2}$ and while the MWI range for DLE and SAC combustors ranges from 40 to 60 Btu/scf-°$R^{1/2}$ [4]. In addition to meeting the allowable limits, the rate of change in MWI should also be discussed with the gas turbines manufacturer. As shown in GE presentation slides, presented at the CORE Symposium, Houston, May 2011, the allowable MWI rate of change for GE Heavy Duty gas turbine is 1% per second [3].

2.3 Pressure

HP fuel gas supply pressure requirements are governed by the gas turbine burner, site minimum ambient temperature, and fuel gas composition. The typical range of HP fuel gas supply pressure is between 25 barg to 45 barg. The fuel gas supply pressure for Aeroderivatives gas turbines is generally higher than Heavy Duty gas turbines.

2.4 Contaminants

Fuel gas to the gas turbine should not contain any significant concentration of liquid droplets or solids. The presence of contaminants that include but not limited to particulates, liquids, and sulfur can damage the combustor and gas turbine hardware. Normally, there is a fuel gas filter located downstream of the fuel gas KO drum, to remove the solid particulates up to 5 microns in size. Stainless steel piping between the fuel gas filter and inlet flange to gas turbines can avoid corrosion products from entering the combustors [6]. Liquid contaminants can cause performance related issues such as firing temperature excursions, re-ignition and flashback, and damages to the hardware [4]. Liquid contaminants can be avoided by meeting the superheat requirements of the fuel gas. The presence of sulfur in the fuel gas stream can cause severe corrosion due to sulfuric acid condensation when used in combination with an HRSG unit [4]. Therefore, the allowable sulfur concentration should be discussed with the gas turbine vendor. The presence of trace amounts of diolefins can cause fouling and should be removed from the fuel gas [4]. H_2S in the presence of oxygen can form elemental sulfur. Elemental sulfur deposition downstream of the control valve can potentially plug the valve and lead to poor plant controllability and possible plant shutdown.

Feed to the liquefaction unit can contain a high mol% of nitrogen. The fuel gas system typically serves as a purge for nitrogen that cannot be condensed in the liquefaction unit. The maximum amount of nitrogen to the fuel gas system is restricted by the gas turbine combustor design. Nitrogen Rejection Units may be required to remove the additional nitrogen that cannot be sent to the fuel gas system. Tables 2 and 3 summarize the GE fuel gas specifications for DLN and DLE combustors.

Table 2. Summary of Fuel Gas Specification of GE DLN Combustor [4]

		Minimum	Maximum
Pressure		(Note 1)	
Temperature		(Note 2)	
LHV	Btu/scf	775	1160
MWI	Btu/(scf . $R^{1/2}$)	40	54
Variation of MWI		- 5%	+ 5%
Constituent Limits, %mole			
Methane		85	100
Ethane		0	15
Propane		0	15
Butane + C_4+		0	5
Hydrogen		0	Trace
Carbon Monoxide		0	Trace
Oxygen		0	Trace
Total Inerts (N_2 + CO_2 + Ar)		0	15

Notes:

1. *Fuel gas supply pressure varies based on the gas turbine model and combustor configuration.*
2. *Fuel gas temperature should meet both hydrocarbon and moisture superheat requirement. Whereas, maximum allowable fuel gas temperature varies with fuel gas pressure and is also limited by metallurgy.*

Table 3. Summary of Fuel Gas Specification of GE DLE Combustor [5]

		Minimum	Maximum
Pressure		(Note 1)	
Temperature		(Note 2)	
LHV	Btu/lb	18000	
MWI	Btu/(scf . $R^{1/2}$)	40	60
Constituent Limits, % mole			
Methane		50	100
Ethane (C2+ paraffins)		0	(Note 3)
Hydrogen		0	5
Diolefins*		0	None allowed
Other Combustibles**		0	Consult GE

* Diolefins - butadiene, propadiene
** Other Combustible - CO, olefin hydrocarbons, etc.

Notes:
1. *Fuel gas supply pressure varies based on the gas turbine model and combustor configuration.*
2. *Fuel gas temperature should meet both hydrocarbon and moisture superheat requirement. Whereas, maximum allowable fuel gas temperature varies with fuel gas pressure and is also limited by metallurgy.*
3. *Maximum allowable C_2+ paraffins for DLE combustor varies based on the type of gas turbine.*

In summary, there are several parameters that are controlled to maintain a stable supply of fuel gas to the high pressure fuel gas users. These parameters are:
1. Fuel gas pressure at the pre-mixing chamber to ensure smooth combustion process.
2. Fuel gas temperature entering the combustor to avoid any liquid droplets in the combustion chamber.
3. Fuel gas system should be designed to ensure that the fuel gas MWI and its rate of change are within the allowable range.

There are several ways to control each parameter in ensuring a continuous and reliable fuel gas supply to the gas turbines. One method is to install a robust control scheme that combines temperature and composition analysis to mitigate the rate of change of MWI. Another method is to install a fuel gas mixing drum, which will provide smooth transition of MWI due to sudden changes in the fuel gas source (i.e. fuel gas composition). The next section discusses a few design features and control schemes that can be used to handle various operating scenarios.

3.0 Fuel Gas System Design Features for Various Operating Scenarios

3.1 Temperature control to meet fuel gas requirements

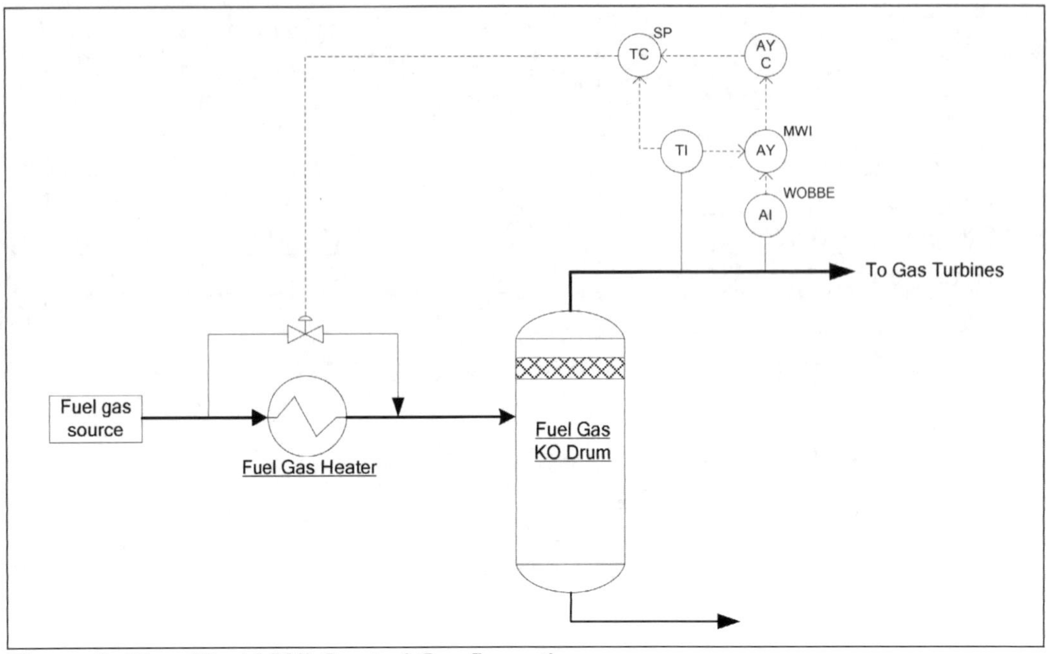

Figure 3. Simplified MWI Control Configuration

Figure 3 shows a simplified MWI control configuration. Fuel gas flows through a fuel gas heater, fuel gas KO drum, and fuel gas filter before it reaches gas turbine skids. Fuel gas temperature is adjusted to minimize the rate of change in MWI and also to ensure that the fuel gas is well above its dew point. A Wobbe analyzer provides the set point to the temperature controller. Any liquid droplets in the fuel gas should be knocked out at the upstream KO drum and there should be no compositional or MWI change between Fuel Gas KO Drum and inlet to the gas turbine skids.

3.1.1. Loss of fuel gas heater

The normal fuel gas source to the gas turbine from the liquefaction unit is dry and at least 28°C superheated. Therefore, the fuel gas heater duty is used primarily to ensure stable temperature to the users and the heating demand can be expected to be minimal. If the heating medium supply valve fails closed, the fuel gas temperature will be reduced. In the absence of fuel gas compositional changes, the MWI usually stays within the allowable limits and significant impact to the gas turbine performance is not anticipated.

If the alternate fuel gas source is instead used and gas temperature drops below its dew point, the fuel gas sources will be shut down to protect downstream gas turbine skids. One of the mitigation measures is to install a standby parallel electric heater to ensure heating duty is always available during plant operation.

3.1.2. Temperature spike from heating medium supply problem

If heat input to the fuel gas heater is accidentally increased, the temperature will rise and the bypass valve on the fuel gas heater will open to maintain a fixed temperature. If the temperature continues to rise, it will raise an alarm on the DCS panel and eventually shut down the fuel gas source to protect the gas turbines.

3.2 Equipment configurations to meet fuel gas requirements

When the fuel gas supply is switched to an alternate source, the gas turbine will most likely experience a MWI change due to compositional variation. The following mitigations is used to meet the rate of change MWI requirements and also to ensure that the MWI is still within the allowable limits:

i) If the compositional variation is significant and falls outside the MWI allowable range, the temperature can be adjusted to meet the target MWI using control configuration shown in Figure 3. A fuel gas mixing drum may be used as an alternate mitigation to dampen the step change of MWI.

ii)If the compositional variation is insignificant provided the superheat requirement is met, installing sufficient piping volume between Fuel Gas KO Drum and Gas Turbine skids can enhance fuel gas mixing and dampen the rate of change of MWI. The control configuration is simpler, in which the Wobbe analyzer is not connected to the temperature controller for this mitigation and is instead sent to the gas turbine controller.

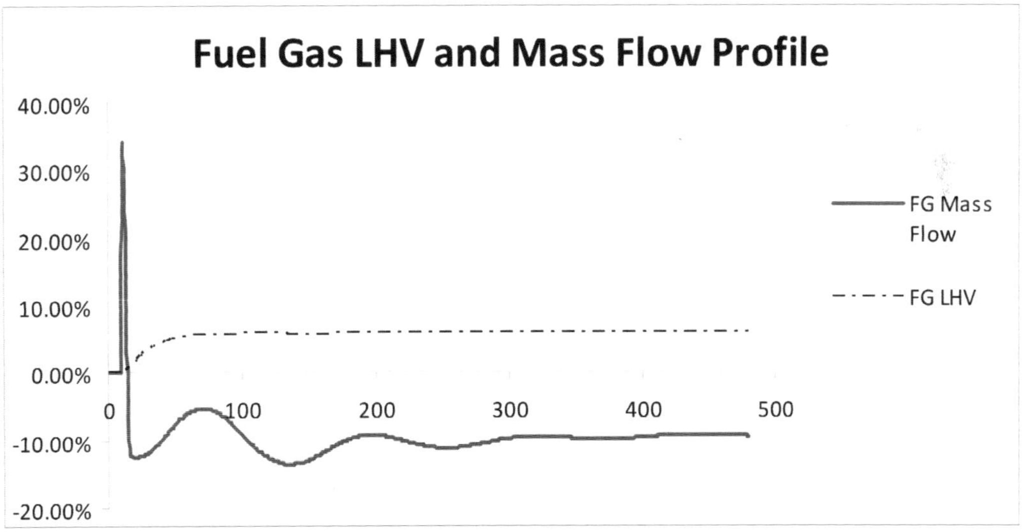

Figure 4. Fuel Gas LHV and Mass Flow Rate Profile

Figure 4 illustrates a typically step change in MWI and composition when switching from a lower LHV fuel source to a higher LHV fuel source. The MWI rate of change is shown to be approximately 0.2% per second, which is less than GE Heavy Duty gas turbine maximum specification of 1% per second [3]. The fuel gas flow is reduced when sending a richer fuel gas source to the gas turbine skids provided all other gas turbine operating parameters (e.g. ambient temperature, gas turbine speed, power output) are constant. An initial spike from the sudden opening of the control valve on the alternate fuel gas source can be noticed in Figure 4. Flow control is however restored in less than 350 seconds.

3.3 Pressure fluctuation in fuel gas system

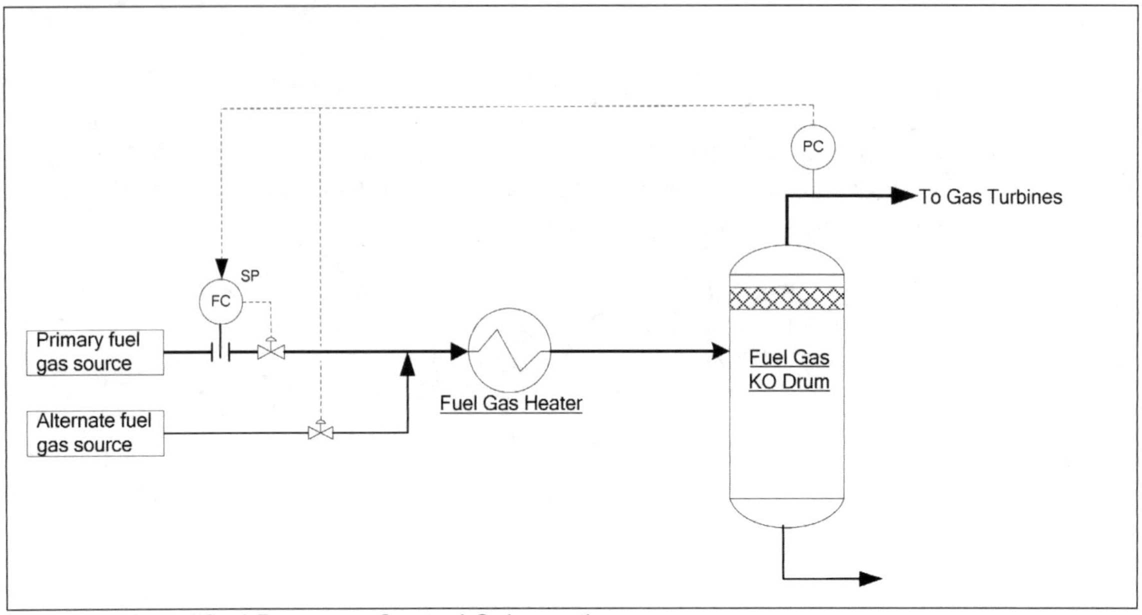

Figure 5. Simplified Pressure Control Schematic

Fuel gas supply for both high and low pressure systems need to be maintained at relatively constant pressure. Any significant pressure fluctuation can affect the fuel gas supply and overall plant performance. If the fuel gas supply losses pressure, the possible consequences are tripping of the gas turbines, the flare system, and the fired heater. In order to ensure stable fuel gas pressure, a combination of pressure and flow controller is used as shown in Figure 5. This figure illustrates a simplified pressure control schematic where a normal bleed from the primary fuel gas source along with a fast acting control valve is used to ensure smooth transfer when fuel sources are switched.

3.4 Fuel gas system design consideration during Start-up / Planned Shutdown

The typical fuel gas source for the GTGs during initial startup is the feed pipeline since most of the units are not ready. The pipeline feed gas may not only be in saturated condition, but the pressure let down from pipeline pressure to the fuel gas system pressure may lead to formation of liquid particles. Therefore, it is very important that the electrical startup fuel gas heater is in operation. During plant start-up, fuel gas will initially supply the flare units and the gas turbine generator to generate sufficient electric power to start the remainder of the facilities.

The fuel gas supply source is switched from pipeline feed gas to the fuel gas from the liquefaction unit once normal operation is established. The composition, MWI, degree of superheat requirement, temperature and pressure conditions are different for these two sources. These factors must be considered in the fuel gas system design to ensure that manual and automated fuel source switching will not trip the gas turbine operability and reliability.

For a planned shut down, the fuel gas demands will gradually reduce. However, a stable and continuous fuel gas purge to the flare unit should be maintained to ensure plant safety at all times.

4. Summary

This paper identifies the fuel gas users in an LNG Liquefaction plant and explains the importance of a well designed fuel gas system to meet all requirements such as, degree of superheat, pressure, MWI, LHV, contaminant particle size, and the composition limits. These factors are used to design a robust fuel gas system and to maintain stable and continuous fuel gas supply during various plant operation modes. A few real operating scenarios are shown to demonstrate how the fuel gas control system can react to meet OEMs requirements during startup or other upset conditions.

5. Acknowledgements

The authors would like to show their gratitude to Vinod Rajkumar, Himanshu Patel, and David Messersmith for their time to give us advice and motivation in preparing this paper. The authors would also like to thank Cyrus Meher-Homji for his technical guidance on gas turbine and combustor fuel requirement and James Kernan for his expertise on LNG plant commissioning and start up. Lastly, the authors would like to thank Dipanjan Bhattacharya for his help and support in developing the simulation for case studies.

Abbreviations

BOG – Boil Off Gas
CO – Carbon monoxide
DCS – Distributed Control System
DLE – Dry Low Emissions
DLN – Dry Low NO_x
GE – General Electric
GTG – Gas Turbine Generator
HP – High Pressure
HRSG – Heat Recovery Steam Generator
KO – Knock Out
LHV – Lower Heating Value
LNG – Liquefied Natural Gas
LP – Low Pressure
LPG - Liquefied Propane Gas
MWI – Modified Wobbe Index
NGL – Natural Gas Liquids
NO_x – Nitrogen oxides
OEM – Original equipment manufacturer
PPMV – Parts per million by volume
SAC – Single Annular Combustor
WHR – Waste Heat Recovery

References

[1] Badeer, G. H. (2005, September). *GE's LM2500+G4 Aeroderivative Gas Turbine for Marine and Industrial Applications*. Retrieved January 4, 2012, from http://site.ge-energy.com/prod_serv/products/tech_docs/en/all_gers.htm

[2] Davis, L. B., & Black, S. H. (2000, October). *Dry Low NOx Combustion Systems for GE Heavy-Duty Gas Turbines*. Retrieved January 4, 2012, from http://site.ge-energy.com/prod_serv/products/tech_docs/en/all_gers.htm

[3] Evulet, A. T. (2011, May). *Gas Turbine Combustion Technologies* [Presentation Slides]. Retrieved December 10, 2011, from http://www.coresymposium.com/2011_COREPresentations/Proven%20Solutions%20GE-Evulet.pdf

[4] GE Energy. (2009, January). *Specification for Fuel Gases for Combustion in Heavy-Duty Gas Turbines* (GEI 41040k).

[5] GE Energy. (2009, September). *Process Specification Fuel Gases for Combustion in AeroDerivative Gas Turbines* (MID-TD-0000-1).

[6] Meher-Homji, C. B., Zachary, J., & Bromley, A. (2010). *Gas Turbine Fuels - System Design, Combustion and Operability*. Proceedings of the 39th Turbomachinery Symposium. Turbomachinery Laboratory, Texas A&M University.

[7] Physical Properties. (n.d.). In *GPSA Engineering Data Book* (12th ed., Vol. 2, pp. 23-5). (Original work published 2004)

[8] Wien, S., Beres, J., & Richani, B. (2005, August). *Air Emissions Terms, Definitions and General Information*. Retrieved January 4, 2012, from http://site.ge-energy.com/prod_serv/products/tech_docs/en/all_gers.htm

Authors Biography

Bindupriya Bandari

Bindu Bandari is a Process Engineering Specialist for Bechtel's LNG. She also has design and field experience in LNG Re-gasification facility; and Safety Relief System for the gas processing facilities and refineries. She holds a Masters of Science, Chemical Engineering degree from Texas A&M University.

Pei-Han (Jason) Yong, P.E.

Pei-Han (Jason) Yong is a Process Engineer for Bechtel with experience on LNG liquefaction, gas processing and treating, and IGCC coal gasification projects. Mr. Yong holds a BS degree in Chemical Engineering from University of Texas at Austin and Master of Chemical Engineering from University of Houston. He is a licensed engineer in the state of Texas.

Christine Teh

Christine Teh is a Process Engineering Specialist for Bechtel working on LNG Liquefaction projects. She holds a Bachelor of Science in Chemical Engineering degree from the University of Texas at Austin.

Robert A. Zinsmeister, P.E.

Robb Zinsmeister is a Senior Control Systems Engineer for Bechtel's LNG Technology Group located in Houston, Texas. He has design experience in LNG, gas processing and refining. He also has field experience in construction, commissioning, startup and plant operation/maintenance. He is responsible for providing support to LNG projects and Control Systems discipline design standards.

Robb has ten years of experience in the Oil and Gas industry. He holds an B.A.Sc. degree in Electrical Engineering from Clarkson University and is a registered Professional Engineer in the state of Texas. He has previous work experience with ExxonMobil at one of their gulf coast refineries.

WASTE HEAT RECOVERY CONSIDERATIONS IN LNG PLANTS

DOUGLAS WILES; STANLEY KWAN; VINOD RAJKUMAR

ABSTRACT

Thermal integration in an LNG plant improves its overall plant thermal efficiency. Thermally integrated plants tend to have lower operating costs and reduced environmental emissions including, CO, CO_2, SO_x and NO_X.

Waste Heat Recovery Unit (WHRU) viability will depend on project specific factors including the availability of waste heat, type, quality, quantity and cost of fuel, type of heating medium and environmental regulations. The scope of the project, for example whether it is a green-field or retrofit construction, will determine the relative importance of many of these factors.

WHRU installation at the gas turbine driver exhaust requires higher initial capital. The economic viability will depend on the magnitude of the accrual of operating cost savings, and their ability to counteract the initial capital outlay. Savings can be in the shape of reduced fuel gas costs and reduced carbon dioxide taxes. Ultimately the impact of these savings will depend on the owner's measurement of the value of fuel gas; whether fuel usage is accounted for as lost feed or lost product. The negative impacts include the reduction in nitrogen rejection that occurs with reduced fuel gas usage and the power restrictions imposed on gas turbine drivers due to the increased back-pressure caused by the presence of the WHRU.

This paper investigates the factors impacting the viability of WHRU in an LNG plant and identifies scenarios where WHRU equipment can add value to a plant. A real world case study will be used to quantify the impact of these variables on the financial considerations of the WHRU.

1. INTRODUCTION

The cost of operating an LNG plant can be shown to be closely related to the plant's overall thermal efficiency, and its impact on the environment. LNG plants have sometimes been criticized for emitting greenhouse gases. The recovery of waste heat can minimize these concerns by integrating plant energy; however the technologies available to recover waste heat can be associated with higher initial capital.

2. WASTE HEAT RECOVERY UNIT DESIGN INTENT

The installation of a waste heat recovery unit (WHRU) enables the overall plant thermal efficiency of an LNG plant to be increased and consequently the use of fuel gas to be reduced. Lower fuel gas consumption enables a reduction in environmental emissions and a reduction of operating costs related to fuel use. The monetary impact of these savings will depend on the owner's fuel gas valuation methodology; whether fuel usage is accounted for as lost feed or lost product.

The commonly accepted source of waste heat in an LNG plant is from the exhaust of a gas turbine driver. Gas turbine drivers are used extensively in LNG plants to drive both refrigeration compressors and power generators. Gas turbine driver exhaust provides a high quantity of waste heat, as typical exhaust temperatures are in excess of 450°C (840°F). The waste heat can be recovered using heat transfer coils, placed in the exhaust stacks. A heat transfer medium such as hot oil or boiler feed water/steam can be circulated within the coil in a closed loop system to recover and distribute the heat to process users

The decision to implement waste heat recovery will be influenced by the Plant Owner's corporate goals, or may be a response to local environmental regulatory emission requirements such as carbon dioxide taxes. The Plant Owner must also consider the slight reduction in overall power available from a gas turbine driver when a waste heat recovery unit (WHRU) is installed on its exhaust stack. The installation of a WHRU also increases the total installed cost (TIC) of the plant and the operating complexity of the process.

3. WASTE HEAT RECOVERABLE

The maximum amount of waste heat available for recovery is dependent on gas turbine driver selection. In general, more waste heat is available from larger gas turbines due to increased exhaust flow rates.

Table 1.0 and Figure 1.0 show the exhaust heat recoverable for a series of gas turbine drivers typically used in the LNG industry. Both hot oil and steam applications have been summarized. Note that the steam analysis does not include co-generation and is purely heat recovered for the process users. Exhaust stack gas temperatures must be kept above a certain temperature to prevent moisture or acid gas condensation and subsequent corrosion. At the same time gas temperatures must be selected such that sufficient heat can be delivered to process users. Hot Oil temperatures in the 200°C to 250°C range are often required.

The data in Table 1 is based on a stack gas temperature of 250°C with the use of waste heat recovery, inlet air conditions of 1.013 bara @ 26°C and fuel gas conditions of 34.47 bara @ 25°C.

Table 1.0: Heat Recoverable from Turbine Exhausts for Hot Oil and Steam Recovery Mediums

	GAS TURBINE DRIVER TYPE			
	GE Frame 6B	GE Frame 7EA	GE LM2500+G4	GE LM6000
Fuel Gas Flow Rate (kg/s)	8,867	17,622	5,436	6,502
	Turbine Operating with Hot Oil Waste Heat Recovery			
Exhaust Temperature (°C)	557	545	527	475
Exhaust Pressure (bara)	1.03	1.03	1.03	1.03
Exhaust Flow (kg/hr)	503,640	1,026,000	294,840	394,200
Heat Recoverable (kW)	50,278	98,056	26,389	28,333
	Turbine Operating With Steam Waste Heat Recovery			
Exhaust Temperature (°C)	558	546	529	476
Exhaust Pressure (bara)	1.04	1.04	1.04	1.04
Exhaust Flow (kg/hr)	503,640	1,026,000	294,876	394,200
Steam Mass Flow (kg/hr) @50 bar, 315°C	82,944	161,064	43,164	44,568
Heat Recoverable (kW)	65,000	126,667	32,000	34,944

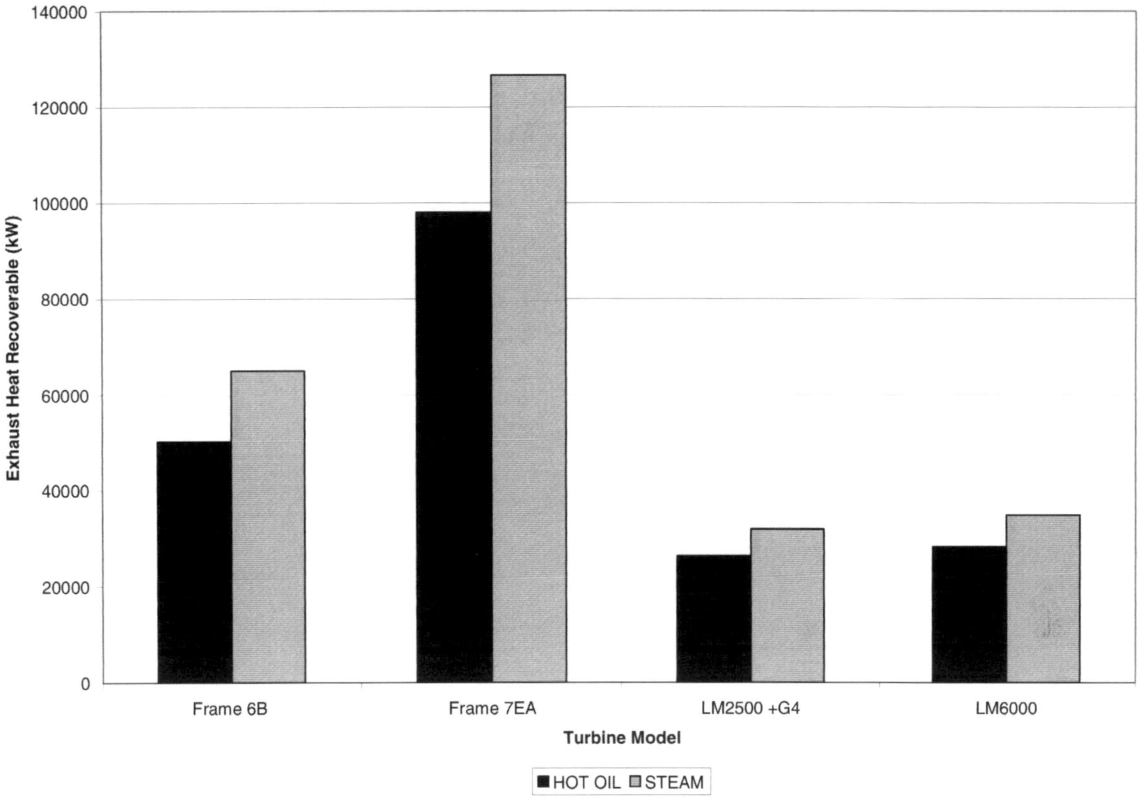

Figure 1.0: Heat Recoverable from Turbine Exhaust for Hot Oil and Steam Recovery Mediums

4. OVERALL PLANT THERMAL EFFICIENCY

Waste heat recovery increases gas turbine driver exhaust pressure loss, causing a slight increase in turbine heat-rate, decreasing available power. This pressure reduction is approximately 10 inches water for a typical WHRU coil arrangement (of which 6 inches account for normal losses in the stack and an additional 4 inches resulting from the waste heat recovery coil). For a Frame 6 gas turbine driver this equates to a loss of about 0.4 % available power.

The installation of a waste heat recovery unit will not significantly affect the overall fuel gas requirements. However elimination of the fired heaters decreases low-pressure fuel gas requirements. This net decrease in fuel gas consumption draw improves the overall plant thermal efficiency. Previous studies have shown a net increase of up to 1.4% in overall plant thermal efficiency is possible.

Overall plant thermal efficiency is defined as the ratio of the product higher heating value (HHV) to the plant feed gas higher heating value (HHV):

$$\eta_{TH} = \left(\frac{HHV_{product}}{HHV_{feed}} \right)$$

The overall plant thermal efficiency improvements for operating LNG facilities with a variety of duty/LNG production (MW/MTPA) ratios are shown in Figure 2.0.

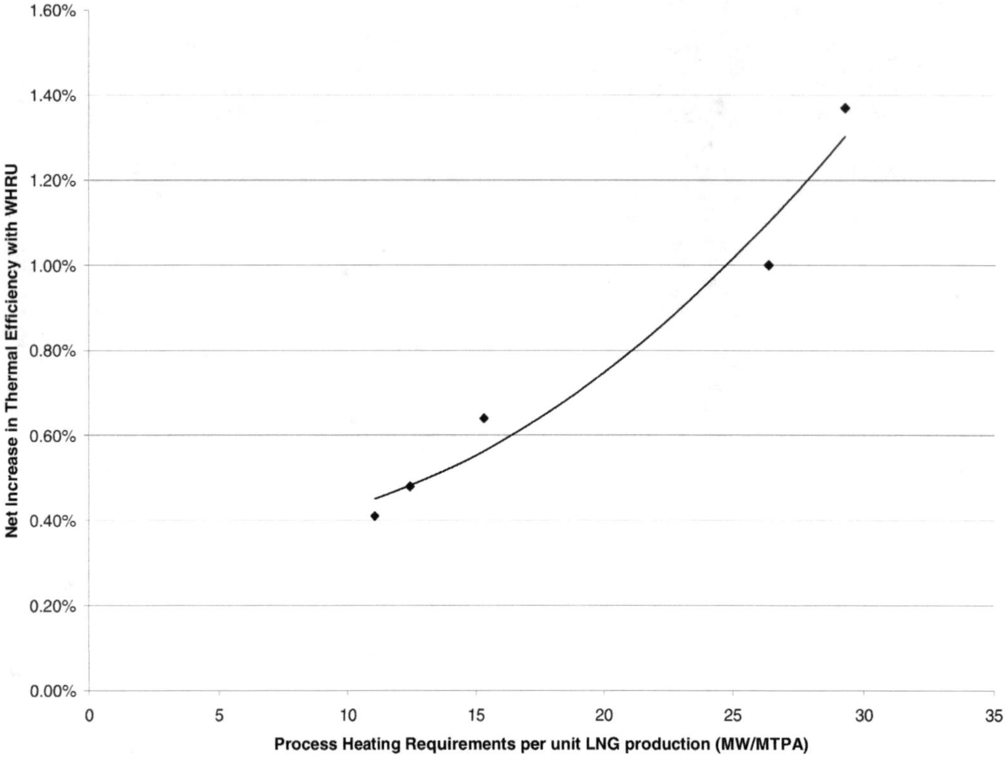

Figure 2.0: Overall Plant Thermal Efficiency improvements with waste heat recovery

Thermal efficiency is a function of many factors such as the feed gas inlet conditions including pressure, temperature, composition and ambient air conditions. Fuel gas can be obtained from both the front and back-end of an LNG facility. The withdrawal of fuel gas from the back-end of the LNG facility can provide a purge for unwanted inert gases, such as nitrogen. Inert gases can build up over time and consume unnecessary compression power, reducing overall plant thermal efficiency. As the fuel gas demand is reduced, so is the amount of nitrogen that can be rejected to the fuel gas system. Therefore plants with a high inert content in the feed gas, a Nitrogen Removal Unit may be required.

5. HEAT TRANSFER MEDIUM CONSIDERATIONS

5.1 HOT OIL

Use of synthetic heat transfer fluids is a common trend in the LNG plant. These are typically hydrocarbon based fluids with properties designed for long life, low degrees of fouling and low temperature pumpability. Modern synthetic heat transfer fluids are most often referred to simply as 'Hot Oil'. LNG plants require Hot Oils that can comfortably operate in a moderate temperature range; Hot Oil is typically supplied to users at approximately 200 – 250°C and returned at 110-130°C. Synthetic fluids resist the effects of oxidation many times better than mineral oils (manufacturers have quoted oxidation resistance up to 10 times that of comparable mineral oils). Less oxidation means less solids formation and much less fouling. Typical design parameters for hot oil include a minimum oil flow through the WHRU coils to keep the temperature below the maximum film and bulk temperature limits.

For a hot oil application, the temperature of the hot oil leaving each WHRU is controlled by dampers in the turbine exhaust stack; excess flue gas is diverted away from the hot oil coil through a bypass stack. Final temperature control is achieved by blending with an unheated bypassed hot oil stream.

In order to ensure availability of hot oil system at all times it is common to install an auxiliary/standby heater. A flow controller at the inlet of the WHRU allows the operator to balance hot oil flows between the WHRU and the auxiliary hot oil heater. Under normal operation, the auxiliary hot oil heater will be operating at minimum duty.

5.2 BOILER FEED WATER/STEAM

Steam as a heat transfer medium has a higher heat transfer coefficient compared to hot oil. As a result, steam requires a lesser surface area compared to hot oil for a given heat demand.

Steam is also fire resistant, so the safety issues associated with hot oil and direct heating are minimized. Steam has an unlimited fluid life, unlike hot oils and can be readily available.

The preparation and production of steam requires significant monitoring and treatment of the boiler feed water to remove minerals which could cause deposition and therefore fouling in heat exchangers and pipes. In addition steam systems with cogeneration require piping and equipment rated for high pressures. In addition there are high maintenance costs for the removal of scale, the repair of steam traps and leaks. The high costs associated with steam have to be balanced with the ability to recover more heat from gas turbine driver exhaust streams than hot oil.

5.3 REGENERATION GAS HEATING

There are specific applications where the circulated fluid can be directly heated in the gas turbine driver exhaust utilizing a WHRU coil. One common example is the heating of the regeneration gas used to dry the molecular sieves in the dehydration beds.

Direct heating does however come with some safety considerations. It is considered possible for the regeneration gas tubing welds to fail within the exhaust duct, resulting in the ignition of the regeneration gas. Such a problem can be resolved by locating all welds on heat recovery coils outside of the WHRU ducting.

The maximum allowable film temperature for regeneration gas is higher than that of hot oil; a typical value is around 400°C.

6. START-UP/CONTROLLABILITY

Increased heat integration makes the process more complex. Waste heat recovery provides a less flexible heat source than fired heaters, since waste heat is only available when the turbine is running.

Replacing fired heaters with waste heat recovery can complicate first time start-up. Most refrigerant gas turbines are not available early in start-up; only waste heat from power generation turbines is available. A start-up fired heater therefore may be beneficial to provide the heat necessary to dry out the dehydration system and start the acid gas removal unit. However during normal operation WHRUs require less maintenance than fired heaters.

The control scheme for a WHRU is often more complicated than that for a fired heater. If the same unit heats the regeneration gas and the hot oil, for example, the fact that the regeneration gas is an intermittent service can cause difficulties in controlling heat supplied to the hot oil.
Waste heat recovery schemes have found extensive application the in Oil and Gas industry, as such safety issues associated with their installation and operation have been well documented and managed. The source of waste heat and the choice of heat transfer medium will determine the potential risk of tube leak and ignition. The degree of risk and the level of safeguards should be addressed every time when waste heat recovery is considered.

Waste heat recovery for gas turbine driver exhaust is considered safer than fired heaters because of indirect heating. A WHRU can be considered as a shell and tube heat exchanger, and must be designed in accordance with API 660.

7. CONSTRUCTABILITY

A sketch of a simplified WHRU arrangement is shown in Figure 3.0.

Plot space must be made available for the sizeable ductwork required. The ductwork will likely be in the main process area, where the large turbines are located providing challenges in the plant layout design.

During start-up of the refrigerant gas turbine, it is possible that the mechanical coils installed in the exhaust ducting can be exposed to high temperatures prior to the introduction of heat transfer medium. Thermal cyclic fatigue can occur in intermittent services, for example, regeneration gas heating requirements are dictated by dehydration bed sequencing. The material of construction of the heat transfer coils for WHRUs must be chosen such that the coils can run dry, withstand thermal cyclic fatigue and avoid moisture and acid gas condensation.

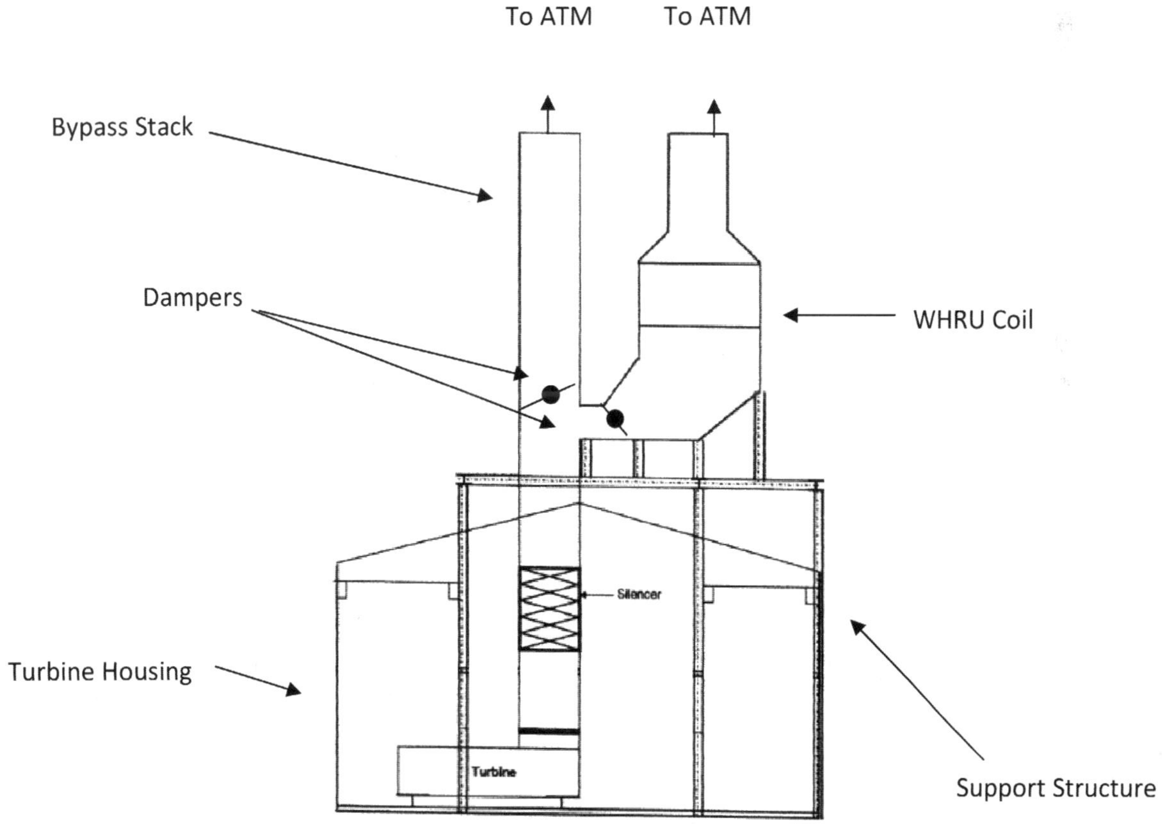

Figure 3.0: A Simplified WHRU Layout

It is possible to design the waste heat recovery unit to include an arrangement of multiple coils, containing different fluids, within one common exhaust. A combination that is generally used includes regeneration gas and hot oil or steam coils in the same gas turbine driver exhaust.

With multiple fluid services included in the same exhaust stack, there are design considerations on how to orientate the coils with respect to one another. The coil that is installed lower in the stack will be exposed to the hottest exhaust gas. Similarly coils further up the stack will be exposed to cooler exhaust gas; the lower coil having absorbed a portion of the exhaust heat before the exhaust comes into contact with the upper coil.

Regeneration gas is typically heated to a much higher temperature than hot oil. An exit temperature of approximately 315°C is generally required for dehydration bed regeneration. Regeneration gas entering the WHRU coil is typically 20°C. Such a large temperature gradient favors a design that locates the regeneration coils closer to the hotter gas or lower in the exhaust stack. On the other hand, hot oil entering the coil at 130°C is heated to approximately 250°C. Therefore, locating the hot oil higher up in the coil bundle is desirable. It can be seen that such an arrangement does not come without its problem. The regeneration gas heating duty is cyclical; hot regeneration gas only being required in the heating stage of the absorption/regeneration cycle of the dehydration beds. During the remaining stages of the heating/regeneration cycle the heat sink provided by the regeneration gas will not be available. As a result hot oil will be exposed to temperature swings. It is important to understand these swings when configuring coil arrangement.

8. FINANCIAL CONSIDERATIONS

An effective economic model will provide the owner/developer of an LNG plant with a guide to evaluate the factors that impact a financial decision as to whether to invest in a WHRU. This section outlines the inputs to such a model.

WHRU total installed costs can include the cost of ductwork, high temperature tubing, trim coolers, back-up fired heaters, plot space inside the battery limit and multiple structures to support the units on the exhaust stacks of multiple turbines. The economic viability will depend on the accrual of operating cost savings, and their ability to counteract the initial capital outlay. Savings can be in the shape of reduced fuel gas costs and reduced carbon dioxide emissions. Ultimately the impact of these savings will depend on the owner's measurement of the value of fuel gas that would otherwise be burnt in a fired heater. The owner could account for fuel gas usage as lost feed or lost product.

In order to determine the attractiveness of a WHRU, it is necessary to perform a comparison of the cash flows associated with waste heat recovery and those associated with a fired heater.

The cumulative cash flows (CCF) for fired heater and waste heat recovery unit can be represented by the model shown below:

$CCF_{\text{Fired heater or WHRU}}$ = -(Capital Cost,$) - (Annual Fuel Gas Expenses, $) – (Annual CO_2 taxes, $)

A typical cash flow model for an LNG plant comparing various waste heat recovery options is presented in Figure 4.0

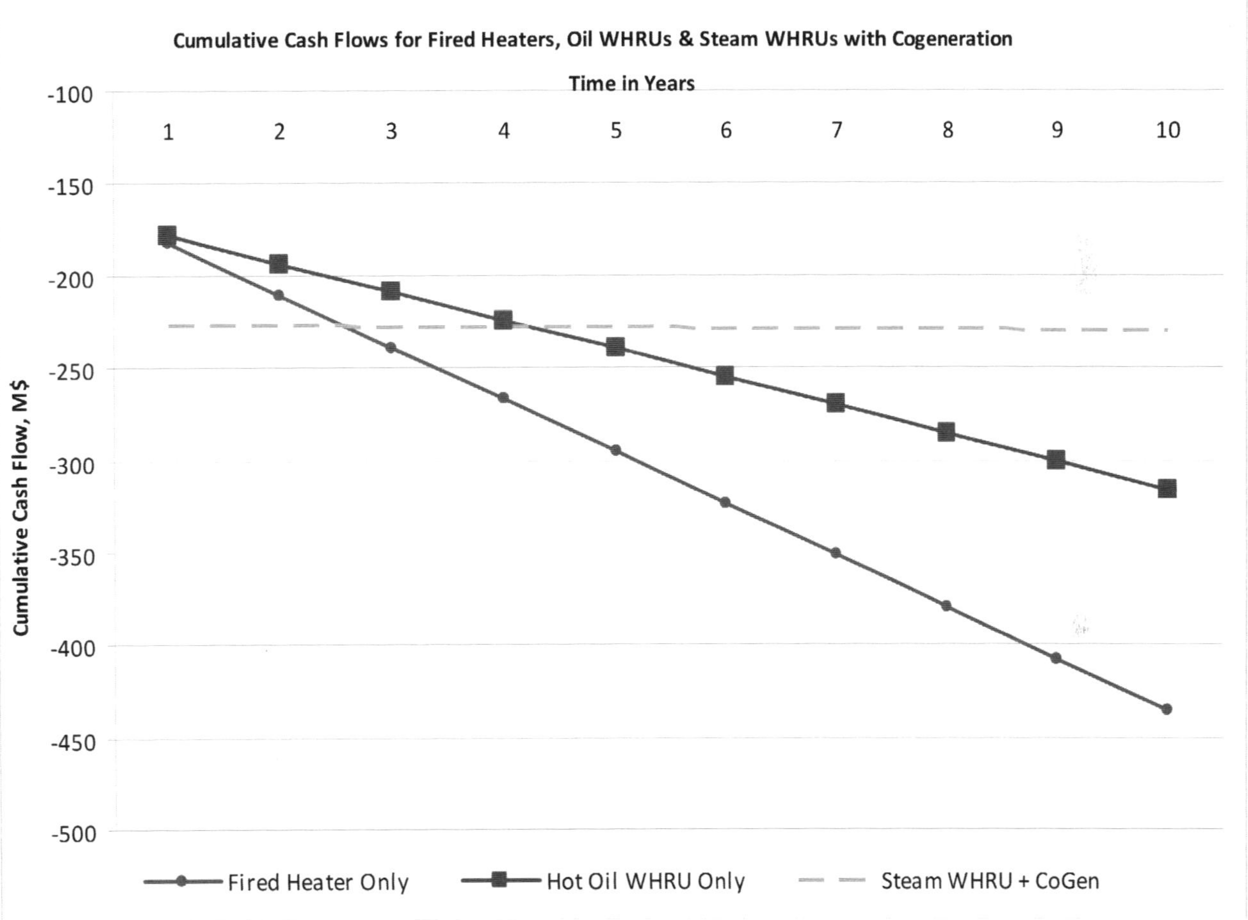

Figure 4.0: Cumulative Cash Flow Chart for Fuel Gas valued at 3$ / MMBTU

The input variables for the model are summarized in Table 2.0

Table 2.0: Total annual savings from Waste Heat Recovery

Description	Fired Heater Only	Hot Oil WHRU Only	Steam WHRU with Cogeneration
Total Fuel for process heat + electric power, MMBtu/hr	800	437	9
Assumed Fuel Cost $/MMBtu	3$	3$	3$
CO_2 tax ($/tonne)	20	20	20
Total Annual Cost for fuel and CO_2 ($ millions/year)	28 M$	15 M$	0.3 M$
Total Annual Savings from Base Case ($ millions/year)	0	13 M$	28 M$
Total Installed Cost ($ millions)	yy M$	1.05 * yy M$	1.45 * yy M$

A similar cash flow model can be developed for fuel gas valued as LNG product, say 8 $/MMBtu.

From figure 4.0 it can be inferred that eliminating fired heaters and providing all process heat from waste heat can provide significant reductions in fuel consumption and emissions with relatively little increase in capital cost and operating and maintenance complexity. The greater the process heat requirements the greater the savings.

Using steam for process heating and electric power generation offers the potential for even greater savings in fuel and emissions, although there could be a significant increase in capital cost and equipment complexity. A similar analysis would be conducted on a case-by-case basis to estimate expenses or cost savings associated with the different waste heat recovery options.

9. EXTERNALITIES

Owner priorities and environmental regulations could both influence the decision as to whether to install waste heat recovery. Future technological developments are likely to make waste heat recovery more attractive to the plant owner. Gas turbine driver suppliers may look at leveraging synergies between the waste heat recovery unit and the turbine, becoming the single point of responsibility for the entire unit.

Some LNG plant owners may have corporate policies that encourage the justification of additional expenditure in order to comply with environmental restrictions or statutes in the country in which they operate.

10. SUMMARY

This paper investigates the factors impacting the viability of WHRU in an LNG plant and identifies scenarios where WHRU equipment can add value to a plant.

11. REFERENCES

1. C. B. Meher-Homji, H. Patel, V. Rajkumar, D. Messersmith; Bechtel Corporation, J. Rockwell; ConocoPhillips Company ***"Thermo-Economic Analysis of Combined Cycle Based Liquefaction Plants"*** GASTECH 2011 Amsterdam, Netherlands 21-24 March 2011

2. Norrazak Hj. Ismail, MLNG Tiga Sdn Bhd, PETRONAS and Thomas Roy Stuart, Foster Wheeler Energy Ltd, Malaysia **"The Train 7 Fire at PETRONAS' LNG Complex, Bintulu, Malaysia"** LNG journal July/August 2005 page 12-14

3. Solutia Inc, audience guide slides **"Heat Transfer System Options and Considerations in Fluid Selection"** Conrad E. Gamble, MTS Fellow.

12. AUTHORS

Douglas Wiles

Douglas Wiles is a Process Engineer for Bechtel Houston, Texas. He has worked on multiple LNG FEED and EPC projects from conceptual design to commissioning and start-up.

Douglas has ten years of experience in LNG and Gas Processing. He holds an MEng degree in Chemical Engineering from the University of Nottingham, United Kingdom, and an MBA from the University of Texas, USA. He is a chartered engineer in the United Kingdom.

Stanley Kwan

Stanley Kwan is a Process Technology Principal Engineer of the LNG Technology Group for Bechtel Houston, Texas. He is responsible for providing project support of LNG Technology and conducting LNG Product Development Studies, and Technical Research.

Stanley has over thirty years of experience in refining and heavy oil upgrading technologies and process design for LNG, refinery and petrochemical projects. He holds a B.A.Sc. degree in Chemical Engineering from the University of Toronto, Ontario, Canada, and is a registered professional engineer in the state of Texas and the province of Ontario, Canada.

Vinod Rajkumar

Vinod Rajkumar is the Manager of the LNG Product Development Center – for Bechtel Houston, Texas. He has served in various roles in LNG Projects for the last 11 years. His LNG experience includes everything from conceptual design through commissioning, start-up and testing. He holds a Bachelor's of Science degree in BioChemical Engineering from National Institute of Technology, Jalandhar.

(25d) Demonstration Test Result of High Pressure Acid-Gas Capture Technology (HiPACT)

Koji Tanaka, Advanced Fuels Research & Development Division, JGC Corporation, Yokohama, Japan, Takehiro Komi, Domestic Project Division, INPEX Corporation, Niigata, Japan, Yasushi Fujimara, Research & Development Division, JGC Corporation, Ibaraki, Japan, Torsten Katz, Gas Treatment Process Technology, BASF SE, Ludwigshafen,, Germany, Oliver Spuhl, Sour Gas Treatment Process Engineering, BASF SE, Ludwigshafen, Germany, Erick Contreras, OASE Gas Treating Excellence, BASF East Asia Regional Headquarters Ltd., Hong Kong, Hong Kong and **Kumar Sugavanam**, Gas Treatment Process Technology, BASF Corporation, Houston, TX

The technology HiPACT (High Pressure Acid-gas Capture Technology), jointly developed by JGC and BASF, is an advanced CO2 capture technology to contribute to economical CCS implementation in natural gas processing plants. The solvent has high stability against thermal degradation, which enables high pressure and elevated temperature operation during solvent regeneration to reduce CO2 compression cost downstream. The solvent also has a higher CO2 absorption capacity than commercially available solvent technologies, which results in decreased solvent circulation rates and lower AGRU (Acid Gas Removal Unit) capital and operating costs.

The demonstration test was conducted in INPEX's Koshijihara natural gas plant in Nagaoka City. INPEX consistently improves the natural gas production efficiency to reduce environmental impact. As for the AGRU technology, DGA was initially selected, followed by aMDEA(TM) then finally by HiPACT for demonstration purposes. A commercial AGRU train was temporarily modified and aMDEA(TM) solvent was swapped with HiPACT. Treated gas from HiPACT was sent to the pipeline via dew point control system during the demonstration test.

HiPACT showed excellent performance in CO2 absorption; 20% reductions of solvent circulation rate and reboiler duty compared with aMDEA(TM) were achieved, keeping CO2 specification at absorber overhead unaffected. The temperature profile in the absorber indicated that the CO2 absorption kinetics was much higher compared to aMDEA(TM). HiPACT showed much sharper and narrower temperature bulge within the lower packing bed in contrast to aMDEA(TM)'s broader bulge. There was no negative indication for operation such as foaming. Through numerous data acquisition, the process modeling and simulation tools were updated.

(25e) Application of Exergy Analysis for LNG Plants and Receiving Terminals

Chen-Hwa Chiu, Chevron Energy Technology Company, Houston, TX and Lixin You, Chevron Technology Ventures LLC, Hosuton, TX

Natural gas liquefaction is an energy-intensive process. New world scale LNG plant usually has multiple trains with each train has a very high capacity (up to 8 million tons per annum of LNG). Increasing the plant efficiency can increase LNG plant output and economics, with reduced greenhouse gas emission.

On the other hand at LNG receiving terminals, LNG needs to be regasified to be put in pipelines to deliver to customers. LNG has a great cold energy potential. This cold energy can be used to generate power. The power generation scheme needs to be optimized to fully utilize the LNG cold energy potential.

Exergy analysis is based on the Second Law of Thermodynamics. It can identify the exergy loss distribution in the whole LNG plant or receiving terminals. Improving exergy efficiency for the unit operations with highest exergy loss will result in increased production, lower power consumption and better overall economics. For power generation using LNG cold energy, exergy analysis can identify the best power generation scheme for different LNG feed, composition and required delivery pressure.

This paper first review applications of exergy analysis in LNG plants and import terminals and then present major exergy analysis results and discoveries. Finally, recommendations for improving exergy efficiency in both LNG plants and receiving terminals are proposed.

(25f) Maximizing LNG Capacity for Liquefaction Processes Utilizing Electric Motors

Jonathan Berg, Air Products and Chemicals, Inc., Allentown, PA and Yu-Nan Liu, Air Products & Chemicals, Allentown, PA

In some locations currently being considered for LNG projects, relatively low cost electricity from the local power grid may be available. For these facilities, refrigeration compressors driven by electric motors drawing power from the grid may be attractive. The fuel requirements for these plants will be relatively low since there will be no need for fuel to drive refrigeration compressors or to generate electricity. This presentation will describe several natural gas liquefaction process configurations that are particularly amenable to LNG facilities without substantial fuel requirement. The applicability of the process configurations based on specific project requirements will also be discussed.

(29a) The Economics of Process Measurements

Douglas C. White, Emerson Process Management, Houston, TX

What is the economic value of a new measuring device in a process plant – a new analyzer, a more accurate flow measurement, an additional temperature or pressure measurement, etc.? Excluding the minimum set of measurements required to safely operate the plant and meet regulatory requirements, how can additional measurements be economically justified? Engineers are often presented with this issue, both for proposed new plants and existing sites. More measurements generally reduce the uncertainty in evaluating the past performance, the current state and the expected future performance of the plant. The reduction in uncertainty should then result in improved decisions. There are a hierarchy of decisions in plants that have different time scales and different magnitudes of possible financial impact. In some cases the decisions are algorithmic, i.e. a control loop. In others they are more complex, involving manual processing. Quantifying the possible improvement in future decisions and their possible impact on plant profitability is used to economically value the measurement. The value can be compared with the cost to determine an expected return on investment. In this paper a framework is presented to quantitatively address this subject and examples from the process industries are used to illustrate the procedures developed.

(29b) Applying Offline Tools to Reduce On-Site Control Implementation Effort

Kenneth V. Allsford[1], Bhaskar V Iyer[2]and Aric Tomlins[2], (1)Honeywell Process Solutions, Honeywell, Houston, TX, (2)Honeywell Process Solutions, Honeywell International, Houston, TX

Process control engineers are continuously challenged to reduce the effort to commission new advanced regulatory control strategies and/or linear/nonlinear multivariable control applications. Reducing on-site controller identification and tuning and the associated "interference" with process operations is especially desired – provided the commissioned application(s) meet(s) performance expectations. This paper will present a refining practitioner's perspective on applying tools to address this challenge.

For multivariable control applications on a number of refining units, it is becoming established practice to run automated closed-loop step testing tools to hone in on the control models that will enable good performance for the tuned/commissioned controller. However, this closed-loop step testing approach requires satisfactory “seed models” to enable control to be maintained during the step testing. For advanced regulatory control, similarly knowledge of approximate relevant process dynamics (effectively also “seed models”) – coupled with an appropriate tuning philosophy – enables good initial estimates for PID and other tuning parameters. As such the focus of this paper is on capturing seed models that can be used for either of these purposes. Approaches, referred to here as “Offline Tools”, that enable reduction of actual plant testing are reviewed.

A general resource in all instances is that (large) engineering organizations have the capacity to maintain, explicitly or implicitly, relevant data libraries – seed models for multivariable controllers or tuning parameters, possibly with associated identified process dynamics, for advanced regulatory control strategies.

For greenfield or significantly modified brownfield sites, operating data may not be available initially. However, operator training (or other (dynamic) process engineering) simulators may be available and their use for generating seed models and/or initial tuning for advanced regulatory control strategies will be reviewed. Process engineering model assumptions that impact the use of the models for these purposes will be highlighted. Of note is that with appropriate dynamic process engineering models, controller performance across a wide range of operations can conceptually be evaluated relatively quickly.

For actual operating plants, a range of operating data exists from the descriptive provided by operations and engineering staff through the very granular stored in historians. Steady state and/or dynamic models may also exist and the operating data can be used to identify the usefulness of such models, potentially initiating a discussion on making such models “current”. A range of approaches (“offline tools”) to utilize the available offline data to estimate seed models exists; it will be discussed that the optimal approach is typically some combination of these tools selected for the particular circumstances.

Useful Tips for Fired Heater Optimization

Nikki Bishop, Emerson Process Management, Atlanta, GA
Barbara Hamilton, Emerson Process Management, Garrison, NY

Introduction

Fired Heaters are used extensively throughout refineries for heating, vaporization and thermal cracking of various process fluids. These heaters are essential for refinery operation. Figure 1 below shows a typical process fired heater.

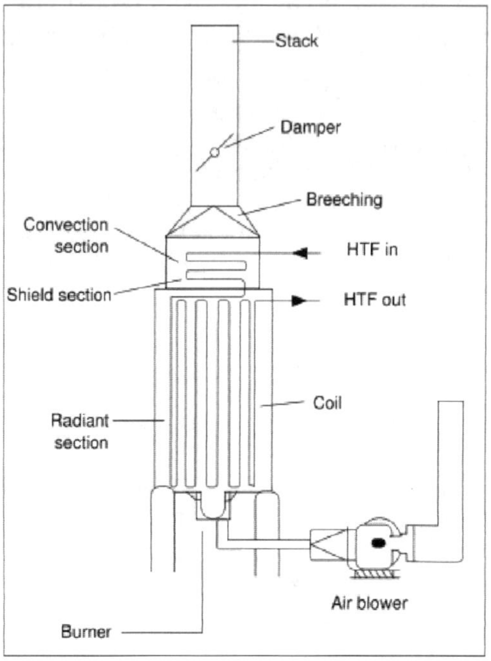

Figure 1. Typical Process Fired Heater

The main objective for fired heater optimization is to safely control to a target combined coil outlet temperature while minimizing energy costs, emissions, and overall variability. To achieve this objective, the heater must be able to safely operate near constraints and demonstrate stable and robust performance. A holistic approach that examines the field devices, basic control strategy and loop tuning is recommended to establish a solid foundation. Advanced process control methods coupled with a solid combustion control strategy provides an effective means of safely operating near constraints. Even further, pre-engineered application solutions offer great benefits in the ease of implementation and support.

Back to Basics: The Instruments

Just as a building is only as good as its foundation, a process control system is only as good as the instruments which provide the measurement data and final control. It's important to address and mitigate process and mechanical issues. Necessary measurement and actuator improvements are made in order to reduce control loop variability. Reducing variability on key loops enables the heater to operate closer to the constraint conditions, which increases throughput, efficiency, and safety. For example, it is essential to tightly control the air and fuel flows for safe and efficient combustion.

Figure 2 below shows typical fired heater measurements.

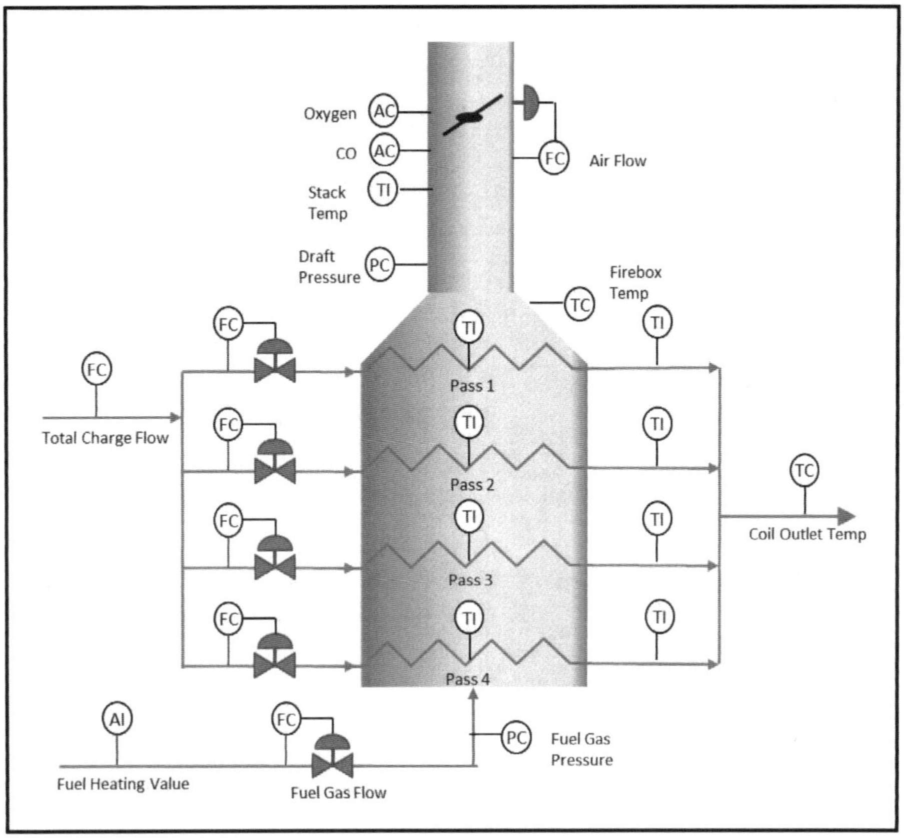

Figure 2. Typical Measurements

Beyond Base Regulatory: The Full Solution

Many fired heaters are instrumented with only the minimum instrumentation necessary for operation. While this strategy allows for operation of the heaters, it lacks the sophistication necessary for optimal performance. A full solution, including a burner management system (BMS) for safety and the necessary instruments, final control elements and damper actuator, is the ideal recipe for optimal heater performance.

The Ultimate Control Strategy

The ultimate goal for a fired heater is to heat a process fluid to a desired temperature. Maintaining a constant outlet temperature is critical for the process. Variations in outlet temperature introduce variability in the overall process. Since the optimum operation will almost always be near constraints (e.g. maximum tube temperatures, minimum excess air), variation in the process requires the operator to stay further away from the actual limit to provide the necessary buffer or safety margin to handle any unexpected process upsets. As a result, manufacturers are not always able to achieve the most efficient operation of their assets.

Figure 4 below shows a representation of the impact of variability on cost.

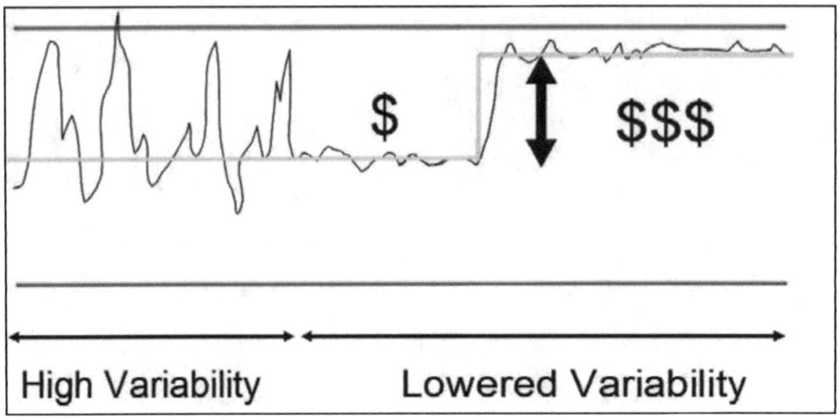

Figure 3. Process Variability Cost Impact

Reducing variability means ensuring a robust and stable control strategy is in place. Implementing a fully automatic, regulatory control system ensures that coil outlet temperature is maintained at the desired setpoint, while simultaneously balancing pass temperatures and air flow. Pass balancing adjusts the flows through each tube such that no pass runs hotter than the others, reducing hot spots within the coils, which translates to longer run lengths and the potential to operate at a higher severity.

Bringing the Heat: Combustion Controls

As previously noted, the ultimate goal for a fired heater is to safely heat a process fluid to a desired temperature. The required energy is provided by combustion of a fuel to heat the coils. The efficiency of the combustion process determines the efficiency of the heater. A "single-knob" combustion control strategy allows for safe, efficient firing at any desired charge rate. With "single-knob" control, the desired coil outlet temperature is maintained while operating within mechanical and thermodynamic heater constraints at all times.

A proven strategy for safe combustion is to employ an air and fuel "cross-limiting" solution. Cross-limiting is the traditional approach of increasing air flow first on an increasing fuel demand and decreasing fuel first on a decreasing demand. This strategy prevents unsafe, fuel-rich environments from being introduced by the automation system. Typically this is done by empirically determining the air, fuel, and stack oxygen levels at various demand points. A linear relationship is typically used between demand and fuel flow. The relationship between demand and air flow follows the empirically derived curve such that the oxygen target at each demand point ensures complete combustion.

Typically, the lowest stack oxygen level that does not produce carbon monoxide (CO) in the stack is best for both efficiency and emissions compliance. CO in the stack is an indication that the fuel is not completely burned.

Advanced combustion controls allow for maximum capture of the available heat in the fuel with less variability. This is important even if the fuel is a refinery waste gas that is "free" because it reduces greenhouse gas emissions, makes the fuel available for other uses like boilers or co-gen plants, and allows for more throughput in a capacity-constrained situation. It is impossible to safely drive stack oxygen levels down without robust combustion controls. When air and fuel are not well coordinated, not only is efficiency affected, but a fuel-rich environment risks an explosion if furnace oxygen levels get too low.

Figure 5 below shows the relationship between excess air and available heat.

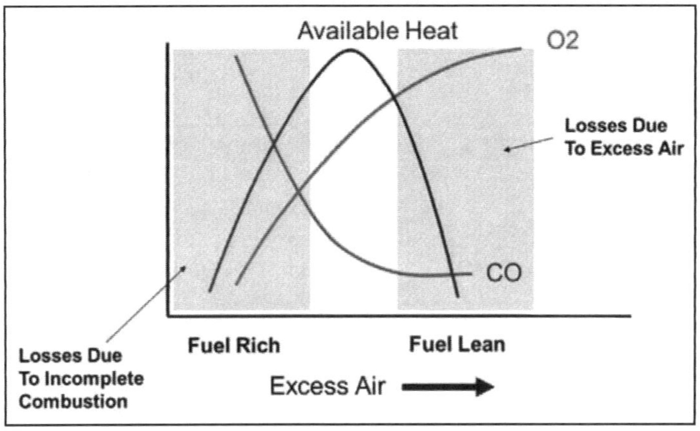

Figure 4. Stack O_2 vs. Available Heat

The economic savings resulting from operating a furnace at a lower excess air can easily be calculated based on the extra energy used to heat the extra air from ambient to the stack exhaust conditions. The stack O_2 concentration is an indication of how much excess air there is. If all the O_2 is perfectly consumed, the O_2 concentration will be 0. This, however, is not safe, so most furnaces operate with O_2 concentrations of 1% or higher. Decreasing the O_2 concentration in the stack from the current level to a lower value results in energy savings that can be substantial, depending on the starting point and the cost of fuel.

Figure 6 illustrates the relationship between reduced stack oxygen, efficiency and energy savings.

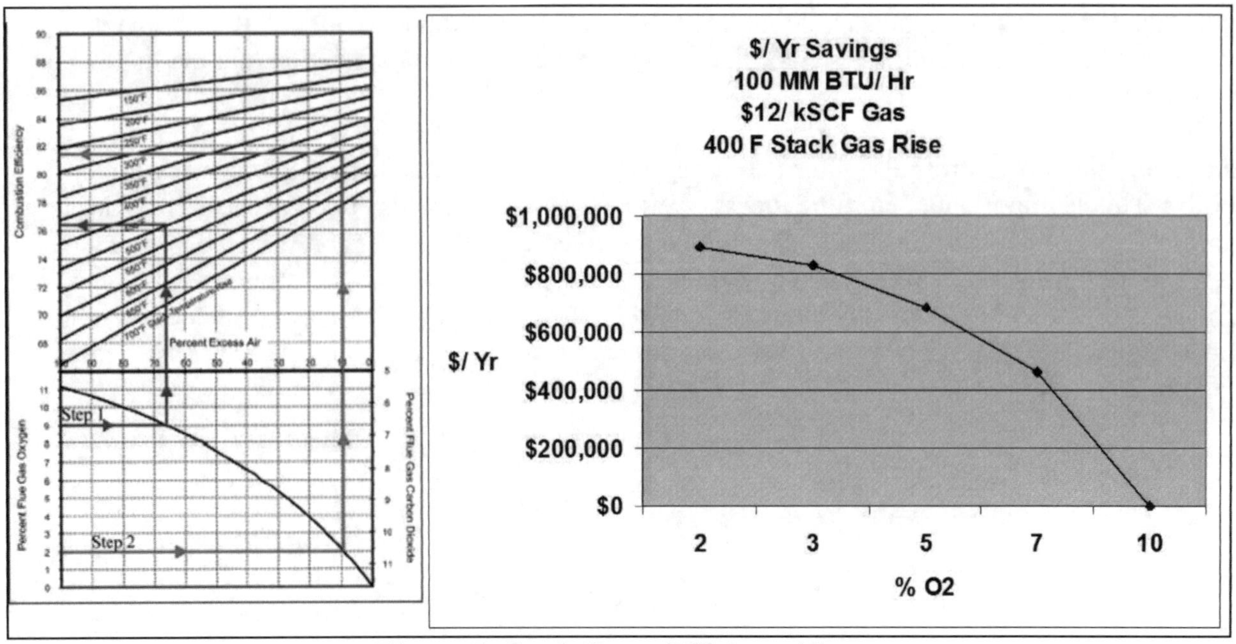

Figure 5. Stack O_2 vs. Efficiency/Cost

Advanced Process Control

To achieve optimal heater operation, the heater must be able to safely operate near constraints and demonstrate stable and robust performance. This means balancing pass temperatures while simultaneously ensuring complete combustion and minimizing energy. Model predictive control (MPC) that is native to the regulatory control processor provides a high speed, redundant, robust control solution.

MPC uses relationships between controlled, manipulated, disturbance, and constraint variables to create dynamic process models that are used with an optimizer to determine the best set of process changes that control all variables to target. Known relationships between variables can be input directly into the MPC control module. Alternatively, an automated "bump test" sequence is used to establish the process models. Constraint violations (such as skin temperatures, fuel pressures, and air/fuel cross limits) are anticipated (predicted) and manipulated variables adjusted accordingly.

A model predictive control strategy also reduces process variability, because for optimal operation, passes are balanced not to achieve equal flow, but to achieve equal temperature rise across all the tubes. Pass balancing based upon temperature rise creates process conditions that favor more efficient overall heat transfer and balance the coke build up, allowing the unit to run longer between de-coking shutdowns.

Figure 7 illustrates an actual example on an ethylene cracking furnace where MPC is used to control the temperature deviation between the passes. This chart demonstrates the significant reduction in pass outlet variability when the model predictive controller is online.

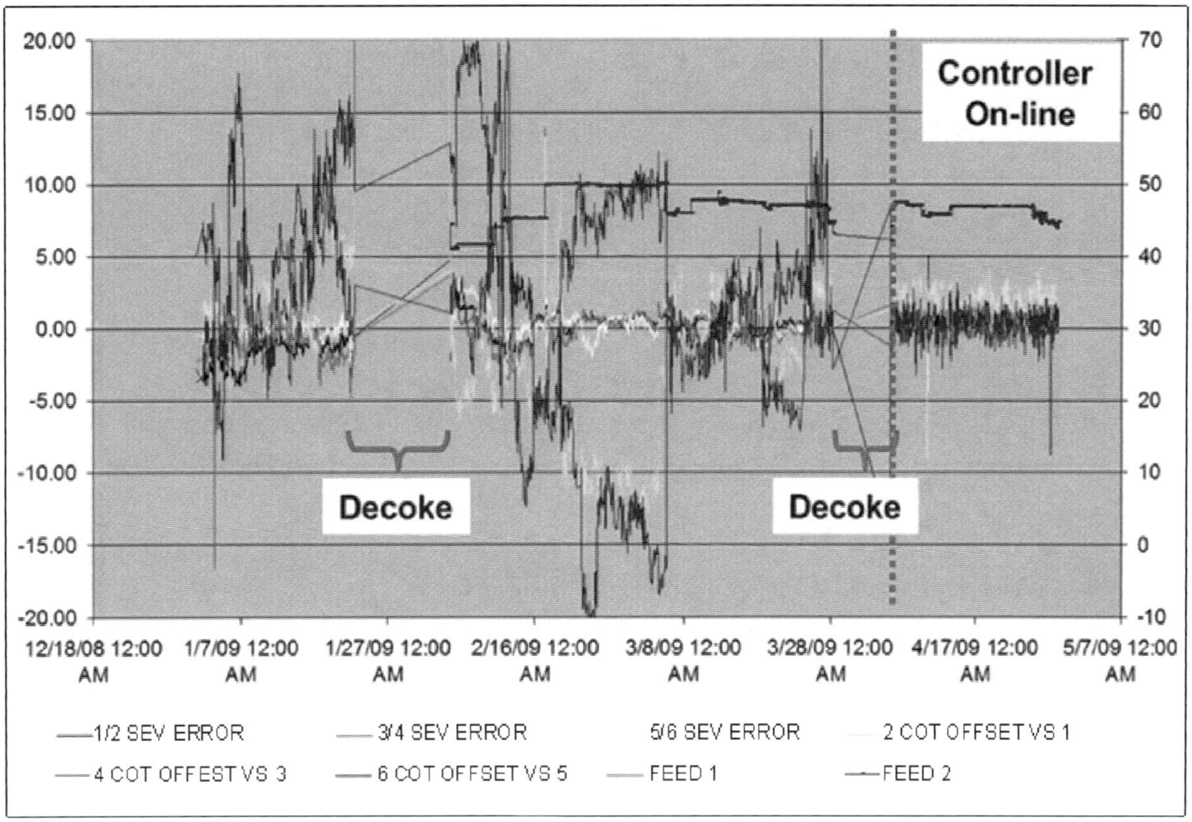

Figure 6. MPC Controller Online Variability

MPC Put to Good Use

One multi-national oil company successfully implemented an advanced process control strategy on its fired heaters as part of a larger APC project on the unit and achieved remarkable business results. Two fired heaters on the atmospheric crude unit were operated in parallel, with poor efficiency. Due to various process and equipment reasons, these heaters operated only with significant manual intervention. Process interactions made conventional, automatic control less than ideal.

The solution to the problem was an MPC strategy which allowed for safe and stable operation of both furnaces simultaneously. These furnaces can now operate near their constraints in spite of significant fuel gas, feed rate, crude composition and feed temperature variations. Implementation of this advanced process control strategy on the heaters resulted in an annual fuel savings of more than $100,000. The heaters now require less manual actions and can manage load changes automatically in closed-loop control. Excess stack oxygen is reduced and efficiency increased in both heaters.

Abqaiq Plant Virtual Plant RTO Implementation

Dr. Kuangyao Brian Peng and Saleh H. AL-Musallam,

Saudi Aramco, Abqaiq, Saudi Arabia

Introduction

Abqaiq Plant is the largest worldwide crude stabilization plant with $200MM energy consumption per year. Abqaiq NGL plant shares 30% of total energy consumption. High NGL recovery rate has been the objective because of undesired loss of hydrocarbon liquid condensate in the offgas pipeline. Recent efforts for process improvements including commissioning of TEG dehydration unit and compression train upgrades have been implemented to enhance NGL recovery and to improve the availability of desired cryogenic cooling for C2/C3 separation.

Abqaiq Natural gas liquid process includes three major processes: compressing, high pressure stripping and deethanizing. Simultaneous optimization for all three major processes under the constraints of NGL product spec. requires fundamental understanding and a robust simulation tool. It has been a dream and a challenge for process and control engineers to develop such a complete model which can map the plant operating conditions in situ and perform a complete plant optimization online.

ROMeo (Rigorous Online Modeling with Equation-based Optimization), a widely used real time optimization (RTO) platform, can be used to achieve our goal. A real time optimizer can take the variations in upstream units such as changes in crude feed flow combinations and stabilization depth in real time, and calculates the most profitable mode of operating condition by using the rigorous plant model with energy consumption equations and process and product constraint equations. Then the calculated optimal set points are submitted to the DCS control room to ramp the plant process.

Abqaiq Plant NGL engineering department has successfully implemented and tested an in-house developed RTO NGL plant model. A potential increase of $200K to $400K revenue increase for each adjustment has been predicted under the current operating mode and unit operation constraints. In addition to the quantifiable revenue advantages, this presentation will also highlight other substantial advantages of the system such as measurement health monitoring, next generation process KPIs, and energy optimization.

Background of NGL process optimization initiatives

To meet the future energy demand and sustainability requirements in the Kingdom of Saudi Arabia, Abqaiq Plant starts several energy efficiency programs for all process units. A rough study with Saudi Aramco P&CSD for NGL recovery enhancement has shown great potentials with online optimization. As a result of this study, by simply adjusting the operating conditions of HP-Stripper columns could increase NGL recovery rate by 3 to 5 MBD. Further study with relaxing product spec. of individual columns and maintain the final product spec has shown an increase of NGL production up to 9 MBD.

A joint effort among corporate optimization specialists and plant operation engineers took off in 2010. A road map was proposed to implement ROMeo RTO at Abqaiq NGL plant by the end of year 2012. The road map includes three major milestones: rigorous model development, PI integration and offline optimization.

Process description

Continuous feed gas from oil stabilization units is fed into the NGL plant with a flow rate of 550MMSCFD at 3psig and 140F. Feed gas is compressed with two stage compressors to 450psig and is cooled by cooling tower for the outlet temperature at 110F-125F.

Feed compositions are studied with the available historical data. The major variation of feed compositions come from crude oil mix containing neat condensate, Arabia Light and Arabia Extra light. As a baseline model, average compositions over the past few years are used for model stability. Typical compositional data for hydrocarbon contains C3+ above 70%. Table 1 shows the average compositional data for a typical feed.

	N2	C1	CO2	C2	H2S	C3	IC4	NC4	IC5	NC5	C6	C7	C8
Overall	0.019706	0.0785301	0.028414	0.224106	0.043989	0.342631	0.056587	0.132501	0.032532	0.032833	0.006334	0.0018	3.4E-05

Table 1

Abqaiq NGL plant contains five major high pressure stripper columns and one deethanizer column. The average operation condition of high pressure column is 145F and 400 psig. Column efficiencies vary from 68% to 85% because of different tray designs. The five high pressure stripping columns were designed with 22 bubble trays for separation of C2 and C3+. Each column equips with steam kettle reboilers and temperature controller at the 7 th stage of the column. The column temperature is the key handle for NGL product specification.

The deethanizer chiller is designed to operate at 65F and the reflux condenser is 60F. Currently, operation requires all columns product spec at 0.4 vol% which is a conservative approach. Overstripping of NGL product to offgas line would raise the liquid condensation in the pipeline, reduction of valuable NGL product recovery and waste of energy.

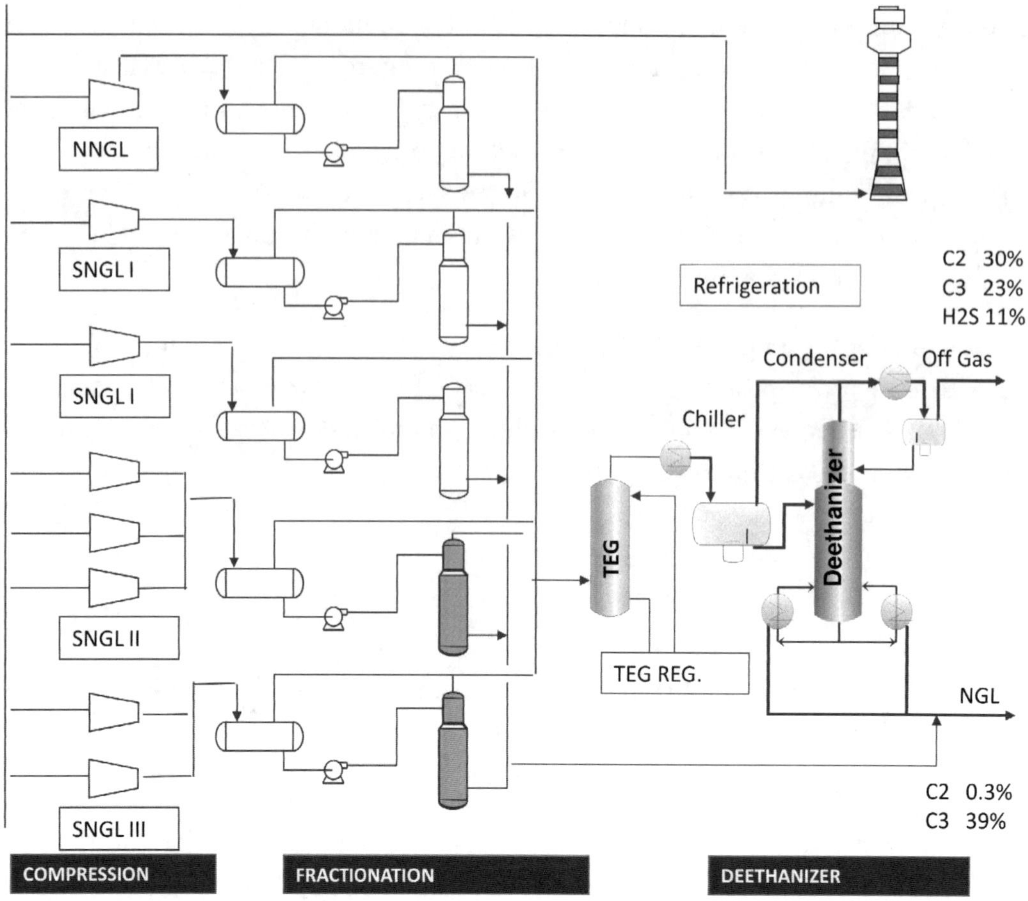

Figure 1: A simplified NGL process flow diagram.

Model development and implementation

Rigorous process model is developed based on the current process conditions in the field. Due to time constraint, the preliminary process model does not integrate with detail performance data of compressors. However, because the phase I of this program is to maximize NGL recovery and to minimize the impact of liquid condensation, the performance functions are not required at this stage.

One of the challenges of the team is to identify all the piping routes of cross feed lines. The cross feed flow rates have to match the field instrumentations. The on-off setting of process units is relied on the macro programs that were developed in house with the engineering support and training from Invensys. A total number of 180 major PI-tags were linked to PI historian for data reconciliation.

Economic variables for the sources include 60# and 625# steam, propane and feed gas. For the product steams, two offgas streams and one liquid NGL stream are included. The environmental variables include the release of CO2 to atmosphere and the amount of sour water steam to downstream facility.

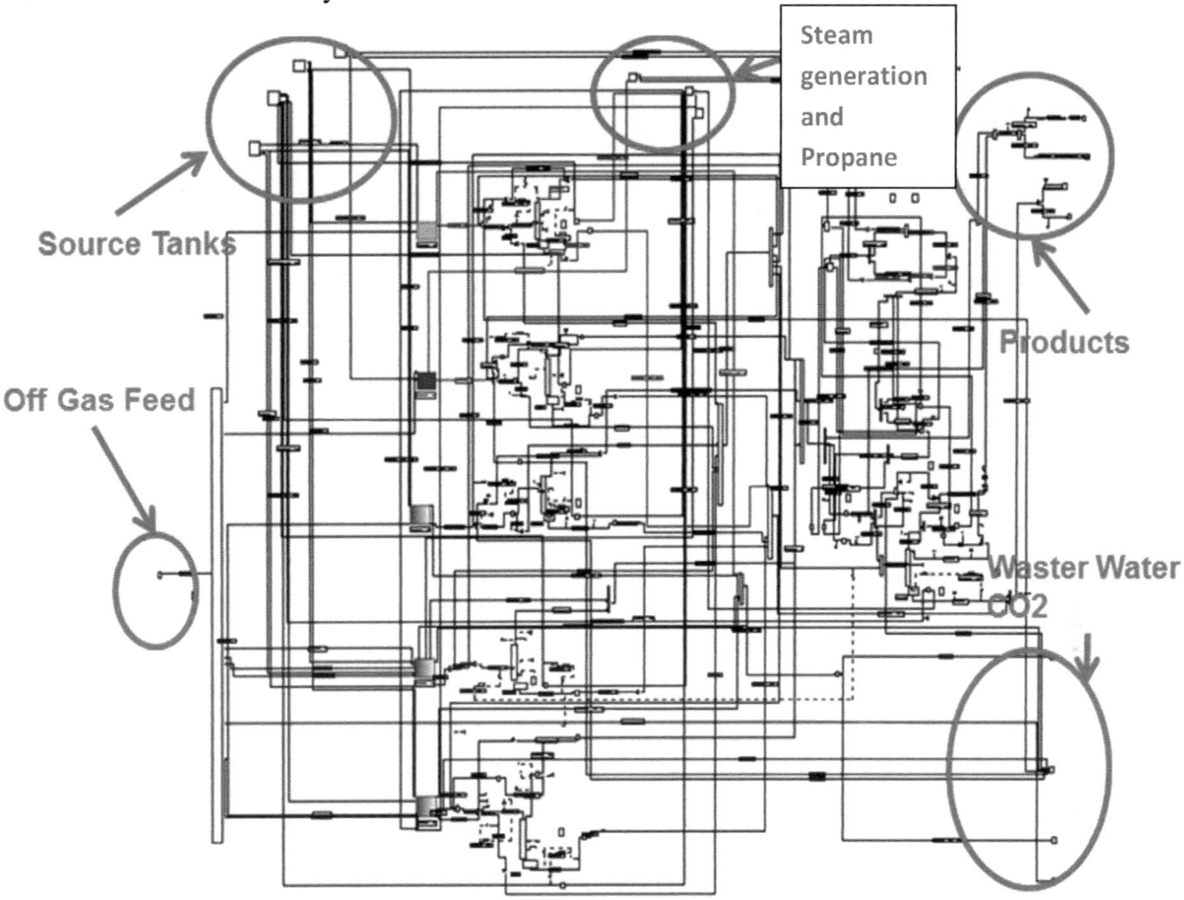

Figure 2

Figure 2 demonstrates the process models with eight compression trains, five high pressure stripping columns, one deethanizer column and a propane refrigeration system. All process units in this project are simulated with rigorous thermodynamic models.

Data Reconciliation

Process model has been integrated with PI historian for data reconciliation purpose. Macro routines with TCL scripts are developed to turn on and turn down major equipment under different operation modes. Data reconciliation results deliver the report for measurement system analysis that indicates major instrumentation issues. Operation, instrumentation engineering and maintenance are able to utilize this report to identify the issue and improve the situation. Furthermore, accuracies of some major KPIs are improved after the improvement of measurement system.

Tags	Unit	Scan	Model	Bias	% ABS(Bias)	Comments
462-FI-310.PV	MMft3/day	114.851	91.047	-23.804	26.14474	
340-Total-Off-Gas.CAL	MMft3/day	393.3	297.296	-96.004	32.2924	
461-FIC-182.PV	MMft3/day	18.471	14.189	-4.282	30.17831	
499-TI-764AI1PV.CV	F	159.65	125.356	-34.294	27.35729	
462-FI-1147.PV	MMft3/day	79.345	61.404	-17.941	29.21797	
334-FI-1005.PV	lb/hr	38297.5	25444.8	-12852.7	50.51209	
461-FIC-12.PV	lb/hr	62396.6	29909.7	-32486.9	108.6166	

Table 2

Table 2 shows an example of data reconciliation results with problematic instrumentations related to the major KPIs. Engineering and operation was able to take immediate actions with the reported warnings and resolved the issue in time.

Online Open Loop Real Time Optimization

The online process model has been up running to provide advisory for operation engineers. The new process set points for high pressure stripping columns such as the feed flow rates and individual column specs are reported on daily basis. A typical adjustment could increase a few MBD of NGL product. When feed composition shift due to variation of oil feed combination and stabilization depth, the NGL recovery can be enhanced and lined out quickly with ROMeo RTO.

Table 3 shows an example of production increase with RTO. The long term benefit is estimated at least $3MM per our study.

Optimization Summary

Variable	UOM	Value	Change
Net Objective	currency/day	5.628E+006	56752.500
Positive Contributors			
+ Products	currency/day	1.119E+007	54915.900
Negative Contributors			
- Steam Feeds	currency/day	5.563E+006	-1836.620

Table 3

ROMeo has also identified a potential increase of 3MBD recovery with simple process modification. As shown in the figure 3, under the optimization mode, ROMeo tends to divert all the overhead gas from deethanizer feed drum to bypass deethanizer overhead condenser. It has been the normal practice for NGL operation since the system was designed. Similar result was concluded from an independent energy optimization study from corporate process optimization group. The online RTO demonstrates capability of evolving process by justifying the bottleneck of current process design.

Other Monitoring System Development

A couple of monitoring indexes have been proposed for the NGL plant and pipeline operations. For NGL process, a fouling index is defined to capture the early stage of fouling. Instead of relying on the steam flow meter, a fouling index is defined with fundamental equations to remove the impact of temperature gradient across the heat exchanger. The fouling index can provide information for unplanned maintenance schedule.

Table 4 shows the examples of fouling indexes for the kettle reboilers. Out temperature at the cold side is assumed to be the bubble point of the hydrocarbon flow. Hot side temperature is assumed to be the dew point of the superheated steam flow. Temperature difference is calculated with the average temperature from both sides. Based on the engineering data, the heat transfer coefficient in red shows the inconsistent dimension which requires further validation during turn around activity.

Fouling Studies

Variable	UOM	Initial	Value
/340_E3_Duty_CU/e3ua	BTU/hr-F	337104.000	317283.000
/340_E3_Duty_CU/e3u	BTU/hr-ft2-F	67.421	63.457
/461_C_101Duty_CU/e103ua	BTU/hr-F	1.900E+006	1.789E+006
/461_C_101Duty_CU/e102u	BTU/hr-ft2-F	383.799	361.470
/462_stripper_duty_cu/e203ua	BTU/hr-F	2.037E+006	1.905E+006
/462_stripper_duty_cu/e203u	BTU/hr-ft2-F	59.916	56.036
/499_E3_Duty_CU/E3UA	BTU/hr-F	1.515E+006	1.515E+006
/499_E3_Duty_CU/E3U	BTU/hr-ft2-F	69.030	69.030

Table 4

Table 5 shows another environmental variable established for carbon dioxide release. The total amount of carbon dioxide can be tracked with online optimization model. With data collection from historian, justification for carbon dioxide capture can be established because supercritical carbon dioxide injection could be the potential technology for depleted oil fields in the Kingdom.

Overview Stream Economics: CO2 Generation

Variable	UOM	Value	Change
Flow	MMft3/day	59.916	0.000

Table 5

Figure 3 shows another important process variable to be monitored is the predicted hydrocarbon liquid condensate. The offgas pressure will drop along the pipeline to the gas plant for C1-C2 separation and sulfur removal. When the initial temperature is higher than the ambient temperature, hydrocarbon liquid condensate is expected in the pipeline. The amount of hydrocarbon liquid condensate can be predicted with online optimization model which has been updated on daily basis. The amount of accumulated liquid can be correlated to the back pressure incidents for future optimization of pipeline scraping frequency and plant design criteria. For instance, the NGL target recovery and frequency of pipeline scrapping shall be determined based on the total cost analysis.

Tracking of hydrocarbon liquid condensate can be achieved with an online model to mitigate back pressure and flaring of valuable products. This new performance index could potentially save the company millions of dollars.

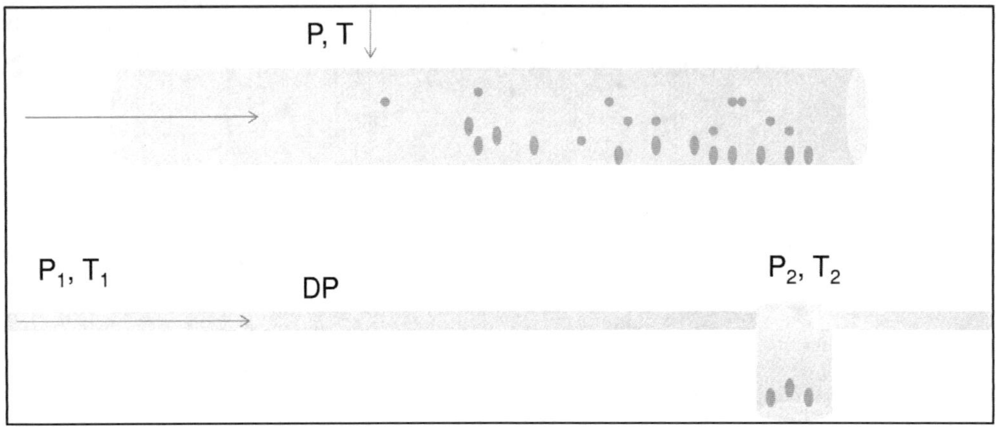

Figure 3

Conclusion

Abqaiq NGL Plant has successfully implemented and tested an in-house developed RTO NGL plant model. A potential increase of $200K to $400K revenue increase for each adjustment has been predicted under the current operating mode and unit operation constraints. In addition to the quantifiable revenue, this presentation also highlights other substantial advantages such as measurement health monitoring, next generation process KPIs, and energy optimization. Future development of this project will focus on supervisory PI Processbook for DCS operators and implementation of engineering details for major equipment.

Acknowledgement

The authors thanks for the supports from the managements of South Area Oil Production Department at Abqaiq and Process simulation and Control System Department at Dhahran under Saudi Aramco Oil Company. This joint technology development has demonstrated a successful story for Saudi Aramco accelerated transformation program.

(29e) Dynamic Simulation of CGC System Under Ethylene Plant Turnarounds

Chuanyu Zhao, Jie Fu and Qiang Xu, Dan F. Smith Department of Chemical Engineering, Lamar University, Beaumont, TX

CGC system (Cracked Gas Compression system), or Multistage compression system, is a vital section during ethylene production. It is located after the cracking furnaces and quench towers, and ahead of the distillation section, which is considered as the "heart" of an ethylene plant because its reliability, stability and sustainability determines the production capacity of that plant. Thus, the CGC system needs to be strictly controlled and operated, especially during plant turnarounds. Generally, plant planned turnarounds include startup and shutdown operations, which have two levels of requirements: i) A feasible and safe startup/shutdown strategy needs to be well developed. Many past turnaround cases failed due to compressor surge, which caused severe equipment damages and unnecessary flaring. ii) An optimal startup/shutdown strategy is better to be pursued if possible in order to minimize startup/shutdown time and reduce flaring emission, raw material usage and energy consumption.

Based on the above consideration, this paper employs rigorous dynamic simulations to help plant turnaround operations, i.e., examine and improve operation strategies of a CGC system for ethylene plant startup and shutdown. For safety consideration, anti-surge process design and control strategies are dynamically evaluated along with alternative turnaround procedures. Dynamic operating trajectories for every stage of CGC are also identified. Assisted by the dynamic simulation, the improvement of plant startup and shutdown performance are demonstrated.

(29f) Dynamic Simulation of Demethanizer and Chilling Train System

Shujing Zhang, Meiqian Wang, Karthik Krishnadevarajan, Qiang Xu and Kuyen Li, Dan F. Smith
Department of Chemical Engineering, Lamar University, Beaumont, TX

The chilling train system is an important part of ethylene plant which is usually combined with demethanizer and a series of flash drums to separate hydrogen and methane from cracked gas, and recover the hydrogen and methane as different products through multi-stream heat transfer with refrigerants. The system of demethanizer and chilling train is the intermediate part of an ethylene plant whose performance can significantly affect the amount of flaring during start-up and thus energy consumption. However, most research efforts are focused on design and steady-state optimization aspects, the dynamic behavior of chilling train is rarely explored. In this study, several control strategies of demethanizer and chilling train system will be simulated and corresponding dynamic responses will be investigated in order to reduce time and flaring during start-up.

General methodology includes the following steps: 1) In the first place, a steady-state model will be built using the sequential modular approach and plant design data; 2) the model will be further validated with normal steady state operating conditions obtained from the DCS (distributed control system) historian; 3) If the steady-state model is satisfied, it will be exported to the dynamic simulation environment. Three types of data will be needed in this transferring, including equipment dimensions, control strategies and P& ID (piping and instrument diagram) parameters, and the heat transfer option; 4) after the dynamic modeling is completed, it will be further validated to match some process upsets in DCS historian; 5) the control strategies will be tested in dynamic simulation to decrease start-up time and energy consumption. A case study is used for the demonstration.

(37a) Impact of Low Levels of Ammonia in Syngas on the Fischer-Tropsch Synthesis Performance of Cobalt and Iron Catalysts in Fixed-Bed Operation

Heinz J. Robota, University of Dayton Research Institute, Dayton, OH, Jhoanna Alger, Energy and Environmental Engineering, University of Dayton Research Institute, Dayton, OH and P. Julius Pretorius, AITF, Edmonton, AB, Canada

The potential for liquid fuel production using the Fischer-Tropsch synthesis with biomass as the carbon source is confronted by a variety of practical issues. Among these is the ability to cost effectively reduce a variety of potential catalyst poisons to acceptably low levels within the capital constraints of a modest capacity biomass-based plant. As noted by Tsakoumis*et al.*in connection with the deactivation cobalt-based FT synthesis catalysts, "Little is known on the effect of nitrogen containing compounds and the available information is mainly found in patents".[1]We have undertaken an investigation into the effects of ammonia on both cobalt and iron FT synthesis catalysts operating under fixed-bed conditions at ammonia levels between 6 ppmv and 300 ppbv. Conversion performance after exposing the Fe catalyst to a steady 6 ppmv ammonia in the syngas remains unaffected relative to the performance in an ammonia-free gas. Conversion performance is stable over several hundred hours of operation without detectable deactivation. In contrast, two different cobalt catalysts both exhibit considerable sensitivity to the presence of ammonia in the syngas. The loss of conversion performance is a linear function of the total ammonia exposure until a N:Surface Cobalt ratio near 0.15 is reached, at which point half of the catalyst's activity is lost. The impact of each adsorbed N entity is quantitatively similar to the impact of S observed during H_2S exposure. Unlike the effect of S, the impact of N entities on cobalt is self limiting. Further exposure beyond N:Surface Cobalt = 0.15 results in no further change in performance. Altered performance persists unchanged for over 100 hours after removal of ammonia from the syngas stream. While the measured α value of the cobalt catalysts is unaffected, the relative selectivity to olefins and alcohols increases.

1) Tsakoumis, Nikolaos E.; Rønning, Magnus; Borg, Øyvind; Rytter, Erling; Holmen, Anders;*Deactivation of cobalt based Fischer-Tropsch Catalysts: A Review*; Catalysis Today 154 (2010) 162-182

(37b) The Synergism of Natural Gas Conversion with Biomass for Maximizing Biofuels Production

Daniel Resasco[1], Lance Lobban[2]and **R. G. Mallinson**[2], (1)School of Chemical, Biological and Materials Engineering and Carbon Nanotube Technology Center, University of Oklahoma, Norman, OK, (2)Chemical, Biological, and Materials Engineering, University of Oklahoma, Norman, OK

Hydrogen from natural gas maximizes the liquid fuel yield from thermochemical biofuel production pathways. This produces the greatest net fossil CO2 reduction compared with biomass produced hydrogen and renewable wind and solar based hydrogen production. This also appears to result in the lowest overall cost per unit of liquid fuel produced.

(37c) Utilization of Supercritical Fluids As Reaction Media for the Synthesis of Higher Alcohols From Syngas

Rui Xu, Sihe Zhang and Christopher B. Roberts, Department of Chemical Engineering, Auburn University, Auburn University, AL

Synthesis of higher alcohols, isobutanol in particular, from synthesis gas (syngas) has been attracting interest due to the potential for producing clean and high octane value fuels. Higher alcohols could be either used directly as fuels, as fuels additives for octane enhancement, or as feed stocks in the chemical industry. The abundance of various carbonaceous sources for producing syngas, including coal, natural gas and biomass, makes this process appealing as an alternative to non-renewable petroleum resources. However, due to the highly exothermic nature of the reactions and the underlying kinetic, this procedure suffers from poor productivity and selectivity towards higher alcohols preventing a large scale and economically feasible commercialized process.

For the last two decades, there has been intensive research centered on shifting the synthesis away from methanol and towards higher alcohols, including; i. the use of modified catalysts, ii. utilizing dual-bed reactors, or iii. slurry phase reactors. In this study, a supercritical medium (i.e. hexanes) was employed to investigate the effect of supercritical solvent on the formation of higher alcohols.

In this work, 0.5 wt% K promoted $CuCo/ZnO/Al_2O_3$was investigated as a catalyst for the synthesis of higher alcohols from syngas. The catalyst was prepared by coprecipitation methods and tested in a fixed bed reactor using a supercritical hexanes as the reaction medium. The results suggest that the presence of supercritical hexanes improved the heat transfer from the catalyst bed. It has also been observed that the formation of $CH_4$was significantly reduced under supercritical hexanes reaction conditions. The results of the catalytic investigations both with and without the supercritical hexane medium demonstrated that supercritical hexane has a significant effect on the selectivity and the productivity towards higher alcohols.

(37d) Cleaner Conversion of Coal to Liquid Fuels with Proven Gasification Technology:

Ravindra Agrawal, Process, KBR, Houston, TX

Gasification is typically conducted at high temperature and high pressure with aid of gasification agents such as oxygen/steam to generate syngas that is rich in CO and H_2. CO and H_2can be used as an intermediate for a wide variety of products including: chemicals, fuels, and metals. The high pressure reduces the volume of the syngas and makes it relatively simpler to remove potential pollutants. Traditionally sulfur compounds from coal are captured with lime and results in a large amount of solid and liquid waste that needs to be disposed off. In gasification process, sulfur compounds can be captured and discarded as sulfur rather than as calcium sulfate, thereby reducing the amount of solid waste to landfill. Also, the high pressure makes it easier to capture CO2 reducing the amount of greenhouse emissions. The high oil price in recent years has increased the demand for CO2 for EOR. This has resulted in a number of projects now seeking to capture CO2. This presentation will discuss some recent projects at KBR that involve proven gasification technologies. The advantages of integrating proven gasification technologies with technologies proven in oil and gas sector to produce liquid fuels from coal will be discussed.

Determining the Rate of Biodegradation in Fuel Impacted Karst Aquifers

Roger Painter[1], Tom Byl[1,2], Lonnie Sharpe[1], Valetta Watson[1] and Tony Patterson[1]

[1] Civil and Environmental Engineering, Tennessee State University, Nashville, TN 37209, USA; rpainter@tnstate.edu, lsharpe@tnstate.edu, ahmadk19@gmail.com, justinvaughnharris@gmail.com

[2] United States Geological Survey, Nashville, TN 37211, USA; tdbyl@yahoo.com,

* Author to whom correspondence should be addressed; E-Mail: Rpainter@Tnstate.edu; 3500 John A. Merritt Blvd., Nashville, TN, 37209 Tel.: (615)963-5388; Fax :(615)963-5902 .

Abstract

Hydrologic and geologic characteristics of fractured rock aquifers have been described as not being suited for natural bioremediation because of small microbial populations. It is also widely perceived that karst groundwater often has insufficient residence time for biodegradation to occur. It is perhaps due to these perceptions that less research has been conducted for quantitative modeling of biodegradation in karst as compared to consolidated aquifers. Modeling biodegradation in karst is in the domain of non-ideal chemical reaction kinetics. For all non-ideal flow reaction zones, the effluent concentration response to an input of tracer must lie between the limiting cases of completely mixed flow and purely plug flow. The residence time distribution function (RTD) for tracer molecules in a single karst conduit or a complex system of conduits is a probability density function which can be interpreted to define the probability that contaminant molecules present at the influent at time equals zero will arrive at the effluent after a particular amount of time. To demonstrate this methodology the biodegradation rate of a contaminant (toluene) in raw karst groundwater from a BTEX impacted site in central Kentucky was quantitatively measured in batch microcosm studies and the extent of biodegradation of toluene in the same groundwater was measured for a complex flow system. The values of the pseudo first order rate constant (k') obtained ranged from 0.017 $(hr)^{-1}$ to 0.0210 $(hr)^{-1}$ compared to 0.0186 $(hr)^{-1}$ for the microcosm experiments. The close agreement between the values of k' obtained from the static microcosms and the ADE model indicate that the ADE does an adequate job of modeling the RTD for modeling biodegradation in karst aquifers. The values of k' obtained correspond to a half-life of less than two days for toluene and this has major implications to issues regarding natural attenuation of fuel impacted karst sites. The RTD approach presented here is applicable to biodegradation in any karst system that has been characterized with a tracer response curve from a conservative tracer study.

Key Words: Bioremediation, Karst, Resident Time Distribution, Advection Dispersion

---

1. Introduction

The high costs and technical difficulties associated with conventional remediation strategies make in situ biodegradation a promising approach for cleaning up contaminated aquifers [1]. The lack of studies examining biodegradation in karst aquifers may be due to the widespread perception that contaminants are rapidly flushed out of karst aquifers [2]. In highly developed and well-connected conduit systems, the rate of contaminant migration is expected to be much faster than the rate of biodegradation [3, 4]. However, the belief that contaminants are rapidly flushed out of karst aquifers is a popular misconception. Tracer studies suggest that large volumes of water may be trapped in fractures along bedding planes and other features isolated from active groundwater flow paths in karst aquifers [5]. In areas isolated from the major conduit flow paths, contaminant migration may be slow enough that biodegradation could significantly reduce contaminant mass if favorable microorganisms, food sources, and geochemical conditions are present [6, 7, and 8]. Researchers have also implied that natural bioremediation in karst or fractured rock is unlikely to occur because of the microbiological characteristics of karst aquifers; small microbial populations and low surface-area-to-volume ratio. Typical microbial numbers for material from unconsolidated aquifers have been reported to range from 1 x 104 to 1 x 107 cells per milliliter (cells/mL) [9].Studies have shown that water from bedrock aquifers also may contain microbial populations within this range. For example, total microbial populations of 9.7 x 105 to 8.5 x 106 cells/mL and heterotrophic bacteria populations of 3.5 x 103 to 5.0 x 105 cells/mL were detected in ground-water samples collected from a gasoline-contaminated karst aquifer in Missouri [10]. Greater than 70 percent of bacteria in consolidated aquifers are attached to solid surfaces. This fact may have led to the assumption that natural bioremediation in karst conduits is negligible because contact between attached bacteria and contaminants would be limited by the smaller surface area to volume ratio for karst.

The literature shows that efforts to mathematically model karst aquifers have followed two different and complementary approaches [11, 12, 13, and 14]:

1. A probabilistic approach that focuses on input events and output responses and takes into account local field observations of flow and transport processes.
2. A determinist approach that takes into account theoretical concepts of aquifer structure, the physics of each mechanism involved in the transfer and flow of water, and from that tries to simulate the hydraulic behavior of the aquifer.

There are four major conceptual approaches for deterministic models for karst groundwater:

1 The equivalent porous medium (MODFLOW);

2 The discrete fracture network;

3 The double porosity continuum;

4 The triple porosity approach – matrix, fracture and conduit.

MODFLOW should be used with caution because the degree of geologic heterogeneity exhibited by karst aquifers precludes the application of Darcy's Law and the use of average model inputs based on the assumed homogeneity of the subsurface [15]. Karst modeling is difficult due to the inability to adequately characterize the heterogeneity of the flow parameters, i.e. determine the geometry and hydraulic parameters of the high-conductivity conduit network [16, 17]. The second problem is related to the development of a computer code based on numerical methods. Strong heterogeneity of the flow medium and different flow regimes in high-conductivity channels and in the low-permeability matrix make the development of appropriate computer modeling codes difficult. The third problem is the transfer of the simulated results to a real system. Because of the higher degree of heterogeneity, uncertainties are much higher for karst systems than for most porous aquifers [18, 19]. Efforts are under way to develop more sophisticated deterministic models for karst and the triple porosity approach shows the most promise going forward since most karst aquifers have the common characteristics of storing most of the groundwater in the primary matrix and secondary fissure system and yet transmitting most of the groundwater through tertiary conduits.

When considering the adequacy and appropriateness of a probabilistic or deterministic modeling approach, it is important to be clear about the objective of the model. Deterministic models work well if the objective is to model hydraulic heads, groundwater fluxes and spring discharge. On the other hand, existing deterministic models for karst aquifers perform poorly for the prediction of flow direction, velocity and solute destination. Consequently a probabilistic modeling approach is appropriate for modeling biodegradation of contaminants in karst aquifers. The release of a contaminant into a karst aquifer and the eventual discharge of the contaminant at a down gradient spring is analogous to the operation of a non-ideal flow chemical reactor. Therefore a further justification for a probabilistic modeling approach is the fact that the state of the art chemical reaction engineering approach for modeling such a contaminant undergoing biodegradation is the application of a probabilistic biodegradation model based on an RTD derived from a field tracer study. Finally, another validation of a probabilistic approach for modeling biodegradation in karst aquifers is the wide use of quantitative field tracer studies to provide information regarding karst flow [20, 21].

Tracer studies are an integral step in determining the RTD of contaminants in a karst system. Much work has been done to develop models for complex groundwater and surface water systems which attempt to describe the RTD in terms of mathematical functions independent of tracer study information. The numerical transient model (OTIS) and the Solute Transport and Multi-rate Transport – Linear Coordinates (STAMMT-L) model are examples of such models [22]. However, a RTD based on tracer study data is often the single most important site specific information for modeling karst conduits. The RTD is not a complete description of a karst system but a given RTD is unique for a particular system. For large complex systems, representative tracer studies of particular conduits can be combined to provide a model of the entire complex karst system. The modeling of complex reaction systems as a number of simpler systems connected in series and parallel is an established method used for reactor design by chemical reaction engineers. In interpreting the results of a tracer study it is assumed that the contaminant experiences the same RTD as the conservative tracer except that the contaminant experiences biodegradation while in transient through the karst aquifer.

USEPA's QTRACER model is a probabilistic model that uses the one dimensional advection dispersion equation (ADE) to model the RTD for karst. In this context, it is a natural extension of chemical reaction engineering to use field tracer data to construct an RTD based on the ADE and couple the RTD to a chemical kinetic model to describe biodegradation in the karst aquifers. It is important to stress that the ADE is not being used here to model the physical characteristics of the aquifer flow. The physical characteristics of the aquifer are subsumed into the RTD and the ADE is modeling the RTD. The appropriateness of this approach is evident in the fit of the model RTD to the experimental RTD derived from the tracer study data.

2. Materials and Methods

Laboratory flow through karst microcosms shown in Figure 1 were constructed using a 20-liter glass reservoir and twelve 1-liter volumetric flasks arranged in three replicate systems of four flasks connected in series.

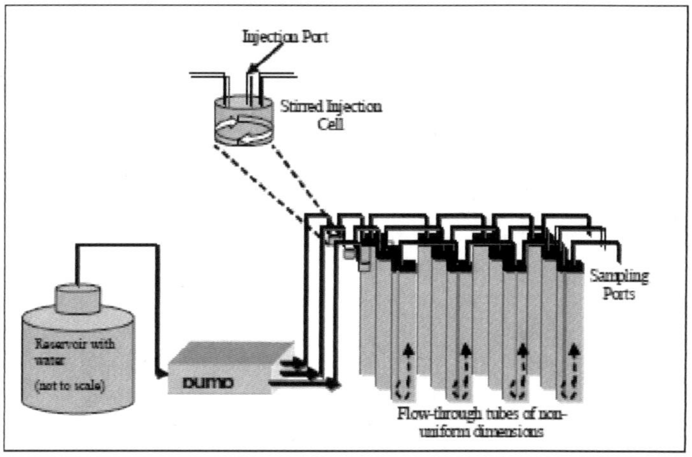

Figure 1: Laboratory flow through karst microcosms.

Water was pumped into all three systems by using a high-performance peristaltic pump. A stirred injection cell (10 mL volume) was placed at the entrance of each replicate system for the injection of tracer or toluene. During the conservative tracer studies Rhodamine dye was simultaneously injected into the stirred injection cells of the packed and unpacked systems. The Rhodamine concentration at the discharge port was monitored through time by collecting samples at 1-2 hour time intervals over a 4 day period. A fluorometer was used to quantify the Rhodamine in the water samples. The lower detection limit on the fluorometer was established at 100 parts per trillion.

Toluene was selected as the experimental contaminant because it is a component in most fuels and because previous work indicated pseudomonad bacteria, which are heterotrophic aerobic bacteria (HAB), from the Kentucky site could grow using toluene as a food source. The biodegradation experiments used water containing live bacteria collected from a 120-foot-deep well completed in a karst aquifer in south-central Kentucky. Bacteria counts ranged from approximately 700,000 bacteria per milliliter to 1.2 million depending on the well and sample collection time. These bacteria counts were derived using two methods, direct counts and BART growth tests, and the results of the two tests were within 20 percent of each other. Bacteria from the fuel contaminated part of the karst aquifer had a 5% lighter buoyant density and a wider range of sizes than the bacteria from the non-contaminated well. Additionally, bacteria isolated from fuel contaminated ground-water samples readily grew with dissolved gasoline as the only source of food. Static microcosms set up using aerated raw karst water and spiked with Toluene at 1 mg/L established a pseudo first order biodegradation rate constant of 0.0186 hr^{-1}. Sterile control microcosms had less than 10% toluene loss over the same time period.

Before the tracer study was initiated, the experimental systems were sterilized with bleach. The bleach was neutralized with sterile sodium thiosulfate. During the conservative tracer study, a constant flow rate of approximately 3 milliliters per minute (mL/min) was established for both systems. The water flowing through the abiotic system was sterile water that had a pH of 10. Previous work indicated that elevating the pH to 10 maintains an abiotic system. At the beginning of the tracer study, 300 micrograms (µg) of Rhodamine dye (1,500 µg of 20% wt/wt solution) was injected into each stirred injection cell. The Rhodamine concentration at the discharge port was monitored through time by collecting samples over a 4-day period.

Before the toluene biodegradation study was initiated toluene was injected into an abiotic system to investigate loss of toluene to vaporization and or adsorption, The experimental systems were sterilized with bleach. The bleach was neutralized with sterile sodium thiosulfate. Filter-sterilized toluene (87 µg) dissolved in 100 micro liters (µL) of methanol was delivered into the injection cell and pushed through the abiotic system with sterile water that had a pH of 10. Previous work indicated that elevating the pH to 10 maintains an abiotic system. Toluene concentration was monitored at the discharge port over the next 5 days. Water samples were collected in clean 40-mL volatile organic compound (VOC) vials every 1 to 4 hours. The water samples were immediately analyzed on a gas chromatograph (GC) equipped with a purge-and-trap system, silica-film capillary column, argon-carrying gas, and micro-argon ionization detector.

The karst groundwater containing live bacteria was then pumped through the system for four days to establish a bio-film on the glass surfaces. An 87-μg aliquot of toluene was then injected into the biotic system in the same manner as for the abiotic test. In order to document the presence of attached bacteria, glass slides were suspended in both the packed and unpacked systems. The suspended slides were removed prior to and at the end of the experiments and viewed using an epifluorescent microscope and the direct count method [23]. Bacteria counts using MPN and microscopic methods were used to confirm that bacteria covered the glass surfaces. Figures 1 and 2 shown below are representative of the photographs of the glass surfaces.

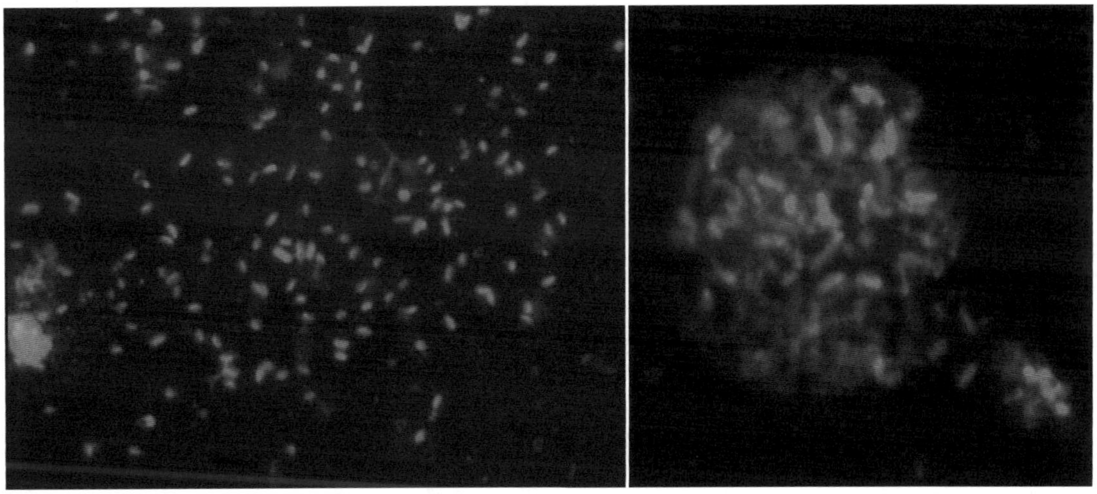

Figure 1. (Left) Bacteria (white objects) attached to the surface of the glass after 3 days of pumping water through the system (400x magnification, epifluorescent), and (**Right**) close up of a bacteria cluster on the surface of the glass (800x magnification, epifluorescent).

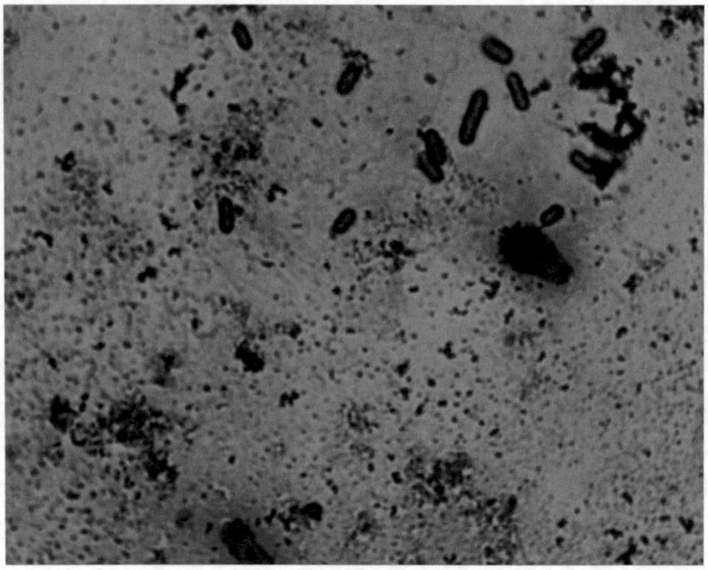

Figure 2. Free-living bacteria (dark objects) collected from the water column after 3 days. Flagella can be observed attached to the rod-shaped bacteria (1,000x magnification, bright field).

For a sparingly soluble contaminant (A) reacting with (B) the reaction may actually be second order or variable order:

$$\frac{dC_A}{dt} = kC_A C_B \qquad \text{Equation 1}$$

Where:
C_A = Concentration of Contaminant;
C_A = Concentration of Microbes;
k = Biodegradation Rate Constant;
and;
t = Time.

But in a case where the microbes are acclimated and at steady state and electron acceptors are not limiting the concentration C_B may not change appreciably while C_A does. Treating C_B like a constant:

$$\frac{dC_A}{dt} = kC_A C_B \Rightarrow k'C_A \qquad \text{Equation 2}$$

Where:

$k' = kC_B$ = pseudo first order rate constant and the biodegradation is modeled as pseudo-first order.

This does not however require that the mechanism of the microscopic chemical reaction be mono-molecular. The actual biodegradation rate is variable order and is a function of electron donor (carbon substrate) and microbial population. However, it is a common practice in chemical engineering to model complex reactions with pseudo first order kinetics. The following application of chemical reaction engineering theory to is strictly valid if the chemical biodegradation kinetics information is liner with respect to intensive fluid properties.

The residence time distribution function (E(t)) for contaminant molecules in a single karst conduit or a complex system of conduits is a probability density function (p.d.f.) which can be interpreted to define the probability that contaminant molecules present at the influent at time equals zero will arrive at the effluent after a particular amount of time. The RTD is depicted as a plot of E(t) versus time as time goes from zero to infinity.

The mathematical approach presented below involves coupling biodegradation kinetics to the RTD by application of the ADE. The transient form of the ADE was solved and fitted to the tracer study data to estimate the systems Peclet number. This Peclet number was incorporated into a second solution of the ADE with a chemical reaction term that was fitted to the toluene degradation data to estimate the pseudo first order rate constant.

E(t) was determined by injecting a pulse of a conservative tracer (rhodamine) into the system shown in Figure 1 at time (t) = 0 and then measuring the tracer concentration in the effluent as a function of time. At some time after injection of the tracer, consider an increment of time dt small enough that the tracer concentration exiting the reactor between t and t + dt is essentially constant. The amount of tracer leaving the reactor between t and t + dt is then in differential form:

$$dN = QC(t)\,dt \qquad \text{Equation 3}$$

Where:
Q = volumetric flow rate;
C(t) = concentration at time t;
and;
N = Mass

Where dN denotes the amount of material that has spent an amount of time between t and t + dt and Q is the volumetric flow rate which is taken to be constant. If N_0 is the amount of tracer injected at t = 0 then the fraction of material leaving the reactor between t and t + dt is

$$\frac{dN}{N_0} = \frac{QC(t)}{N_0}dt \qquad \text{Equation 4}$$

And for a pulse input the residence time distribution function is defined as

$$E(t) = \frac{QC(t)}{N_0} \qquad \text{Equation 5}$$

So that

$$\frac{dN}{N_0} = E(t)\,dt \qquad \text{Equation 6}$$

Noting that N_0 is the sum of all the tracer leaving the system

$$N_0 = \int_0^{\infty} QC(t)\,dt \qquad \text{Equation 7}$$

And combining Equations 3, 6 and 7 gives

$$E(t) = \frac{C(t)}{\int_0^{\infty} C(t)\,dt} \qquad \text{Equation 8}$$

E(t) is the pdf for the random variable C(t) and the first two moments of E(t), define the mean residence time and variance of the distribution. The mean residence time (t_m) is:

$$t_m = \int_0^\infty t\, E(t)\, dt \qquad \text{Equation 9}$$

And the variance (σ^2) is:

$$\sigma^2 = \int_0^\infty (t - t_m)^2\, E(t)\, dt \qquad \text{Equation 10}$$

For the non-ideal flow characteristics of karst E(t) and its moments are obtained by numerical integration of the tracer breakthrough data.

The advection, dispersion equation for the conservative tracer is:

$$\frac{\partial C(t)}{\partial t} = D_a \frac{\partial^2 C(t)}{\partial z^2} - U \frac{\partial C(t)}{\partial z} \qquad \text{Equation 11}$$

Where:
D_a = dispersion coefficient;
z = distance
and;
U = velocity.

D_a is a measure of the degree of mixing in the fluid due to molecular and eddy diffusion as well an indication of the nature of the fluids velocity profile due to viscous effects. This equation effectively superimposes mass transfer of the tracer/contaminant due to axial dispersion equal to $-D_aA_c(dc/dz)$ onto the convective mass transfer which is equal to UA_c. It may appear that this model is capable of accounting only for axial mixing but it can be shown that the model accounts for radial mixing as well.

Equation 11 can be put in dimensionless form to facilitate its solution as:

$$\frac{1}{P_e}\frac{\partial^2 \psi}{d\lambda^2} - \frac{\partial \psi}{d\lambda} = \frac{\partial \psi}{\partial \lambda} \qquad \text{Equation 12}$$

Where,

$$\psi = \frac{C}{C_0};$$
$$C_0 = concentration\ at\ time = 0$$
$$\lambda = \frac{z}{L};$$
$$\theta = \frac{tU}{L};$$
$$and;$$
$$P_e = \frac{UL}{\lambda}$$

After a narrow pulse of tracer is injected, molecular diffusion and eddy diffusion cause the pulse to widen as the tracer molecules disperse randomly in all directions. This is depicted below in Figure 3. Consequently D_a and hence Pe are directly related to the resulting mean residence time and variance of the RTD obtained from the tracer concentration versus time data.

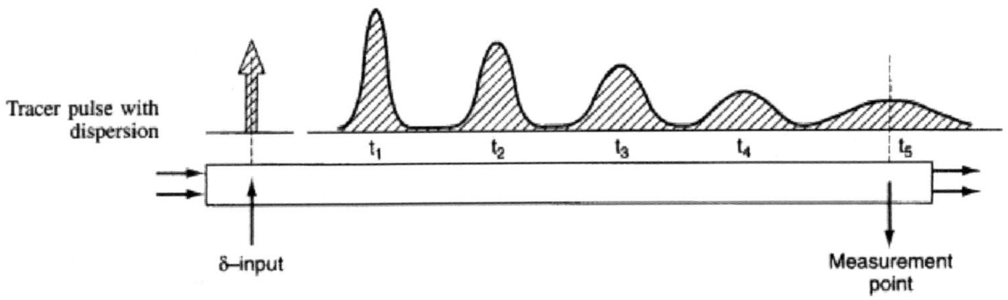

Consequently the solution of Equation 12 for Dankwerts boundary conditions for an open-open system gives the following relationship between the Peclet number (P_e) for the non-ideal flow system, the mean residence time (t_m), and the variance (σ^2) [24, 25]:

$$\psi = \frac{1}{2\sqrt{\frac{\pi\theta}{Pe}}} e^{\frac{-(1-\theta)^2}{\frac{4\pi}{Pe}}} \qquad \text{Equation 13}$$

$$t_m = \left(1 + \frac{2}{P_e}\right)\tau \qquad \text{Equation 14}$$

$$\frac{\sigma^2}{\tau^2} = \frac{2}{P_e} + \frac{8}{P_e^2}$$ Equation 15

Equations 14 and 15 often appear in the literature as equalities, in reality, the equations result from truncation of infinite series and are strictly valid only for large P_e. This is of practical importance since the whole concept of an apparent diffusivity in a dispersion-like equation breaks down for very low values of P_e where the flow approaches mixed flow. Mixed flow on the macroscopic scale is not achieved in typical karst. Fortunately P_e values for karst aquifers are relatively large due to their long effective path lengths. The Peclet number is a measure of the relative contribution of mechanical dispersion and diffusion to solute transport. It relates the effectiveness of mass transport by advection to the effectiveness of mass transport by either dispersion or diffusion. Peclet numbers below 0 4 indicate diffusion control; 0.4 – 6.0 suggest that diffusion and advection are in transition and thus approximately equal to each other; and greater than 6 0 indicates advection control . The Peclet numbers for karst are typically many times greater than 6.0. [26, 27]. In this study, P_e ranged from 12.01 to 14.80.

Values of σ^2 and t_m from the tracer study are used to estimate P_e and τ and these parameters are used to estimate the extent to which the toluene flowing through the system is biodegraded. The experimental value for the fraction of toluene biodegraded was obtained by numerically integrating the toluene concentration versus time data. The conversion as a function of time (X(t)) is:

$$X(t) = 1 - \frac{C_A(t)}{C_{A_0}(0)}$$ Equation 16

X(t) is the fraction of the initial contaminant biodegraded after some time (t) and since the RTD function (E(t)) is the probability density function (p.d.f.) for C(t), the mean or expected value of X(t) is:

$$X_m = \int_0^\infty X(t)\, E(t)\, dt$$ Equation 17

And noting that the rate of change of X(t) is independent of concentration and hence independent of the degree of mixing it follows that for first order kinetics:

$$X(t) = 1 - e^{-k't} \qquad \text{Equation 18}$$

And the mean conversion is obtained by combining equations 17 and 18 [28]:

$$X_m = 1 - \int_0^\infty e^{-k't} E(t)dt \qquad \text{Equation 19}$$

The steady state advection dispersion equation for the toluene undergoing biodegradation as it flows through the system is:

$$\frac{D_a}{U}\frac{dC}{dz^2} - \frac{dC}{dz} + \frac{r_A}{U} = 0 \qquad \text{Equation 20}$$

Where;
r_A = volumetric reaction rate.

This equation is linear for zero or first order kinetics and thus can be solved analytically. For higher order kinetics the equation must be solved numerically. As with the advection dispersion equation for the conservative tracer the equation for toluene undergoing degradation can be written in dimensionless terms as:

$$\frac{1}{P_e}\frac{d^2\psi}{d\lambda^2} - \frac{d\psi}{d\lambda} - D_A\psi = 0 \qquad \text{Equation 21}$$

Where D is the Damkohler number defined as:

$$D = C_{A0}^{n-1}\tau;$$

and for first order reactions (or assuming the reaction can be modeled by pseudo-first order kinetics):

$$D = k'\tau$$

Thus Equation 17 couples information from the RTD obtained from the tracer study to the chemical reaction kinetics term. In steady state mass transfer, the transient terms in the "Open-Open System" boundary conditions disappear and reduce to the Dankwerts "Closed-Closed" Boundary Condition. Furthermore, this equation is linear for first order kinetics and thus can be solved analytically. The solution for assumed pseudo first order biodegradation kinetics is [29]:

$$X_m = (1-\psi_m) = 1 - \frac{4qe^{(P_e/2)}}{(1+q)^2\, e^{(P_e q/2)} - (1-q)^2\, e^{(-P_e q/2)}} \qquad \text{Equation 22}$$

Where;
X_m = the fraction of toluene biodegraded;
and;
$q = \sqrt{1 + 4\, D/P_e}$

3. Results

The experimental values for the conversion (X_m) of toluene from the flow microcosms and the Peclet numbers from the conservative tracer studies allows for the calculation of the values for k'. The results from numerical integration of the concentration versus time data for the three tracer studies conducted are shown in Table 1.

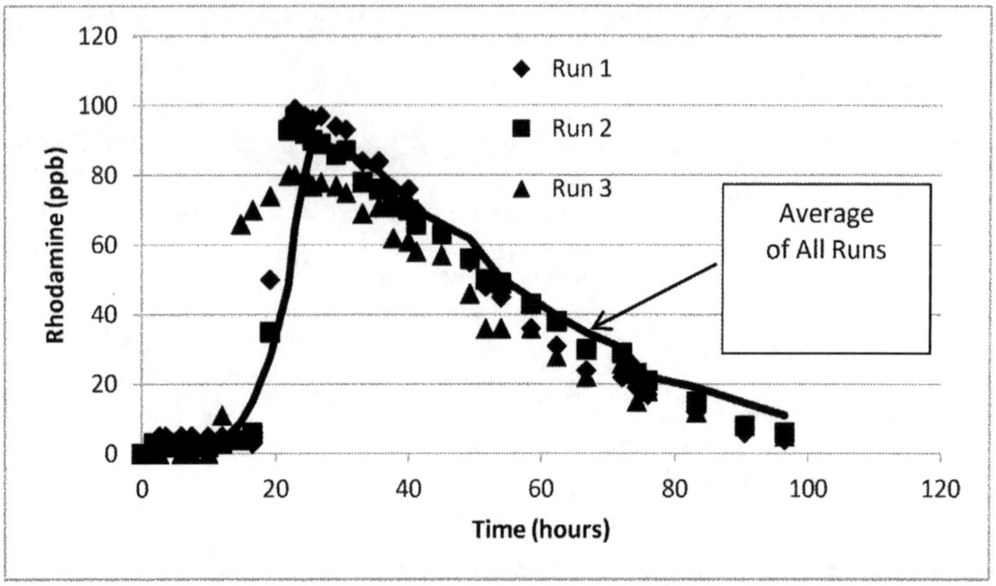

Figure 3. Concentration versus Time data for the conservative tracer test runs. This data was used to determine the mean, variance, and Peclet number for the laboratory karst system.

Table 1 Experimental Results for Tracer Studies			
	Run 1	Run 2	Run 3
σ^2 (hr)^2	297.7	319.4	348.7
t_m (hr)	41.13	43.13	39.65
Space Time (hr)	36.15	38.00	33.99
Peclet Number	14.50	14.80	12.01
Mass in (μg)	476.0	476.0	476.0
Mass out (μg)	461.2	428.8	441.9
Recovery (%)	96.0	90.0	93.0

The toluene data shown in Figure 3 was numerically integrated to determine the experimental values for X_m and the values of k' shown in Table 2 were calculated with Equation 22.

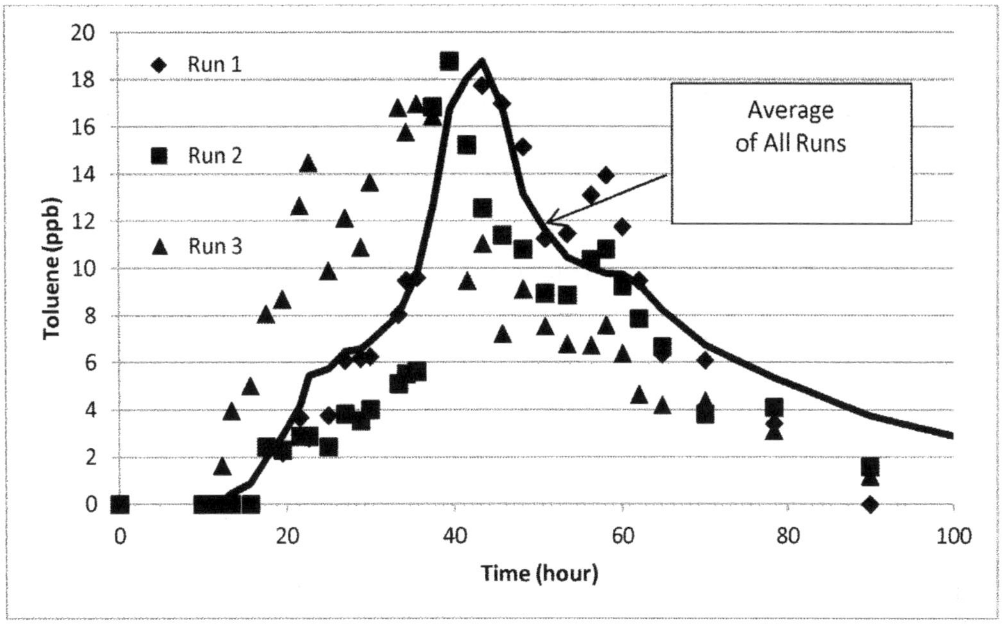

Figure 4. Concentration versus Time data for toluene undergoing biodegradation. This data was used to determine the pseudo first order reaction rate constant for the laboratory karst system.

Table 2 Experimental Results for Biodegradation Studies			
	Run 1	Run 2	Run 3
Toluene Injected (μg)	87.0	87.0	87.0
Recovery Abiotic (μg)	84.6	62.6	75.6
Recovery Biotic (μg)	41.5	36.5	41.6
Conversion (Exp.)	0.44	0.48	0.52
k' $(hr)^{-1}$	.0170	.0180	.0210

To facilitate the comparison of the experimental RTD to the model RTD a dimensionless RTD function was calculated as:

$$E(\theta) = \frac{\psi}{\int_0^{\infty} \psi \, d\theta} \qquad \text{Equation 23}$$

The experimental dimensionless RTDs are compared to the model RTD in Figure 5.

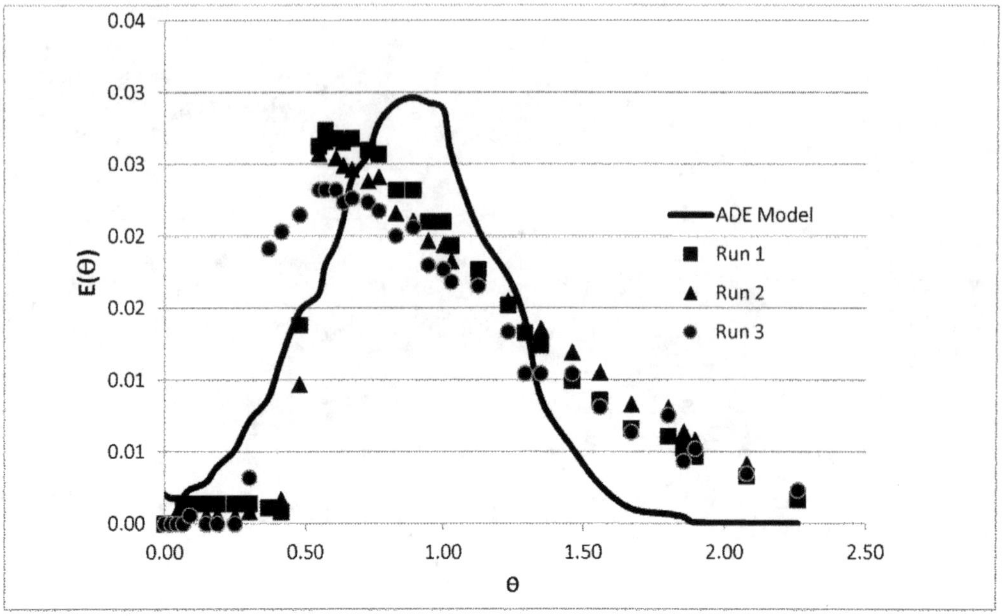

Figure 5. Dimensionless experimental RTDs obtained from numerical integration of the tracer data are compared to the RTD from the ADE.

4. Discussion and Conclusions

The scope of the results of this research is not limited to model development; it also addresses the fundamental issue of whether natural aerobic biodegradation occurs to a significant degree in karst. The perception that contaminants are quickly flushed out of karst aquifers is largely incorrect. Tracer studies for karst aquifers often indicate that these aquifers are characterized by diverse flow regimes and storage capabilities [30, 31]. The range of flow velocities between the least permeable and the most permeable portions of the aquifer is typically over 6 to 10 orders of magnitude [32].

An RTD approach for modeling biodegradation in karst aquifers is validated by the wide use of quantitative field tracer tests to provide information regarding karst flow. It is a natural extension of chemical reaction engineering to use field tracer data to construct an RTD and to model biodegradation in the system. The biodegradation of toluene in groundwater from a karst aquifer in central Kentucky was adequately described by pseudo first-order reaction kinetics. The values of k' obtained from the model ranged from 0.017 $(hr)^{-1}$ to 0.0210 $(hr)^{-1}$. The close agreement between the values of k' obtained from the static microcosms and the ADE model indicate that the ADE does an adequate job of modeling the RTD with regard to the area beneath the RTD curve. This area is proportional to the mass of the contaminant in the effluent. Comparison of the experimental RTDs to the RTD from the ADE in Figure 5 reveal that the solution of the ADE tends to be more symmetric than typical tracer RTD and fails to reflect the long upper tail that is typical of tracer study data. Methanol was used to promote dissolution of toluene to allow the delivery of a true concentration impulse input to the system. This poses a concern as the presence of methanol as a food source could augment biodegradation of toluene artificially. The close agreement of the reaction rate constants for the flow microcosm with the batch microcosm which did not incorporate methanol addresses this concern to some extent. However, this is a legitimate concern and future work may address this concern by conducting the experiment for a stepwise input instead of an impulse. This experimental approach would not require the injection of methanol.

The values obtained for k' correspond to a half-life of less than two days for toluene. This result has major implications for issues regarding the natural attenuation of fuel-impacted sites. This result indicates that the natural aerobic biodegradation of toluene will proceed to a significant degree in karst aquifers despite shorter residence times. This methodology is an adequate and appropriate approach for modeling biodegradation of contaminants traversing any karst system that is characterized by tracer study breakthrough data. The literature reports values of P_e ranging from 6.3 to 22 and in this study P_e ranged from 12.01 to 14.80. The appropriateness of Equation 12 for P_e in this range is directly verifiable by comparing the RTD obtained from the ADE to the experimental RTD from the tracer study data.

References

1. Bombach P, R. H., Kästner M, Fischer A. (2010). "Current approaches for the assessment of in situ biodegradation." *Applied Microbiology Biotechnology*, 86: 839-852.

2. Schwarz, K., T. Gocht, et al. (2011). "Transport of polycyclic aromatic hydrocarbons in highly vulnerable karst systems." *Environmental Pollution* **159**: (1): 133-139.

3. J.-C., Poty, F., Mahler, B. and Drogue, C. (2004), Colonization by Aerobic Bacteria in Karst: Laboratory and In Situ Experiments. *Ground Water*, 42: 526–533. doi: 10.1111/j.1745-6584.2004.tb02621.x

4. Ryan, M., and J. Meiman. 1996. An examination of short-term variations in water quality at a karst spring in Kentucky. *Ground Water* 34, no. 1: 23-30.

5. Field, M.S. Karst hydrology and chemical contamination. *Journal of Environmental Systems* **1993**, *22*, 1-26.

6. Byl, T.D.; Painter, R. Microbial Adaptations to Karst Aquifers with Contaminants. In Proceedings of American Water Resources Association (AWRA) Annual Water Resources Conference, Seattle, WA, USA, 9-12 November 2009.

7. Harvey, R.W.; Barber, L.B, II. Association of free-living bacteria and dissolved organic compounds in a plume of contaminated groundwater. *Journal of Contaminant Hydrology* **1992**, *9*, 91-103.

8. Harvey, R.W.; Smith, R.L.; George, L. Effect of organic contamination upon microbial distributions and heterotrophic uptake in a Cape Cod, Mass., aquifer. *Applied and Environmental Microbiology* **1984**, *48*, 1197-1202.

9. Ghiorse, W.C., and Wilson, J.T., 1988, Microbial ecology of the terrestrial subsurface: *Advances in Applied Microbiology*, v. 33, p. 107-172.

10. O'Connor, J.T.; Brazos, B.J. The Response of Natural Ground Water Bacteria to Ground Water Contamination by Gasoline in a Karst Region. In Proceedings of the Conference on Hazardous Waste Research, Kansas State University, Manhattan, Kansas, USA, 1991; Erickson, L.E. Ed.; Publisher: City, Country, Year; pp 81-293

11. Kiraly, L. 1998. Modeling karst aquifers by the combined discrete channel and continuum approach. *Bulletin Hydrogeology* 16, 77–98.

12. Teutsch, G. and Sauter, M., 1998.Distributed parameter modeling approaches in karst-hydrological investigations. Bulletin d'Hydrogéologie (Neuchâtel), 16, 99-109

13. Smart, C. and Worthington, S.R.H., 2004a. Groundwater in karst. In: J. Gunn (Editor), Encyclopedia of Cave and Karst Science. Fitzroy Dearborn.

14. Domenico, P.A. & Schwartz, W., 1998. *Physical and Chemical Hydrogeology* Second Edition, Wiley.

15. Quinlin, JF. 1989, Ground-water monitoring in karst terrains: recommended protocols and implicit assumptions: Las Vegas, Nev., U.S. Environmental Protection Agency, Environmental Monitoring Systems Laboratory, EPA/600/X-89/050, 100 p. Also available at http://www.karstwaters.org/files/gwkarstr.pdf

16. Wolfe, W.J.; Haugh, C.J.; Webbers, Ank; Diehl, T.H. Preliminary conceptual models of the occurrence, fate, and transport of chlorinated solvents in karst U.S. Geological Survey Water-Resources Investigations Report ; U.S. Geological Survey, 1997

17. Liedl, R. and Sauter, M.: Modelling of aquifer genesis and heat transport in karst system, Bulletin d'Hydrog´eology, 16, 185–200,1998

18. Adams, R. and Parkin, G. 2002. Development of a coupled surface – groundwater – pipe network model for the sustainable management of karstic groundwater. Env. Geol., 42, 4 513-517

19. Field, MS, 1997, Risk assessment methodology for karst aquifers: (2) Solute-transport modeling, *Environmental Monitoring and Assessment* 47, 23–37.

20. Field, MS 2002. The QTRACER2 Program for Tracer Break-through Curve Analysis For Tracer Tests in Karst Aquifers and Other Hydrologic Systems. U.S. Environmental Protection Agency, Office of Research and Development, EPA/600/R-02/001, 194 p

21. Lenore S Clesceri; Andrew D Eaton; Arnold E Greenberg; Mary Ann H Franson; (editors) *Standard Methods for the Examination of Water and Wastewater* : 19th edition supplement, (Prepared and published jointly by:) American Public Health Association, American Water Works Association, Water Environment Federation; Washington, DC, ©1996.

22. Gooseff, M. N., S. M. Wondzell, R. Haggerty, and J. Anderson, Comparing transient storage modeling and residence time distribution (RTD) analysis in geomorphically varied reaches in the Lookout Creek basin, Oregon, USA. Advances in Water Resources, 26, 925-93, 2003.

23. Danckwerts, P.V. Continuous Flow Systems. Distribution of Residence Times. *Chemical Engineering Science* **1952**, *50*, 3857-3866.

24. Shinnar, R.; Naor, P. Residence time distribution in systems with internal reflux. *Chemical Engineering Science* **1967**, *22*, 1369-1381.

25. Fetter, C. W.: 1992, *Contaminant Hydrogeology*, Macmillan Publishing Co. New York. 458 pp

26. Florea, L. J. and Carol M. Wicks, C. M.. "Solute Transport through Laboratory-Scale Karstic Aquifers" *Journal of Cave and Karst Studies* 63.2 (2001): 59-66.

27. Fogler, H. Scott (2006). *Elements of Chemical Reaction Engineering* (4th Edition ed.). Prentice Hall. ISBN 0-13-047394-4

28. Tom D. Byl and Roger Painter, Microbial Adaptations to Karst Aquifers with Contaminants, Tennessee AWRA,2009 Proceedings, Session 2C.9

29. Vogel, T.M. Handbook of bioremediation: Boca Raton, Fla. In Natural bioremediation of chlorinated solvents Norris, R.D.; Matthew, J.E., Eds.; Lewis Publishers, 1994; pp. 201-224.

30. White, WB. 2002. Karst hydrology: Recent developments and open questions. *Engineering Geology* 65 (2002) 85–105 1–21.

31. Bischoff, K.B., and Levenspiel, O., Fluid Dispersion: Generalization and Comparison of Mathematical Models: *Chemical Engineering Science*, **1962**,17, p. 245-255.

32. Tchobanoglous, G.; Burton, F.L.; Stensel, D.H. *Wastewater Engineering, Treatment and Reuse*, 4th ed.; McGraw Hill: City, State, USA, 2003; p. 279.

(54b) Prediction of Condenser Fouling in Thermoelectric Power Plant Cooling Systems

Iman Safari[1], Michael E. Walker[1], Javad Abbasian[1], Hamid Arastoopour[2], Ming-Kai Hsieh[3], Ranjani B. Theregowda[3], David A. Dzombak[3]and David C. Miller[4], (1)Chemical and Biological Engineering, Illinois Institute of Technology, Chicago, IL, (2)Department of Chemical and Biological Engineering, Wanger Institute for Sustainable Energy Research (WISER), Illinois Institute of Technology, Chicago, IL, (3)Department of Civil and Environmental Engineering, Carnegie Mellon University, Pittsburgh, PA, (4)National Energy Technology Laboratory

The thermoelectric power industry in the U.S. uses a large amount of fresh water. As available freshwater for use in thermoelectric power production becomes increasingly limited, use of nontraditional water sources is of growing interest. Utilization of nontraditional water, in cooling systems increases the potential for mineral precipitation on heat exchanger surfaces. In that regard, predicting the accelerated rate of scaling and fouling in condenser is crucial to evaluate the condenser performance. To achieve this goal, water chemistry should be incorporated in cooling system modeling and simulation.

This paper provides a model to estimate the rate of fouling in a recirculating cooling loop as a function of makeup water quality and operational conditions. Detail modeling of salts precipitation in condenser and volatile species desorption in cooling tower is presented. The model developed herein is used to perform case studies of various makeup water quality and treatment strategies to provide insight into the ramifications of degraded water use in the cooling loops of thermoelectric power plants.

(54c) A Combined Cost Model to Evaluate the Economic Impact of Degraded Water Use in Thermoelectric Cooling Loops

Michael E. Walker[1], Ranjani B. Theregowda[2], **Iman Safari**[1], Ming-Kai Hsieh[2], Javad Abbasian[1], Hamid Arastoopour[3], David A. Dzombak[2] and David C. Miller[4], (1)Chemical and Biological Engineering, Illinois Institute of Technology, Chicago, IL, (2)Department of Civil and Environmental Engineering, Carnegie Mellon University, Pittsburgh, PA, (3)Department of Chemical and Biological Engineering, Wanger Institute for Sustainable Energy Research (WISER), Illinois Institute of Technology, Chicago, IL, (4)National Energy Technology Laboratory

The thermoelectric power industry in the U.S. uses a large amount of freshwater. The large water demand is increasingly a problem, especially for new power plant development, as the availability of freshwater diminishes in the United States. Reusing non-traditional water sources, such as treated municipal wastewater, provides one option to mitigate freshwater usage in the thermoelectric power industry. This paper presents the development of an integrated costing model to assess the implications of degraded water use, including treatment requirements and the impact of increased tube scaling in the main condenser. A case study of various treatment scenarios is provided for three plants in regions of differing water availability, to provide insight into the ramifications and benefits of degraded water use as makeup to thermoelectric recirculating cooling loops.

(54d) Can Monochloramine Be Used to Control Biofouling in Recirculating Power Plant Cooling System Using Treated Municipal Wastewater?

Shih-Hsiang Chien[1], David A. Dzombak[2] and Radisav Vidic[1], (1)Department of Civil and Environmental Engineering, University of Pittsburgh, Pittsburgh, PA, (2)Department of Civil and Environmental Engineering, Carnegie Mellon University, Pittsburgh, PA

Biofouling, which involves biological growth in both suspended and sessile form, reduces both heat exchanger and cooling tower efficiency in a thermoelectric power plant. This problem is further aggravated when treated municipal effluent, which contains high concentration of organic matter and nutrients compared to freshwater, is used in a power plant recirculating cooling system as an alternative to freshwater. Although secondary treatment and nitrification remove large fraction of biodegradable components, significant biocide addition is still required to suppress biological activity in cooling systems employing treated municipal wastewater.

The key objective of this study was to evaluate the effectiveness of chlorine-based biocides (sodium hypochlorite, pre-formed monochloramine, and chlorine dioxide) and determine their optimal chemical treatment regimen when using secondary and tertiary treated municipal wastewater as power plant cooling makeup. These three biocides are widely used in water and wastewater disinfection but their effectiveness for biofouling control in highly concentrated (4-6 cycles of concentration, as in recirculating cooling systems) and aerated wastewater has not yet been established. Optimal chemical regimens for achieving successful biofouling control were determined by monitoring biocide usage and heterotrophic planktonic and sessile bacterial populations in both laboratory- and pilot-scale studies. In addition, the cost and degree of environmental hazard associated with each biocide when used under optimal conditions were determined using Life Cycle Costing and Life Cycle Assessment tools.

Laboratory experiments indicated that a pre-formed monochloramine (MCA) dose of 2 mg/L was required to achieve industry biofouling control criteria (below 10^4CFU/mL for planktonic bacteria and below 10^4CFU/cm^{2}for sessile bacteria) at 40 ºC, while sodium hypochlorite and chlorine dioxide required slightly lower initial doses to achieve the same biocidal efficiency. However, pilot-scale studies revealed that less pre-formed monochloramine is needed to maintain sufficient residual and suppress biological growth. Although laboratory studies demonstrated the ability to form MCA in treated municipal wastewater through the reaction between added free chlorine and inorganic ammonia present in the treated wastewater, the addition of free chlorine in pilot-scale studies to achieve a total chlorine residual of 3-4 mg/L resulted in only 23% conversion to MCA. It was observed that 91% of inorganic ammonia in the incoming secondary effluent was stripped out in the cooling tower, leading to unstable in-situ MCA formation. A cost analysis for these biocides was performed based on the results of pilot scale experiments to provide relevant practical information for wastewater reuse. This analysis confirmed that the addition of pre-formed MCA for biocidal control in power plant recirculating cooling systems using treated municipal wastewater is effective and deployable at modest cost. Results of Life Cycle Assessment also suggested that pre-formed MCA has the least environmental and human health impact for this application.

A Statistical Approach to Prediction of Transmembrane Pressure in Membrane Bioreactors

Hiromasa Kaneko, Kimito Finatsu, The University of Tokyo, Tokyo, Japan

Introduction

In water treatment fields such as sewage treatment and industrial liquid waste treatment, the membrane bioreactor (MBR) has been widely used to purify wastewater for reuse [1,2]. MBRs combine biological treatment with membrane filtration. First, bacteria within activated sludge metabolize the organic pollutants and produce environmentally-acceptable metabolites, then a microfiltration (MF) or ultrafiltration membrane separates liquids from solids, i.e. the clean water from the sludge.

However, MBRs suffer from some practical difficulties. One of the critical difficulties is membrane fouling [3,4]. Membrane fouling is a phenomenon wherein foulants, such as activated sludge, sparingly-soluble compounds, high molecular weight solutes and colloids, absorb or deposit on the membrane surface and absorb onto and block the membrane pores. For example, in cases where the MBR is operated under constant-rate filtration, significant energy is required to achieve constant permeate flow rate due to membrane fouling.

To reduce membrane fouling, physical cleaning, for example, aeration, backwashing, or back-pulsing, is performed periodically during MBR operations. Furthermore, chemical cleaning must be carried out with chemical reagents after a given period of processing time, when the transmembrane pressure (TMP) exceeds a given value, because some foulants cannot be removed by physical cleaning and these residual foulants will prevent the recovery of membrane performance with time. Frequent chemical cleaning and replacement of membranes are both expensive.

Hence, to enable chemical cleaning to be performed at an appropriate time [5], membrane fouling must be predicted in the long term. When an MBR plant is operated under constant-rate filtration, this corresponds to prediction of the TMP. Moreover, to enable the distributed MBR systems mentioned above, each MBR must be operated automatically and controlled remotely. Therefore, the TMP must be predicted *a priori* and a schedule of chemical cleaning must be created in advance.

Geng et al. assumed that fouling that could not be reduced by aeration and backwashing was deep pore clogging and that an increase of TMP could be represented as an exponential function of processing time, then fitted parameters of the exponential equation with experimental data [6]. Kim et al. attempted to adapt changes of states in an MBR plant by recursively identifying the parameters of the exponential equation [7]. Additionally, Chen et al. focused on a critical flux [8], below which membrane fouling can be neglected in the short term, and a TMP jump [9], which is a rapid increase in TMP after a period of processing, and constructed two models that predicted the initial increase and rapid increase of TMP, respectively [10]. However, the predictive performance of these models for new data were not confirmed, although the fitting accuracy of the models was high. Furthermore, water quality variables were not considered in the models and hence it is conceivable that they cannot predict TMP well in other MBR plants and lack broad utility.

On the other hand, Meng et al. studied the relationships between membrane fouling resistance and water quality variables [11]. By investigating correlations among the water quality variables and membrane fouling resistance, the results suggested that there should be a direct influence of MLSS concentration, sludge particle size distribution and extracellular polymeric substances and an indirect influence of the other variables on membrane fouling. Thus, water quality variables have multiple effects on MBR fouling phenomena and the simultaneous consideration of these effects is necessary to predict membrane fouling resistance and TMP in the long term in various MBR plants.

Therefore, in this study, we use chemoinformatics [12] to construct statistical models to predict TMP from other MBR parameters, **X**, such as water quality variables and operating conditions. Chemoinformatics uses informatics methods to solve chemistry problems and is used to study a variety of topics, such as quantitative structure–property relationships, quantitative structure–activity relationships, reaction design, drug design, and process systems.

In this paper, we target MBR plants operated under constant-rate filtration, construct prediction models for the increase in TMP per unit time, ΔTMP, from **X**, and predict TMP using the models. Figure 1 shows the basic concept of this TMP prediction. First, a statistical model is constructed between **X** and ΔTMP by using databases of variables that have been measured in the past. Next, the constructed model predicts the increase in TMP from time n to time n+1, $\Delta TMP_{pred}(n+1)$, by inputting MBR parameters in new data **x**(n). The target value of TMP at time n+1, $TMP_{pred}(n+1)$, can be calculated by adding $\Delta TMP_{pred}(n+1)$ to TMP(n), which is measured previously.

When long term prediction of TMP is carried out, the constructed model predicts $\Delta TMP_{pred}(n+2)$, $\Delta TMP_{pred}(n+3)$, ... by inputting expected values or setting values of MBR parameters at times n+1, n+2, ..., respectively. We then use these predicted values as the baseline and can then predict $TMP_{pred}(n+2)$, $TMP_{pred}(n+3)$, ... by adding $\Delta TMP_{pred}(n+2)$, $\Delta TMP_{pred}(n+3)$, ... to $TMP_{pred}(n+1)$, $TMP_{pred}(n+2)$, ..., respectively.

In our study, two methods are used to construct regression models, the partial least-squares (PLS) [13] method and the support vector regression (SVR) [14] method. PLS models are based on linear regression and we can obtain semi-quantitative regression coefficients for each considered parameter to predict the increase in TMP. On the other hand, SVR models are based on nonlinear regression and are reported to have more predictive accuracy than that of neural networks in some cases.

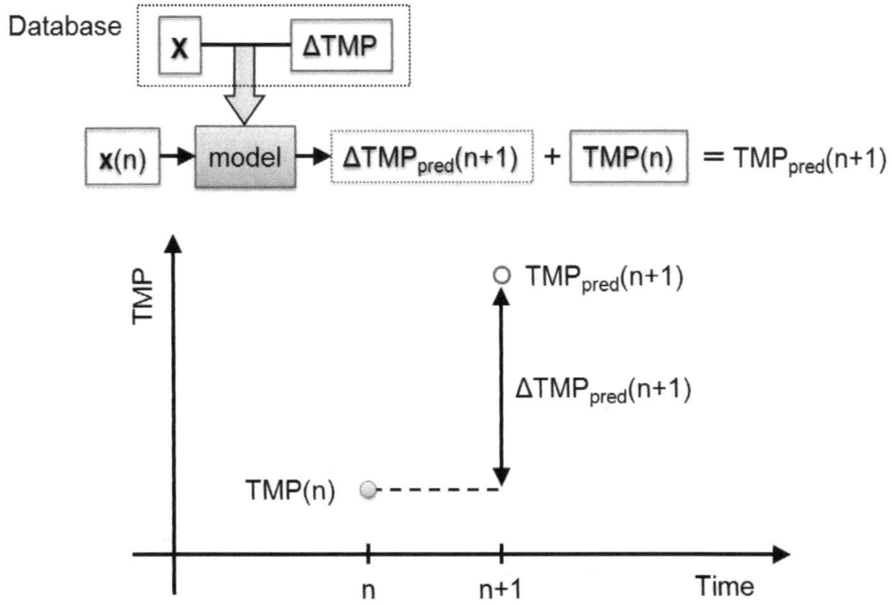

Figure 1. Basic concept of TMP prediction. **X** contains MBR parameters such as water quality variables and operating conditions of a training data set; ΔTMP is an increase in TMP for a training data set; n represents time; **x** represents MBR parameters of new data; ΔTMP_{pred} represents a predicted value of ΔTMP; and TMP_{pred} represents a predicted value of TMP.

Method

PLS

PLS is a method for relating explanatory variables, **X**, and an objective variable, **y**. Using a linear multivariate model, it goes beyond traditional regression methods in that it also models the structures of **X** and **y**. In PLS modeling, the covariance between **y** and the score vector $\mathbf{t}_i$ is maximized.
A PLS model consists of the following two equations:

$$\mathbf{X} = \mathbf{TP}' + \mathbf{E} \tag{1}$$

$$\mathbf{y} = \mathbf{Tq} + \mathbf{f} \tag{2}$$

where **T** is a score matrix, **P** is an **X**-loading matrix, **q** is a **y**-loading vector, **E** is a matrix of **X** residuals, and **f** is the vector of **y** residuals.
The PLS regression model is as follows:

$$\mathbf{y} = \mathbf{Xb} + \mathbf{const} \tag{3}$$

$$\mathbf{b} = \mathbf{W}(\mathbf{P}'\mathbf{W})^{-1}\mathbf{q} \tag{4}$$

where **W** is an **X**-weight matrix and **b** is a vector of regression coefficients. The number of components must be decided appropriately to construct a highly predictive model. r^2 and q^2 values are used as the measure of fit. The details of these statistics are shown in Appendix B. The optimum number of components is determined in this paper by the first local maximum of q^2.

SVR

The SVR method applies a support vector machine (SVM) to a regression analysis and can be used to construct non-linear models by applying a kernel trick as well as a SVM. The primal form of SVR can be shown to be the following optimization problem:
Minimize

$$\frac{1}{2}\|\mathbf{w}\|^2 + C\sum_i |y_i - f(\mathbf{x}_i)|_e \tag{5}$$

subject to

$$|y_i - f(\mathbf{x}_i)|_e = \max(0, |y_i - f(\mathbf{x}_i)| - e) \tag{6}$$

where y_i and $\mathbf{x}_i$ are training data, **w** is a weight vector, *e* is a threshold, and *C* is a penalizing factor that controls the trade-off between a training error and a margin.
By minimization of Equation (5), we can construct a regression model that has a good balance between generalization capabilities and the ability to adapt to the training data. The kernel function in our application is a radial basis function:

$$K(x, x') = \exp\left(-\gamma \left|x - x'\right|^2\right) \tag{7}$$

where γ is a tuning parameter controlling the width of the kernel function. In this study, LIBSVM [15] is used as the machine learning software.

Case Study

We verified the usefulness of the proposed TMP prediction method with real industrial data obtained from operation of MBRs.

Data

The data were obtained from a MBR plant operated under constant-rate filtration. A brief summary of the target MBR plant is shown in Table 1. In this MBR plant, sewage water flows into two lines of the MBR, A and B. MBR system A comprises anoxic and aerobic tanks and B comprises anaerobic, anoxic and aerobic tanks. A membrane module is immersed in each aerobic tank. Flocculants are added in MBR A.

Values of process variables measured online, such as TMP and temperature, were averaged daily while those of other variables, measured offline, such as MLSS concentration and soluble total organic carbon (TOC), could be obtained only once every few days. Time plots of TMP for A and B are shown in Figure 3. At the points where values of TMP decreased rapidly in Figure 2(a) and (b), plant operators had carried out chemical membrane cleaning.

For training data in MBR A, we used data from monitoring that took place from 5th April to 12th September 2003, and for test data we used data from 13th September 2003 to 4th February 2004. The numbers of training and test data points were 119 and 96, respectively. For training data in MBR B, we used data from monitoring that took place from 2nd June to 17th October 2004, and for test data we used data from 18th October 2004 to 12th April 2005. The numbers of training and test data points were 126 and 153, respectively. Data measured during chemical membrane cleaning and shutdown of each MBR were eliminated in advance.

Table 1. Brief Summary of the Target MBR Plant.

Operation	Constant-rate filtration
Treatment object	Urban wastewater
Treatment scale	25–32 m^3/day × 2 lines (A and B)
Residence time	4.3–6.0 hours
MLSS concentration	8–20 g/l
Membrane	Immersion type flat sheet membrane (MF)
Material	Organic polymer
Pore diameter	~0.4 μm
Membrane area	40 m^2/line

Process Variables

We modeled the relationship between an increase of TMP over a period of one day, ΔTMP, and values of other process variables for the day, **X** (Figure 1). If a new value of **X** is obtained, we can estimate the corresponding ΔTMP and predict TMP for the next day by adding this to TMP for the day. Moreover, by using future values of **X**, we can predict future values of TMP. **X**-variables include temperature, inlet flow, variables relevant to water quality and so on, and the measurement frequency for these variables differ. Variables that are rarely measured have little information and therefore the variables measured at a frequency of less than 30% of the TMP measurement were deleted in advance. Table 2 shows the selected process variables. Values of **X** that were not measured were interpolated from measured data. We used a simple interpolation method, where the newest value was used as a missing value, and the Hermite interpolation method [16].

Results and Discussion

We constructed PLS and SVR models predicting TMP through the intermediary of ΔTMP, with training data. The results for MBR A are shown in Table 3. r^2 represents the fitting accuracy of the constructed models and q^2 represents the predictive accuracy of the constructed models. Values close to unity for both r^2 and q^2 are favorable. The low values of root mean squared error (*RMSE*) and $RMSE_{CV}$

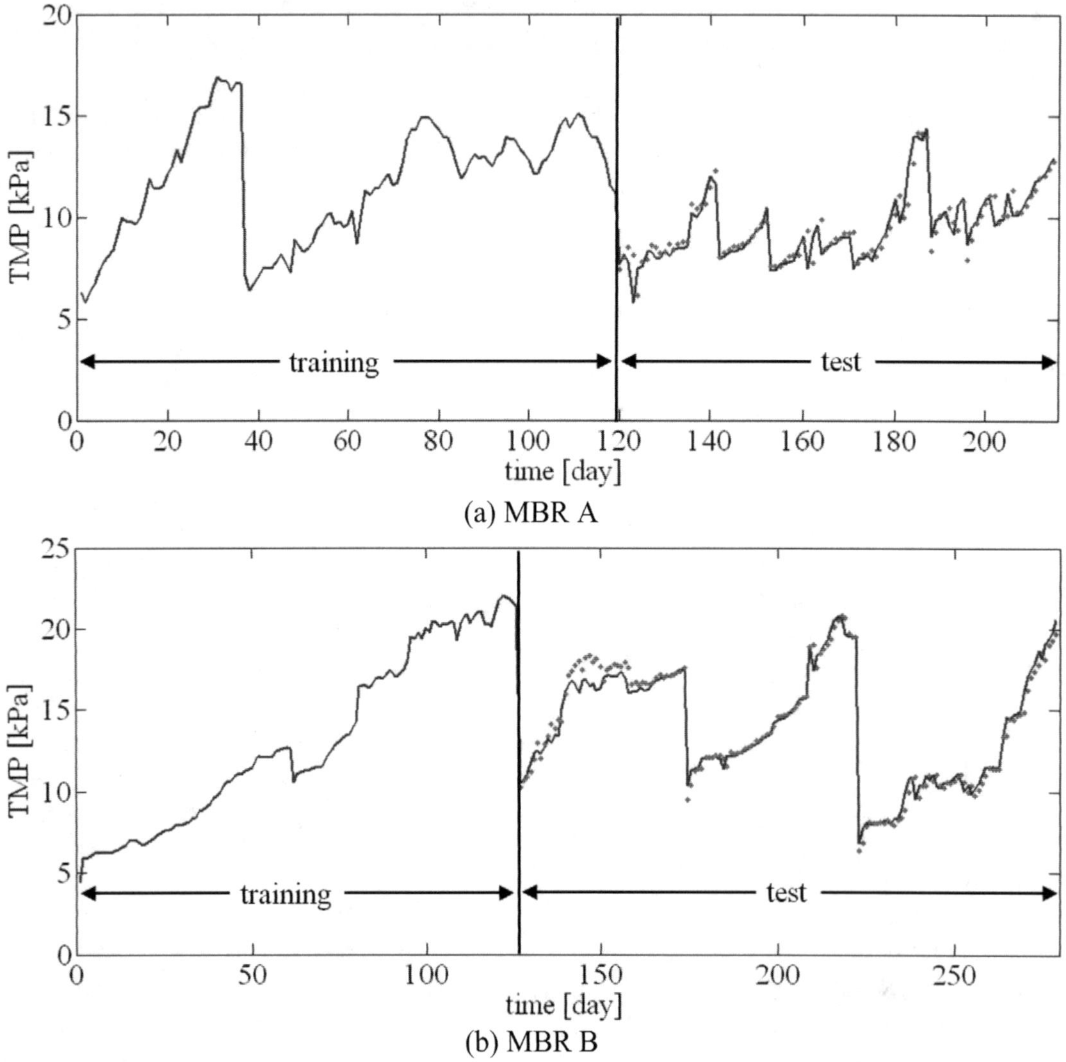

Figure 2. Time plot of TMP. The lines represent the observed values of TMP and the points represent the predicted values of TMP.

indicated that calculation and validation error of the models were small. The predictive model could be constructed with only water temperature, soluble TOC in the aerobic reactor, and MLSS concentration in the aerobic reactor. In addition, there was little difference between r^2, *RMSE*, q^2 and $RMSE_{CV}$ values after applying the simple interpolation technique and those after the Hermite interpolation technique. While nonlinear functions are used in the Hermite interpolation technique, in this case the simple interpolation technique could satisfactorily deal with the data.

Table 2. Process Variables.

Line	Process variable
A	Water temperature Soluble TOC in aerobic reactor MLSS concentration in aerobic reactor
B	Water temperature Treatment flow rate Viscosity in aerobic reactor Filtration rate using a filter paper

For the SVR method, interpolation techniques did not affect the results as much as when using the PLS method, as shown in Table 3.

The modeling results for MBR B are shown in Table 4. All the r^2 and q^2 values exceeded 0.9, reflecting that the TMP models could be constructed with high accuracy. The *RMSE* and $RMSE_{CV}$ values were small. For MBR B, there was little difference between the results using nine and four **X**-variables, similar to MBR A. We concluded that the predictive model could be constructed using water temperature, treatment flow rate, viscosity in the aerobic reactor, and the filtration rate through a filter paper.

The prediction results for MBRs A and B are shown in Tables 3 and 4, respectively. The r_{pred}^2 and the $RMSE_P$ are the r^2 and the *RMSE* for test data, respectively. The r_{pred}^2 value was low and the $RMSE_P$ value was exceptionally high when the data B and the Hermite interpolation are used to construct the TMP model with nine **X**-variables from Table 4. This is because the interval between the day when the variables measured with low frequency were finally measured and the final day of the test data were large,

Table 3. Modeling and Prediction Results for MBR of A.

		Modeling				Prediction	
Reg.[a]	Interp.[b]	r^2	*RMSE*	q^2	$RMSE_{CV}$	r_{pred}^2	$RMSE_P$
PLS	Simple	0.971	0.474	0.970	0.484	0.843	0.665
	Hermite	0.971	0.473	0.970	0.483	0.842	0.668
SVR	Simple	0.990	0.280	0.977	0.422	0.840	0.672
	Hermite	0.985	0.345	0.975	0.447	0.833	0.686

Table 4. Modeling and Prediction Results for MBR of B.

		Modeling				Prediction	
Reg.[a]	Interp.[b]	r^2	*RMSE*	q^2	$RMSE_{CV}$	r_{pred}^2	$RMSE_P$
PLS	Simple	0.997	0.284	0.997	0.291	0.981	0.477
	Hermite	0.997	0.284	0.997	0.291	0.981	0.480
SVR	Simple	0.997	0.285	0.996	0.328	0.983	0.447
	Hermite	1.00	1.27×10^{-4}	0.997	0.298	0.983	0.447

[a] The regression method.
[b] The interpolation technique.

and therefore the values interpolated by the Hermite interpolation became too large. Careful attention is required when applying Hermite interpolation to variables having low measurement frequency. However, even in this situation, prediction can be achieved by using simple interpolation, only **X**-variables with sufficiently high measurement frequencies, or the SVR method, as shown in Table 4.

The $RMSE_P$ values were low, though these are slightly higher than the *RMSE* and $RMSE_{CV}$ values of training data. Almost all the r_{pred}^2 values exceeded 0.8, indicating that the constructed models could predict TMP of test data with high accuracy. In Figure 2, the points represent the predicted values of TMP. As shown in Figure 2(a) and (b), the models could predict the values of TMP, following the observed TMP values with high accuracy. We can therefore conclude that highly predictive models for TMP from other process variables can be constructed by making full use of chemoinformatic methods.

Conclusion

To predict TMP in MBR plants, we proposed statistical models relating the increase in TMP and other process variables with a database of their historical values. The analysis of two sets of real industrial data showed that the measurement frequency of the process variables, interpolation techniques, and regression methods did not affect the modeling and prediction results. However, it is necessary to pay close attention to the use of Hermite interpolation for variables with low measurement frequency. A linear relationship between the increase of TMP and MBR parameters was indicated because there was little difference between the results of PLS and SVR approaches. Furthermore, the effects of the water quality variables and operating conditions on TMP were analyzed using standard regression coefficients obtained by the PLS method.

Variables measured at a low frequency were not be used because they have little information. However, we will be able to use the deleted variables if the frequency of their measurement increases or the variables are predicted by ASM models in the future. In addition, the dynamics of a MBR plant is important and therefore, the predictive accuracy of the prediction models will increase by incorporating the dynamics into the model.

In this paper, we predicted values of TMP over only a one day period. However, we plan to predict TMP values over longer terms. Of course, the constructed models can perform long-term prediction of TMP by inputting expected values or setting values of MBR parameters. However, an improvement of prediction accuracy will be achieved by combining statistical methods with the concepts of critical flux and TMP jump i.e. we will consider water quality, operating conditions, and the phenomena of adsorption and deposition of foulants on the membrane simultaneously.

Our results suggest that the statistical approach used in this study can be a practical tool for the analysis and prediction of TMP. In addition, we believe that by applying our method to process control, MBR plants could be operated effectively and stably.

Acknowledgments

The authors acknowledge the support of the Core Research for Evolutional Science and Technology (CREST) project 'Application of Integrated Intelligent Satellite System (IISS) to construct regional water resources utilization system' of the Japan Science and Technology Agency (JST) and the financial support of Japan Society for the Promotion of Science.

References

1. W.B. Yang, N. Cicek, J. Ilg, State-of-the-art of membrane bioreactors: Worldwide research and commercial applications in North America, J. Membr. Sci. 270 (2006) 201–211.
2. F. G. Meng, S.R. Chae, A. Drews, M. Kraume, H.S. Shin, F.L. Yang, Recent advances in membrane bioreactors (MBRs): Membrane fouling and membrane material, Water Res. 43 (2009)1489–1512.
3. P.L. Clech, V. Chen, T.A.G. Fane, Fouling in membrane bioreactors used in wastewater treatment, J. Membr. Sci. 48 (2008) 534–541.
4. M. Kraume, D. Wedi, J. Schaller, V. Iversen, A. Drews, Fouling in MBR: What use are lab investigations for full scale operation?, Desalination 236 (2009) 94–103.
5. M.J. Kim, B. Sankararao, C.K. Yoo, Determination of MBR fouling and chemical cleaning interval using statistical methods applied on dynamic index data, J. Membr. Sci. 375 (2011) 345–353.
6. Z. Geng, E.R. Hall, P.R. Berube, Membrane fouling mechanisms of a membrane enhanced biological phosphorus removal process, J. Membr. Sci. 296 (2007) 93–101.
7. M.J. Kim, B. Sankararao, C.K. Yoo, Determination of MBR fouling and chemical cleaning interval using statistical methods applied on dynamic index data, J. Membr. Sci. 375 (2011) 345–353.
8. R.W. Field, D. Wu, J.A. Howell, B.B. Gupta, Critical flux concept for microfiltration fouling, J. Membr. Sci. 100 (1995) 259–272.
9. B.D. Cho, A.G. Fane, Fouling transients in nominally sub-critical flux operation of a membrane bioreactor, J. Membr. Sci. 209 (2002) 391–403.
10. Y. Ye, V. Chen, A.G. Fane, Modeling long-term subcritical filtration of model EPS solutions, Desalination 191 (2006) 318–327.
11. F.G. Meng, H.M. Zhang, F.L. Yang, S.T. Zhang, Y.S. Li, X.W. Zhang, Identification of activated sludge properties affecting membrane fouling in submerged membrane bioreactors, Sep. Purif. Technol. 51 (2006) 95–103.
12. J. Gasteiger, T. Engel, Chemoinformatics – A Textbook, first ed., Wiley-VCH, Weinheim, 2003.
13. S. Wold, M. Sjöström, L. Eriksson, PLS-regression: a basic tool of chemometrics, Chemom. Intell. Lab. Syst. 58 (2001) 109–130.
14. V.N. Vapnik, The Nature of Statistical Learning Theory, second ed., Springer, New York, 1999.
15. C. C. Chang, C.J. Lin, LIBSVM: a Library for Support Vector Machines, 2001. Software available at http://www.csie.ntu.edu.tw/~cjlin/libsvm.
16. J.H. Bramble, S.R. Hilbert, Bounds for a class of linear functionals with applications to Hermite interpolation, Numer. Math. 16 (1971) 362–369.

(59a) Sodium Plus Sulfur Promoted Supported Iron Catalysts for the Selective Production of Lower Olefins From Synthesis Gas

Hirsa M. Torres Galvis, Johannes H. Bitter and Krijn P. de Jong, Inorganic Chemistry and Catalysis, Debye Institute for NanoMaterials Science, Utrecht University, Utrecht, Netherlands

The Fischer-Tropsch to olefins (FTO) process is a direct route for the conversion of synthesis gas into light olefins ($C_2^=$-$C_4^=$). The FTO process has to be carried out at high temperatures in order to drive product selectivity to short chain hydrocarbons. Under these conditions, the methanation reaction is kinetically favored resulting in a detrimental effect on product selectivity.

Many studies have been carried out to find a promoter or a combination of promoters that allow an important decrease in methane formation and that could improve the selectivity towards lower olefins simultaneously[1]. It has been reported that promoted bulk iron catalysts modified with sodium and/or sulfur can achieve methane selectivities as low as 10%C and C_2-C_4olefins selectivities of about 35%C[2]. However, up to now, there is no industrial application of these catalysts probably due to their low mechanical stability.

Bulk iron catalysts that are subjected to the very demanding conditions of FTO tend to disintegrate as a result of carbon lay-down[3]. An improvement in the mechanical stability of iron-based catalysts has been achieved by dispersing iron on an inert support that will provide mechanical anchoring for the promoted Fe nanoparticles while allowing the formation of the carbidic active phase[4].

Catalysts with high selectivity towards lower olefins, low methane production and excellent mechanical stability have been synthesized by incipient wetness impregnation of ammonium iron citrate or iron nitrate on alpha Al_2O_3. The sodium and sulfur promoters were incorporated by co-impregnation of aqueous solutions of sodium citrate (or nitrate, depending on the iron precursor) and iron or ammonium sulphate.

The supported and promoted catalysts were tested in the Fischer-Tropsch reaction at 1 bar, 350°C and H_2/CO=1 v/v. The sodium and sulfur promoted catalysts displayed a low methane selectivity (~15%C) while having a high selectivity towards lower olefins (~60%C). The promoted samples prepared with ammonium iron citrate showed a catalytic activity ten times higher than the catalysts prepared with iron nitrate. This behavior is attributed to a higher extent of aggregation of Fe nanoparticles on the samples synthesized with iron nitrate.

[1] Büssemeier, B. et*al.*, U.S. Patent No. 4564642 (1986)

[2] Crous, R.*et al.*, Int. Pat. Appl. No. WO 2010066386 A1(2010)

[3] Kalakkad, D. S.*et al.*, Appl. Catal. A 133 (1995) 335.

[4] Torres Galvis, H. M.*et al.*, Int. Pat. Appl. No. WO2011049456 A1(2011)

E-mail:h.m.torresgalvis@uu.nl. Financial support from NWO/ACTS-ASPECT is gratefully acknowledged

(59b) Hydroisomerization of n-Hexadecane Over Anion Modified Pt/HfO2 Catalysts

Muthu Kumaran Gnanamani, Gary Jacobs, Wilson D. Shafer, Dennis E. Sparks and Burtron H. Davis, Center for Applied Energy Research, University of Kentucky, Lexington, KY

Abstract

Nowadays, isomerization of n-paraffins plays an important role in the petroleum industry [1]. The most widely appliedisomerization catalysts include toxic and corrosive liquid acids, such as H_2SO_4 and HF, metal halides, and metal oxides promoted withCl or F. These reagents are highly corrosive and hence, environmentally hazardous." Many solid acids, such aszeolites, modified metal oxides, and clays, evade these difficulties, and offers to tune not only the number of acid sites but also their strength [2-5]. Among the solid acid studied, ZrO_2 modified by anions such astungstate, sulfate, or phosphate, has attracted considerable attention. ZrO_2promoted with sulfate groups have first been proposed by Holm and Bailey [6] which are active forisomerizationof light straight-chainalkanes at low temperatures (303-423 K)." Further, the stability of sulfatedzirconia catalysts was improved by adding noble metals and other transition metal oxides.

Hino and Arata [7,8], the pioneer in this field reported thatzirconia-tungstatecan be used as a strong solid acid catalysts. Althoughtungstated-zirconiacatalysts are markedly less active than sulfatedzirconia, they showed superior stability and selectivity towardsisomerization by promoted with platinum for largeralkanessuch as n-heptane[9,10]. The reason for the appearance of strong acidity intungstate-containing ZrO_2 remains unclear in many cases and continues to be discussed in literature [11]. Iglesiaet al. [11,12] proposed that platinum particles dissociate molecular hydrogen into hydrogen atoms which spill over to acid sites on the catalyst surface increases n-alkaneconversion rates and isomerization selectivities.

In the present work, the effect ofhafnia on acidic and hydroisomerizationactivity of n-hexadecane was studied using anions such astungstateand sulfate promoted with Pt. Diffuse FT-IR pyridine adsorption technique was employed to estimate the relative amount ofBr nsted and Lewis acids present in the catalysts. Activity and selectivity ofhafniacatalysts were compared withzirconia-tungstate catalysts.

References

1. R.A. Meyers, Handbook of Petroleum Refining Processes, McGraw-Hill, New York, 1996.

2. R.J. Taylor, R.H. Perry, Appl.Catal. A 119 (1994) 121.

3. S. Ernst, J.Weitkamp, J.A. Martens, P.A. Jacobs, Appl.Catal. A 48 (1989) 137.

4. J. Weitkamp, Appl.Catal. 8 (1983) 123.

5. R. Molina, S. Moreno, A.V.-Coelho, J.A. Martens, P.A. Jacobs, G.Poncelet, J.Catal. 148 (1994) 304.

6. V.C.F. Holm, and G.C. Bailey, (Phillips Petroleum Co.), US Patent 3,032,599 (1962)

7. M. Hino, and K.Arata, J. Chem. Soc., Chem.Commun., (1987) 1259.

8. M. Hino, and K.Arata, in Proc. 9thInt.Congr.Catal.,(1988) 1727.

9. E. Iglesia, D.G. Barton, S.L. Soled, S.Miseo, J.E. Baumgartner, W.E. Gates, G.A. Fuentes, G.D.Meitzner, in Proc. 11thInt.Congr.Catal.,(1996) 533.

10.S.L. Soled, E. Gates, E.Iglesia, U.S. Patent 5,422,327 (1995)

11.R.A.Boyse and E.I.Ko, J.Catal. 171 (1997) 191.

12.D.G. Barton, S.L. Soled, and E.Iglesia, Top.Catal. 6 (1998) 87.

(59c) Development of a Fischer-Tropsch Gasoline Process for the Steam Hydrogasification Technology

Yang Li, Chemical and Environmental Engineering, University of California, Riverside, Riverside, CA

Development of a Fischer-Tropsch Gasoline Process for the Steam Hydrogasification Technology

By

Yang Li, J. M. Norbeck, C. S. Park

Chemical and Environmental Engineering

Bourns College of Engineering

Center for Environmental Research and Technology

University of California, Riverside CA

92521

Abstract This paper will present preliminary results from a novel process being developed in our laboratory for the production of high quality F-T gasoline. The system is a combination of a conventional F-T catalyst combined with a zeolite-shell catalyst. This system is being designed for the Steam Hydrogasification process using various carbonaceous feedstocks. The chemical speciation and carbon distribution of the liquid fuel produced with the conventional F-T system will be compared to the new system using the zeolite-shell catalyst. It is found that the zeolite based system shifts the carbon distribution to the lower carbon number species (i.e., C6-C12 range) from the more typical diesel fuel fraction (C12-C20 range) found with FT catalyst systems. Chemical speciation results will be presented along with a assessment of the total carbon conversion expected for this process within the steam hydrogasification technology.

(59d) Direct Conversion of Gasified Biomass to Long Alcohols in Solid Hybrid-Nanoparticles At the Liquid-Liquid Interface of Water/Oil Emulsions

Jimmy Faria[1], Daniel E. Resasco[1], Ali A. Rownaghi[2]and Raymond L. Huhnke[2], (1)University of Oklahoma, Norman, OK, (2)Biosystems and Agricultural Engineering, Oklahoma State University, Stillwater, OK

We have developed a new reaction/separation technology based on a family of recoverable nanohybrid catalysts that simultaneously stabilize emulsions in biphasic systems[1]. These nanostructured solid particles exhibit a unique advantage in the streamlining of biomass syngas to fuels, where the presence of tars greatly complicates purification procedures. These novel catalyst/emulsifier hybrids can catalyze reactions with high "phase-selectivity" either in the aqueous or organic phases. The amphiphilic-catalysts are obtained by fusing carbon nanotubes to metal-oxide particles, which results in a "Janus" like nanoparticle that is able to stabilize water/oil emulsions by forming a rigid film at liquid-liquid interface of the droplets, increasing the apparent viscosity of the system[2]. In the aqueous phase, the tar molecules and small oxygenates can be extracted. Upon phase migration, these species react on the inorganic oxide, which may act both as the hydrophylic side of the emulsifier and as a condensation catalyst. Hence, it is able to catalyze condensation reactions in the aqueous phase, by which small oxygenates soluble in water, with low fuel value, condense via aldol-condensation, ketonization, or etherification. The resulting products are no longer water-soluble molecules and therefore migrate to the organic phase. The oxide used can vary in acid/base characteristics (MgO, and SiO_2).

Ruthenium clusters were anchored onto the hydrophobic carbon nanotubes of the nanohybrids to catalyze Fischer-Tropsch reactions to produce a significant fraction of long alcohols (C5-C15).

The direct conversion of the entire gasified biomass in a biphasic liquid system could greatly simplify the processing of the syngas obtained from biomass gasification, as it does not require the conditioning of the gas prior reaction.

[[1]] S. Crossley, J. Faria, M. Shen and D. E. Resasco,*Science*,**327**, 68 (2009).

[[2]] M. Shen, D. E. Resasco,*Langmuir*, 25, 10843 (2009).

(59e) Comprehensive Kinetic Model for Fischer-Tropsch Synthesis Over a Re-Promoted Co/Al_2O_3Catalyst

Branislav S. Todic[1], Tejas J. Bhatelia[1], Wenping Ma[2], Gary Jacobs[2], Burtron H. Davis[2]and Dragomir B. Bukur[1,3], (1)Chemical Engineering Program, Texas A& M University at Qatar, Doha, Qatar, (2)Center for Applied Energy Research, University of Kentucky, Lexington, KY, (3)Artie Mc Ferrin Department of Chemical Engineering, Texas A& M University, College Station, TX

Abstract

In spite of considerable effort to develop comprehensive kinetic models for Fischer-Tropsch synthesis (FTS), there is still a debate about the reaction mechanism and possible reasons for deviations from Anderson-Schulz-Flory (ASF) distribution. The goal of this work is to develop a comprehensive kinetic model for FTS on promoted cobalt-alumina catalyst and provide an insight into this topic.

Experiments were conducted over 25% Co/0.48%Re/Al_2O_3catalyst in 1L stirred tank slurry reactor over a range of operating conditions (i.e. temperature 220 °C, pressures 200 and 350 psig, H_2/CO feed ratio 1.4 and 2.1 and gas space velocity from 3.6 to 15 NL/g-cat/h). Rate equations were derived on the basis of Langmuir-Hinshelwood-Hougen-Watson approach using well known carbide mechanism for FTS[1]. In order to explain non-ASF behavior it was assumed that the desorption rate of an olefin precursor is a function of carbon number[2].

Kinetic parameters were estimated by minimizing multiple response objective function using Genetic algorithm and Levenberg-Marquardt method, as a global and local optimization tool, respectively. Model evaluation was based on physical meaningfulness of estimated parameters as well as on statistical relevance of the fit.

Model yielded physically meaningful values of parameters. Furthermore, it can be seen from Figure 1 (a-c) that it was also able to correctly predict deviations from the ASF distribution and olefin to paraffin ratio dependence on carbon number.

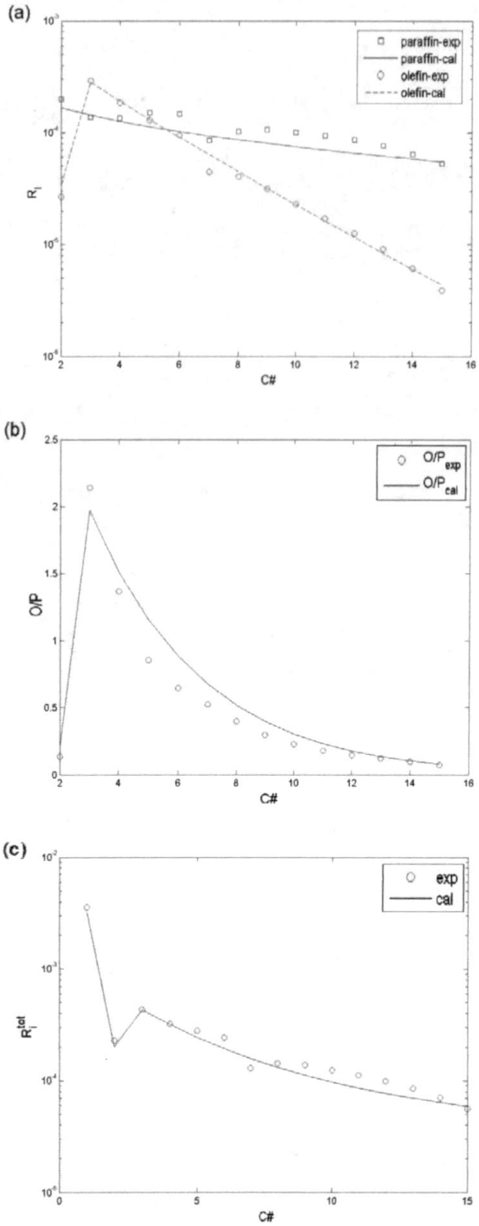

Figure 1 - Comparison of experimental data and model predictions for C_{1-15}hydrocarbons at T=220 °C, P=203.3 psig, H_2/CO=2.1, SV=8.0 NL/g-cat/h (a) rates of formation for n-paraffin and 1-olefin; (b) O/P ratios (1-olefin/n-paraffin); (c) ASF plot (total rates vs. carbon number)

References

[1] LOX E. S., FROMENT G. F. (1993) Kinetics of the Fischer-Tropsch Reaction on a Precipitated Promoted Iron Catalyst. 2. Kinetic Modeling. Ind. Eng. Chem. Res.*32*,71-82.

[2] BOTES G. F. (2007) Proposal of a New Product Characterization Model for the Iron-Based Low-Temperature Fischer-Tropsch Synthesis.*Energy & Fuels*,*21*,1379-1389.

Acknowledgement

This paper was made possible by NPRP grant 08-173-2-050 from the Qatar National Research Fund (a member of Qatar Foundation). The statements made herein are solely the responsibility of the authors.

**Corresponding author: Dragomir Bukur, dragomir.bukur@qatar.tamu.edu*

(59f) Progress on a Comprehensive and Flexible Fischer-Tropsch Synthesis Reactor Model

Joseph W. Pratt, Energy Systems Engineering and Analysis, Sandia National Laboratories, Livermore, CA

Liquid fuels synthesized from renewable CO and H_2could displace petroleum-based fuels to provide clean, renewable energy for the future. Small scale (<20 bbl/day) reactors for synthetic liquid fuels production are an emerging development area that may enable mobile biomass-to-liquids plants suited for on-demand liquid fuel production from diverse, underutilized, local resources. A system compatible with these size requirements may require novel reactor configurations which cannot be efficiently explored without the aid of computer models. Such a model must be general enough to explore non-traditional designs while also being powerful enough to address the known issues of (1) non-uniform temperature distributions, (2) wax buildup, (3) conversion and selectivity optimization, and (4) lifecycle degradation.

Over the last 90 years, the determination of optimal Fischer-Tropsch synthesis reactor conditions has evolved starting with primarily experimentally determined inputs, to increased use of empirical modeling, to incorporation of more theory so that optimal conditions can be predicted and laboratory and pilot plant tests are no longer needed. However, although reactor size and capacity has increased over the years, the same general reactor design today is not much different than that used in the 1940's, especially in the case of fixed-bed reactors. This has led, rightly so, to models that explicitly or implicitly take advantage of this wealth of experience and knowledge with some degree of empiricism. Although they give very accurate results for modeling traditional reactors and/or a specific catalyst, the existing models are unable to examine the significant design changes and innovations that may be needed to solve the challenges described above.

The goal of this work is to create a Fischer-Tropsch synthesis model with the minimum amount of empiricism and assumptions allowed by the current state of the theory, enabling design of efficient, durable, and flexible small scale and/or novel reactors. Accomplishing this requires multiscale heat and mass transport models accounting for inter- and intra-particle gradients, coupled with generalized chemical kinetics models and dynamic (time-variant) degradation phenomena. There is no evidence in the open literature of a model of this type being successfully developed.

This presentation presents our work to-date on the physics component models and the combination of these components into a single comprehensive model, including implementation of schemes for efficient simulation across length scales varying by more than six orders of magnitude. The six component models are:

1. Two-phase interparticle (bulk) flow
2. Interparticle (bulk) heat transfer
3. Intraparticle mass transfer (within pore)
4. Intraparticle heat transfer (within solid and pore)
5. FTS chemical kinetics
6. Time-dependent catalyst deactivation by coke deposition, hydrothermal sintering, liquid/wax accumulation, and poisoning.

The successful, validated model of this type should enable accurate simulation of innovative and non-traditional reactors.

Maximizing Benefits in Selection of Gas Monetization Methodology: Fixed FLNG

Joseph H. Cho, Ph.D., P.E.
Vice President, LNG Plant Unit

Henry Ha
Process Engineer Lead

Wan-Jae Lee
Sr. Vice President, LNG Business Div.

SK Energy & Services
Seoul, Korea

AIChE Spring Meeting 2012
10th Topical Conference on Gas Utilization
Houston, Texas, April 1-5, 2012

ABSTRACT

When Shell announced FID of the Prelude FLNG project, even uncertainties have not been totally cleared up, it is true to get a bright picture in developing stranded wells, which cannot be economically commercialized with the conventional development methodology. However, there are key concerns affecting decision process of an FLNG project development. These may include reserve size with its characteristics (location, water depth, gas to liquid ratio, etc.), technical/commercial risks involved in entire ranges of the project development from evaluating to execution stage.

This paper presents selection criteria and process of natural gas development options in conjunction with technical & commercial issues and risks in developing either a conventional LNG plant or an FLNG option. Newly transformed Fixed FLNG options have been proposed for developing an LNG value chain in areas of environmentally sensitive, remote fields, extremely high labor cost countries, and labor forces shortage areas.

INTRODUCTION

The concept of FLNG, originally known as LNG FPSO (Floating Production and Storage and Offtaking), is to mount the treatment and liquefaction facilities on to a barge-like floating structure, so that processing and distribution can be undertaken at the offshore gas field rather than installing pipelines back to a shore based treatment, liquefaction and export facility. The concept avoids the environmental and social impacts associated with a pipeline to deliver the raw gas onshore from the offshore field, and the coastal modifications and land take that is required for a shore based facility.

FLNG is a niche technology suitable for developing stranded or associated gas fields with midscale production potential. It may be a better commercial solution than conventional LNG value chains, if the size of the field is between 1 and 3 trillion cubic foot (Tcf). Such production potential requires production capacity of 1.0-2.5 MTPA.

The FLNG era has been materialized by Shell's FID in May 2011. It is expected the first FLNG unit will produce LNG by 2017. Since there are many available candidates of wells to be developed with FLNG, the FLNG development concept will be adopted for gas monetization of small or stranded wells in near future.

The FLNG concept can also be transformed for gas monetization at remote areas (no infrastructure is available for construction of a conventional LNG Plant) or specific areas where labor costs are extremely high (or labor productivity is quite low because of both labor costs and labor union). The barge type FLNG can be moored at the fixed mooring facility near shore. This transformed FLNG called F-FLNG (Fixed mooring FLNG) can reduce total installed costs and construction period because the FFLNG can be built at a shipbuilding yard where the working conditions are better than an expected plant job site.

This paper highlights application of F-FLNG and its associated economic and technical benefits compared to both FLNG and conventional LNG plant.

TECHNICAL DRIVERS OF FLNG

FLNG Concept

The FLNG concept provides a technically innovative solution for developing small to medium stranded gas field in a cost-effective manner. This avoids the need for a conventional development scenario comprising off shore platforms, subsea pipelines, onshore liquefaction plant and export jetty. Conceptual outline of an FLNG is illustrated in Fig. 1.

Since an FLNG can be built at a shipbuilder's yard where working conditions are better, higher working productivity can be achieved than other remote areas where less infrastructure is available. This makes FLNG cost competitive against a conventional LNG plant if the development costs are to be compared based on a typical FLNG capacity ranges between 2-2.5 MTPA.

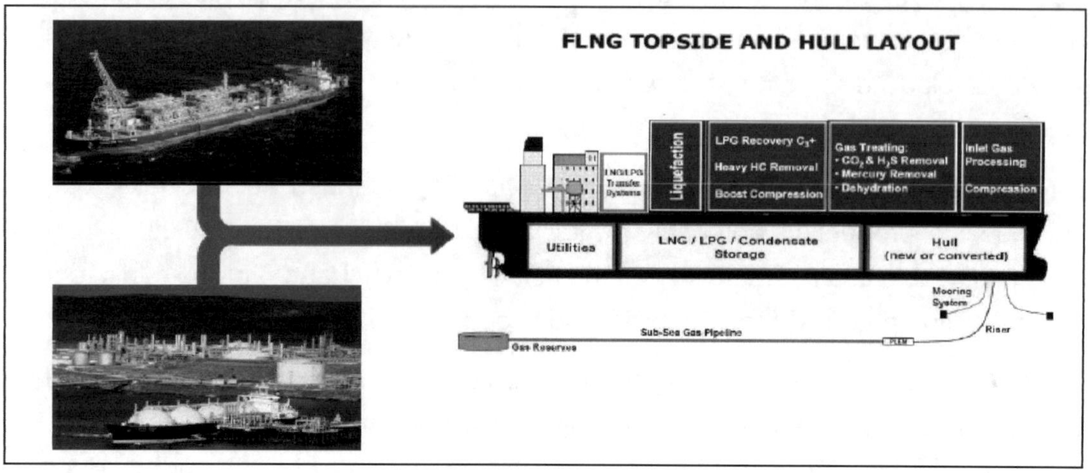

Fig. 1- Conceptual Outline of FLNG

Selection Criteria of FLNG

There are many screening processes and selection criteria before any project developer decides to adopt FLNG concept in developing a gas field. Simple selection criteria comprises of well location and its recoverable reserve size. If subsea pipeline and its associated production platform costs are considerably high, FLNG development can become favorable option. However, in depth feasibility study has to be performed to determine the best development option and an alternative, which identifies technical risks of each feasible development options, economic analysis and other issues. This includes environmental permit, operation permit, geopolitical issues, etc.

The following summarizes selection criteria for FLNG development option [1]:

- Technical - feasibility and safety;
- Environmental - impact during construction, operation and decommissioning, and assessment against compliance with an environmental regulation;
- Commercial - cost, schedule and business case;
- Organizational - do-ability and logistics; and
- Political - access to site and approvals process.

Commercial and Environmental Advantages of FLNG

With a reserve size of about 1 – 3 Tcf, long distance of subsea pipeline and its production platform make the conventional LNG development uneconomical. Elimination of a production platform and considerable distance of subsea pipeline reduces overall development capex and schedule, as such the project makes more economical and less complex.

For the F-FLNG development option, infrastructure that is required for conventional land based LNG plant, such as land acquisition, land development, an export pipeline, trestle and dredging, can be eliminated.

Environmental permit is relatively easier and faster than that of land based case. Opposition from residents, which can be frequently occurred in development of a land based conventional LNG plant, will be insignificant. Moreover, the facility can maintain security and safety against from sabotage at politically unstable areas.

Chemical injection lines contain storage fluid which will be displaced into the flowlines with Monoethylene glycol (MEG) for the MEG lines or compatible buffer fluids for the remaining chemical injection lines. These fluids will be displaced into the flowlines and routed back to the FLNG facility for handling and recycling or onshore disposal.

Technical Advantages of FLNG

FLNG can accommodate facilities for flow assurance line and well control. This includes, gas hydration control (MEG facility), formation water treatment, and umbilicals to control the wells and associated facilities.

In order to minimize the environmental impact, a closed loop freshwater cooling system can be employed. The cooling water can be taken from a depth of around 100 - 200 m, where the seawater temperature is stable and significant lower than that of subsurface. The cold seawater reduces cooling water volumes and increases LNG production.

Modularization with safety distances of each process can allow work safety and easy fabrication at a shipyard, resulting in expedite of overall construction schedule.

Consideration Points in FLNG Option

It is noted that the following should be taken into account in selection of FLNG for offshore gas fields:

- The production capacity of an FLNG facility is limited, due to the deck space required for process equipment, constraining throughput and economics, especially for projects with substantial gas assets. The FLNG sizes are mainly restricted by process configuration and compressor driver output. Two (or less than three) production modules are preferable in view of deck equipment arrangement. The gas turbine drivers currently available at the market have limits of power production. This is main reason of many studies have considered 2-2.5

MTPA FLNG. In order to increase LNG output beyond this capacity, different drivers rather than gas turbines may be considered.

- Expansion for FLNG is limited. It is only achievable by installation of an additional facility. However, expansion of onshore plants can be achieved with less capital costs as a brown field. Onshore expansion requires only incremental, as such saves materials and minimizes environmental impacts because new train uses common infrastructure and utilities of existing train(s).
- FLNG also has higher operating and maintenance costs compared to onshore plants, with very limited operational synergies available for multiple FLNG facilities compared to an onshore plant operating multiple gas trains [1].
- Risks associated with FLNG development have to be taken into consideration. Risks have been well identified and mitigation methodology can be found in the Reference 2.

CONCEPT OF F-FLNG

The basic concept of F-FLNG is almost identical to FLNG except that the F-FLNG is moored at the jetty without weathervaning facility (Turret). The feed gas is supplied by pipelines on a trestle jetty from near shore. Some of facilities, such as quarters, canteen, warehouse, administration offices can be reduced their capacity installed on F-FLNG and moved their capacities to near shore location.

LNG Storage containment system can also be located to onshore. The F-FLNG can then be barge type FLNG. However, the relocation of LNG storage tank to near shore may require vigorous studies in view of land acquisition, ground improvement, construction schedule, construction logistics, and overall costs. Under any situation, the barge requires buoyant system to support the topside gas processing and LNG production facilities.

Figure 2 illustrates F-FLNG compared to FLNG.

.

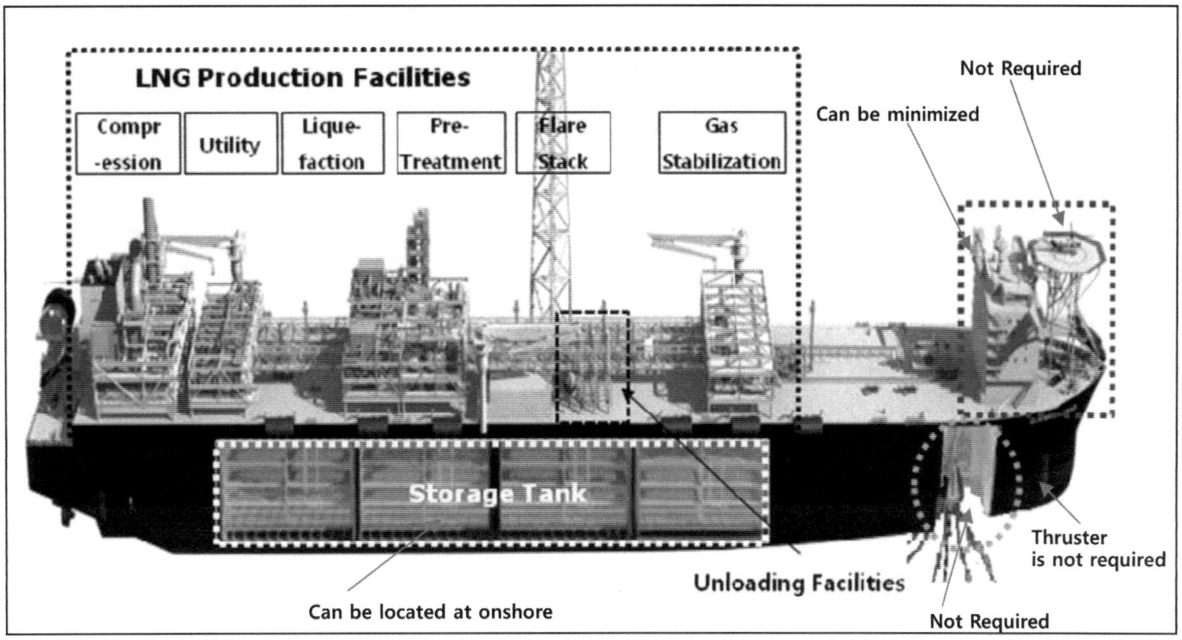

Fig. 2- Typical F-FLNG Outline

APPLICATION OF F-FLNG

Application and Advantages

If many projects are ongoing in any region, acquiring skilled labor is quite difficult, which results in an increase in labor costs. This may cause cost overrun and significant delay in execution of a project. F-FLNG can be applied to near shore or onshore gas monetization where infrastructure has not been well constructed or labor costs are high. Since the F-FLNG needs to be moored on the jetty facility, the sea conditions shall be calm or a manmade breakwater is required.

Because F-FLNG is moored and protected by the breakwater, it also reduces several concerns on even liquid distribution in the high towers, such as the main cryogenic heat exchanger, Natural Gas Liquid (NGL) extraction towers and acid gas removal unit, due to rolling.

F-FLNG needs to make sure available water depth is enough to allow berthing LNG carriers for LNG transfer from the F-FLNG. If it is not enough, there are two possible options may be considered:

- Extension of jetty until the enough water depth can be secured
- Dredging the ship lane so that the carriers can voyage without restriction

F-FLNG achieves high quality workmanship from the well controlled working conditions. The total construction schedule can be reduced compared to stick build conventional LNG plant.

F-FLNG can also apply to a place where getting LNG plant permit is difficult because of environmentally sensitiveness. Different variations of F-FLNG can be designed fit-for-purpose of the site conditions.

Side-by-side LNG transfer technology can be applied to LNG offloading from the F-FLNG because the F-FLNG is moored. The side-by-side technology using the flexible hose has been commercially proven by LNG transfer operation to Floating Storage and Regasification Units (FSRU) from LNG Carriers (LNGC) [3].

Type of Mooring

There are three types of jetty mooring as follows;

1) Jetty moored F-FLNG with side-by-side offloading (Fig.3)

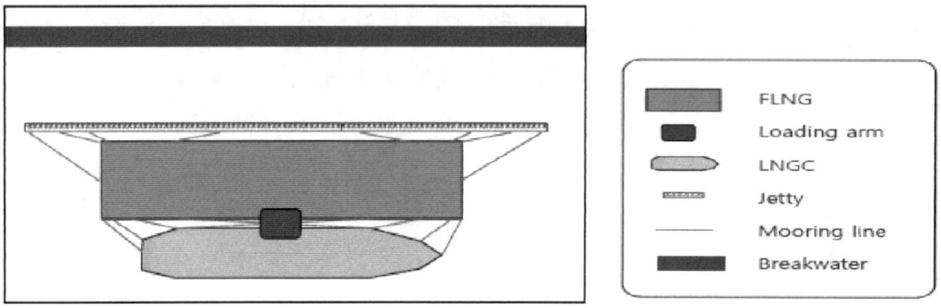

Fig. 3 - Jetty Moored F-FLNG

- No thruster is required
- LNG offload directly to LNGC (Loading arms installed at FLNG)
- Breakwater may be required
- Downtime may be less than turret moored pattern
- Feed gas to be supplied by piping along jetty
- Applicable to shallow water

2) Parallel berthing (Fig. 4)

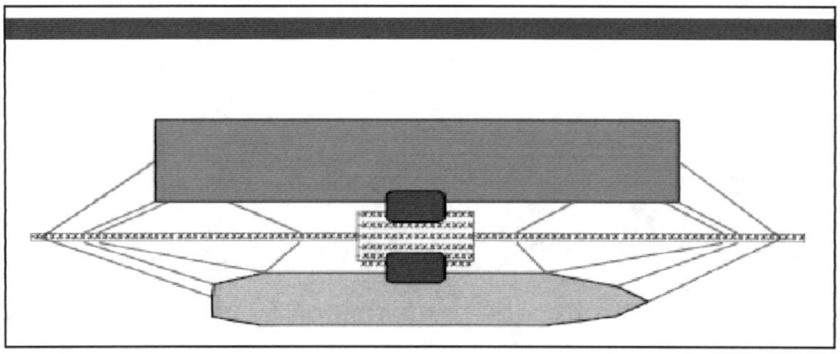

Fig. 4 - Parallel Berthing

- No thruster is required

- LNG offloaded through jetty (FLNG→ Jetty → LNGC)
- Breakwater may be required
- Less downtime (less relative motion)
- Feed gas to be supplied by piping along jetty
- Applicable to shallow water

3) Longitudinal berthing (Fig. 5)

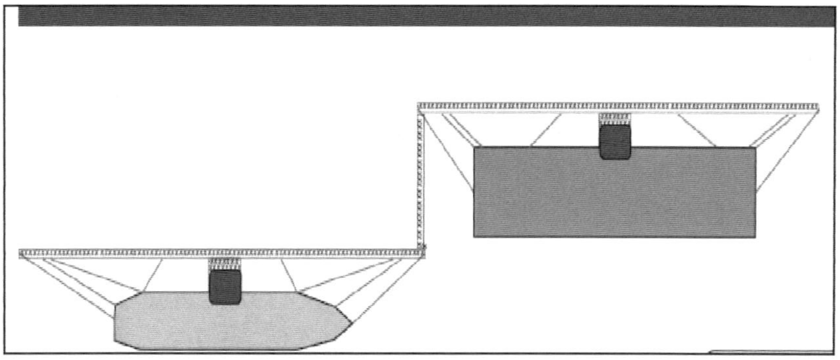

Fig. 5 – Longitudinal Berthing

- No thruster is required
- LNG offloaded through jetty (FLNG→ Jetty → LNGC)
- Longer jetty and breakwater are required
- Less downtime (less relative motion)
- Feed gas to be supplied by piping along jetty
- Applicable to shallow water

ECONO-TECHNICAL ADVANTAGES OF F-FLNG

F-FLNG Facilities

Since F-FLNG is moored and partial function of the following can be moved from FLNG to onshore, overall construction costs can be reduced:

- Quarters
- Warerhouse
- Administration building
- Heliport
- Azimuth Thruster
- Turret mooring system

One of the expensive equipment of the FLNG is a turret mooring system, which is defined as a mooring system where lines are connected to the turret which via bearings allows the vessel to rotate around the anchor legs. This turret can be mounted either internally or externally. An external turret is fixed, with appropriate reinforcements, to bow or stern of the ship. In the internal case the turret is placed within the hull, in a so called moon pool. The chain table, connecting the mooring lines to the turret, can be either above or below the waterline.

External Turret Mooring Systems provide an excellent solution for a wide range of FLNG and oil FPSO applications. These proven single-point mooring systems permit the vessel to freely "weathervane" 360 degrees, allowing normal operations in moderate to extreme sea conditions. External turret systems are less expensive than internal turret designs and can be delivered in a shorter period of time [4]. External turret mooring systems can be mounted at either the bow or stern of converted tankers or new-built vessels.

Internal permanent turret mooring systems provide FLNG/FPSO operators with maximum mooring and fluid transfer capabilities in remote, deepwater and/or harsh environmental conditions. For locations with severe cyclonic weather and harsh sea conditions, the internal turret mooring system provides full weathervaning and enhanced mooring load capacities that enable the FLNG/FPSO to remain permanently on location under the most demanding conditions [4].

The manufactures of the turret mooring system are also limited. Figure 6 presents Internal turret and External turret mooring system.

Fig. 6 – Internal and External Turret (courtesy of Offshore mooring.org)

Since the F-FLNG can be classified as non-vessel, the following safety facility can also be reduced:

- Life boat and Survival raft
- Heli-Deck

<u>Economic Advantages</u>

Majority of quarters can be located to onshore and minimum number of quarters will be equipped in the F-FLNG. Canteen is also minimized on the board. Instead, onshore cafeteria can serve meals for onshore support team and small onboard cafeteria can serve F-FLNG crews only.

An azimuth thruster is a 360 degrees rotatable thruster that is often used for main propulsion. The control system comes with an interface for the auto pilot system. An interface for dynamic positioning systems is optional when enhanced maneuverability is required in applications such as dynamic positioning, ship escort and ship docking services. However, this system can be eliminated because F-FLNG is moored and protected by the breakwater.

For the F-FLNG, mandatory safety facilities for a vessel can be reduced. Inspection and design requirements are less severe than those of vessel classification.

F-FLNG crews' wages and paid time off are different from FLNG. It is intuitively believed that F-FLNG operating costs are less than FLNG.

Construction costs of a breakwater is quite site specific depending on water wave direction and strength, water depth, and its required length is major factors in the cost estimate.

Following table 1 present relative costs comparison of FLNG and F-FLNG. The main factor impacting CAPEX of F-FLNG is seawater condition. If a breakwater is required, the F-FLNG project will increase by about 5%~10% of the CAPEX of the F-FLNG.

Table 2 shows the CAPEX distribution of FLNG and F-FLNG. The basis of CAPEX estimate is as following LNG production.

- Capacity : 2 MTPA
- Sea State : Benign
- Breakwater : Not required (for F-FLNG)
- LNG Storage : Equipped F-FLNG

Table 1 - Relative Costs Comparison between FLNG and F-FLNG (Unit, %)

Item	FLNG	F-FLNG	Remark
CAPEX	100	85~95	5~10% Difference by Breakwater
OPEX	100	75~85	

Table 2 - CAPEX Distribution of FLNG and F-FLNG (Unit, %)

Item	FLNG	F-FLNG	Remark
Pre-FFED/FEED/PMC	2.5~3.0	2.5~3.0	
Topside	60~65	65~70	Fire fighting and emergency gen. costs for F-FLNG can be reduced
Hull	25~30	25~30	For F-FLNG, sloshing is not considered in selecting cargo containment system. The thruster and survival craft is not required.
Turret	10~15	0	
Jetty	0	2.5~4.0	
Onshore Office	0	2.5~4.0	
Logistics Base	2.5~4.0	2.5~4.0	Additional warehouse for emergency supply is not required for F-FLNG
Total	100	100	

CONCLUSIONS

The following conclusions were obtained from this study:

- If many projects are ongoing in any region, acquiring skilled labor is quite difficult, which results in an increase in labor costs. The F-FLNG can be economically and effectively applied to near shore or onshore gas monetization where infrastructure has not been well constructed or labor costs are high.
- F-FLNG can also apply to a place where getting LNG plant permit is difficult because of environmentally sensitiveness.
- Because F-FLNG is moored, it also reduces several concerns on even liquid distribution in the high towers, such as the main cryogenic heat exchanger, NGL extraction towers and acid gas removal unit, due to seawater moving.
- Overall F-FLNG CAPEX and OPEX can save by 5~15% and 15~25% of the FLNG case, respectively.

REFERENCE

1. Shell, Project Description of Prelude FLNG Project, 2011.
2. Cho, J., Ha, H. and Kim D.Y.: "Critical FLNG Project Enablers: Identified Risks and Mitigation", proceeding of AIChE 2011, Chicago, Illinois, March 13-16, 2011.
3. Bokhorst E.V.: "Progress in the Qualification of Components and Systems for LNG Transfer", presented at LNG 16 conference, Oran, Algeria, 17-21 April 2010.
4. Web data: http://www.offshoremoorings.org/moorings/2006/groep4/, 2012.

Author Biography

First Author:
Name: Joseph H. Cho
Position: Vice President
Company: SK Energy & Services

Biography:
Joseph Cho is a Vice President in the LNG Biz Development Div. at SK E&S. He is a registered professional engineer and has been in the LNG industry for twenty six years. He was involved in process design of LNG liquefaction plants and import terminals for various regions as a senior engineer at top tier EPC companies. His scope of work includes managing LNG/Oil/Gas technology, feasibility and optimization studies, economic evaluations, computer simulations, process equipment specifications, control philosophy and review of output from general engineering. He graduated from the University of Oklahoma with a Ph.D. degree in Petroleum Engineering.

ADVANCED TECHNOLOGY HORIZON OF FLNG

Kiil Nam, Jinsang Park, T.J. Kim and Yoon Choon Kim, Hyundai Heavy Industries, Ulsan, Korea

Introduction

With depletion of conventional oil and gas fields, offshore installations have become complex and capital extensive, requiring development of new processes that should be able to operate satisfactorily and economically even for non-conventional environments. The FLNG (Floating Liquefied Natural Gas Plant) is introduced to develop stranded gas fields that have been considered as uneconomical in the past. The FLNG enables natural gas to be produced and liquefied on a floating vessel and offloaded directly onto a carrier without touching onshore facilities. However, there are many challenges that have interrupted the realization of the FLNG project. The installation needs gas purification and liquefaction processes which are completely new in the offshore industry though they are proven technology in the onshore LNG sector. During the conceptual design of such a new process, the selection of the best among the candidates is a formidable task due to the intrinsic features of offshore plants which are characterized by being uncertain, safety-critical, independent and motional.

The FLNG is uncertain for not only adopting onshore LNG process to fit with limited space but also achieving performance and high safety level as per offshore regulation. Safety-criticality stems from the hazards due to treating the flammable hydrocarbon materials in the limited and congested space. A list of common hazards in offshore plants is fire and explosion. The FLNG should additionally offer cryogenic spill protection. Being independent implies that the offshore installations should produce such required utility as electric power, cooling medium and instrument air for themselves. Moreover, they should be able to protect themselves and the people onboard without external helps. Motion comes from offshore environment. The design of the FLNG should consider the mechanical fatigue on structures and the dynamic impact on liquid motions in process facilities including sloshing effect in the tanks.

HHI (Hyundai Heavy Industries), the world's biggest shipbuilder and offshore facilities contractor developed its own Hyundai FLNG model in association with Linde AG, an international industrial gases and engineering company of Germany. The FLNG production capacity is 2.5 MTPA (Million Ton Per Annum) for generic design approach.

The paper presents advanced design features developed by HHI and Linde AG for materialization of FLNG as a market dynamic driver.

Design Approach of FLNG

Selection criteria for liquefaction process

The selection criteria for the liquefaction process are classified into two categories: safety and performance.

The safety requirements should align with the risk management standards and methodologies recommended by ISO 31000 and ISO 17776. The offshore requirements in material selection and layout should be considered in order to prevent the potential risk caused by loss of containment of flammable fluids. A safety and loss prevention management should implement a risk management process including safety studies for the identification of all health and safety related hazards, the assessment of risks to personnel, facility and environment and the reduction of risks by suitable control measures.

The requirements for performance are as follows;

- The process should be economical. The economic evaluations should be performed in a global sense, not in a local sense. The optimized design of a partial system does not necessarily guarantee the optimum of the whole process. As one of global economic performance parameters, the process availability should be considered in the economic evaluation. Complicated process configuration may give better energy efficiency and simultaneously less availability.
- The capacity and configuration of the liquefaction process should be inclusive of the flashed gas and BOG (Boil-Off Gas). The flashed gas occurs when the pressurized LNG is introduced into the cargo tanks while the BOG is generated due to the heat ingress into the cryogenic cargo containment system.

The candidate processes proposed for FLNG are between two fundamentally different processes; refrigerants and nitrogen expander types. The nitrogen expander process uses gaseous nitrogen as refrigerant. Most large LNG plants apply the refrigerants type, either as single or dual mixed refrigerant. For the capacity of less than 2 MTPA, the nitrogen expander process is preferred since the nitrogen expander process is simple and safe in small capacity. For the range capacity of 2~3 MTPA, the SMR (Single Mixed Refrigerant) is suitable and for the capacity of more than 3 MTPA, the DMR (Dual Mixed Refrigerant) is selected since the nitrogen expander and the SMR processes need the more trains due to the limitation in capacity per train.

Based on 2.5 MTPA, the SMR is a good selection due to higher efficiency and lower rotating equipment count than the nitrogen expander.

The results are given in Table 1, showing advantages of refrigerants type for the selection criteria.

Table 1 Comparison of Liquefaction Processes for FLNG

Criteria	N2 Expander	SMR
Energy Consumption	144%	100%
Safety	Inherent Safe	• Flammable Refrigerant • Safety gap required
No. of Refrigerant Train (Based on 25MW Gas Turbine)	5 trains	4 trains
Min. required rotating equipment count	• 1 N2 Compressor • 1 warm expender/compressor • 1 cold expender/compressor Total count : 15	• 1 refrigerant compressor Total count : 4
Plot space requirement	High	Medium

Operation	• Simple process • High No. of rotating equipment • One phase (Gas) flow of Refrigerant	• Relative simple process • Two phase flow of Refrigerant

Selection criteria for power generation system

The main power consumers of an FLNG are the liquefaction unit. The compressors of the refrigerant cycle of the liquefaction unit consume a huge amount of power, which requires a massive quantity of fuel gas. For the power generation of the compressors, various alternatives are considered, each of which is based on electrically driven types and mechanically driven types by steam turbines and gas turbines. The selection criteria are set up to answer the following questions.

- What kind of power generation systems is applicable? What are the basic characteristics of them, including efficiency, required fuel gas properties, physical dimensions, reliability, etc.?
- What configurations are conceivable based on the power generation systems?
- What configuration is economical? What is the implication of the economical evaluation?
- Which is the optimum solution?

The requirements for power generation system should be economical and flexible to fuel gas. The fuel gas consumption is not important in terms of the overall economic evaluation. The process availability should be considered in the overall economic evaluation. The power generation system is capable of utilizing the following gas as the fuel gas.

- Regeneration gas from the columns of pre-treatment Unit
- Non-condensable gas from liquefaction unit
- Flash gas from depressurized LNG
- LNG boil-off gas (BOG) from LNG storage tanks

The combination of the power generation systems gives five alternatives as shown in Table 2. These systems are commonly capable of utilizing gas fuels, which are abundant in the FLNG. The use of the fuel oil as the fuel source is not considered in the economic evaluation.

Table 2 Alternatives for Power Generation Systems

Alternatives	Compressor Drivers of Liquefaction Unit	Others Power Consumers
Alt. 1	Steam Turbine Generator	
Alt. 2	Gas Turbine Generator	
Alt. 3	Mechanically driven by GasTurbine	Gas Turbine Generators
Alt. 4	Mechanically driven by GasTurbine	Steam Turbine Generator
Alt. 5	Mechanically driven by SteamTurbine	Steam Turbine Generator

Table 3 shows efficiency, failure rate and repair time of the power generation systems. The efficiency of boiler and steam turbine is inferior by about 30 %. However, the failure rate of boiler and steam turbine is about 85% lower. Consequently, boiler and steam turbine is less efficient but more reliable and less maintenance-demanding.

Table 3 Comparison of Alternatives for Power Generation Systems

Power Generation System	Thermal Efficiency, %	Failure rate, per 10^6 hr	Repair time, hr
Boiler + Steam Turbine	32	106.5	39.5
Gas Turbine	40	756.8	23.7

The production availability is the measure of the production loss. It is caused by the failure, repair and maintenance of equipment. It is defined as the ratio of actual production rate to planned production rate over a specified period of time.

The production availability was estimated with five alternative configurations as shown Fig. 1. Alt. 3 and 4 show lower production availability since the mechanically driven type by a gas turbine not only has lower availability than that by a steam turbine but also cannot be considered as redundancy.

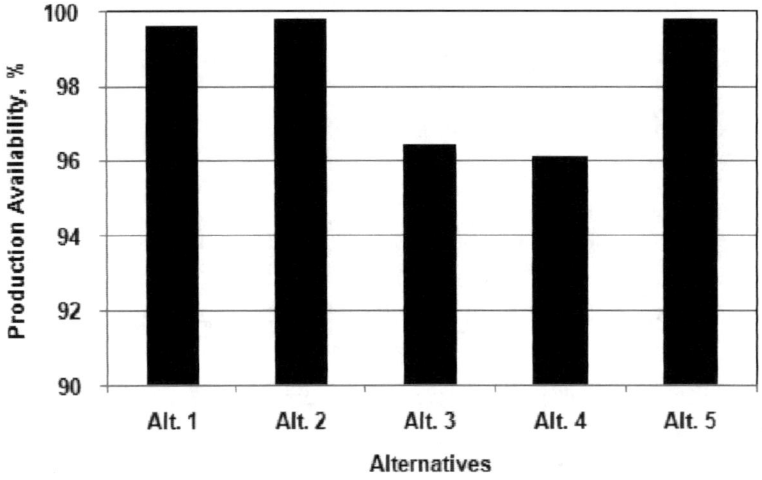

Fig. 1 Production Availability of Alternatives for Power Generation Systems

An increase in the production availability means a decrease in the production loss. The production availability entails the whole examination of the operability of the target plant since it describes how much the plant is capable of meeting the production demand. Fig. 2 represents the fuel gas cost including the production loss cost. The steam turbine driver and generator are high availability systems, while they have low efficiency. In addition, very large plants such as boilers, steam pipes and cooling systems are required with high seawater demand. The electrically driven compressors for the liquefaction unit should install a generator, motor, high voltage switchgear, etc. An increase in equipment count is accompanied by the overall cost and space escalation. The mechanically driven type is able to reduce the risk of the electrical system instability due to trip of large motors. Even though expected low availability of the gas turbine, Alt. 3 was applied due to good efficiency and the above disadvantages of other systems.

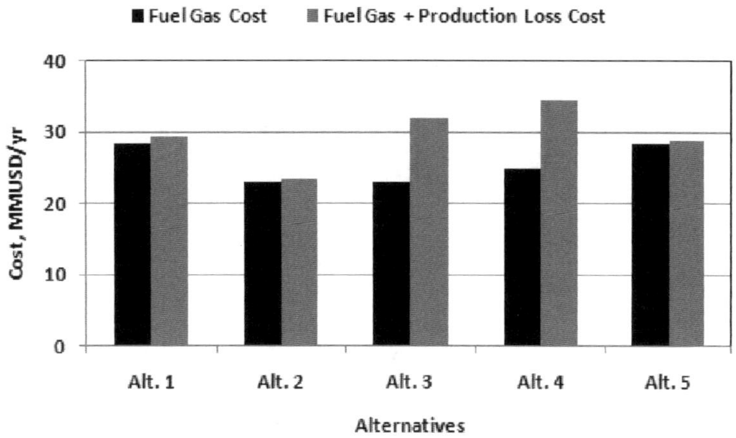

Fig. 2 Fuel Gas Cost including Production Loss Cost

Topside Process Design of FLNG

Fig. 3 shows a block diagram of the topside process for the FLNG. The topside processes are located on a deck platform 6 m above the hull deck. The storage tanks for LNG, LPG and Condensate are located below the hull deck.

Gas treatment process flow

Natural gas, as a multi-phase stream such as gaseous, condensate and water, is taken from the gas reservoir. The three-phase flow is fed to the Inlet facilities and is separated in them. The purpose of the gas pre-treatment units is to remove any of substances which become solid at liquefaction conditions. CO_2 and H_2S in the gas phase stream are normally removed by the amine absorption process. The dehydration system consists of three molecular sieve vessels for removal of water. The mercury removal is fixed bed adsorption unit.

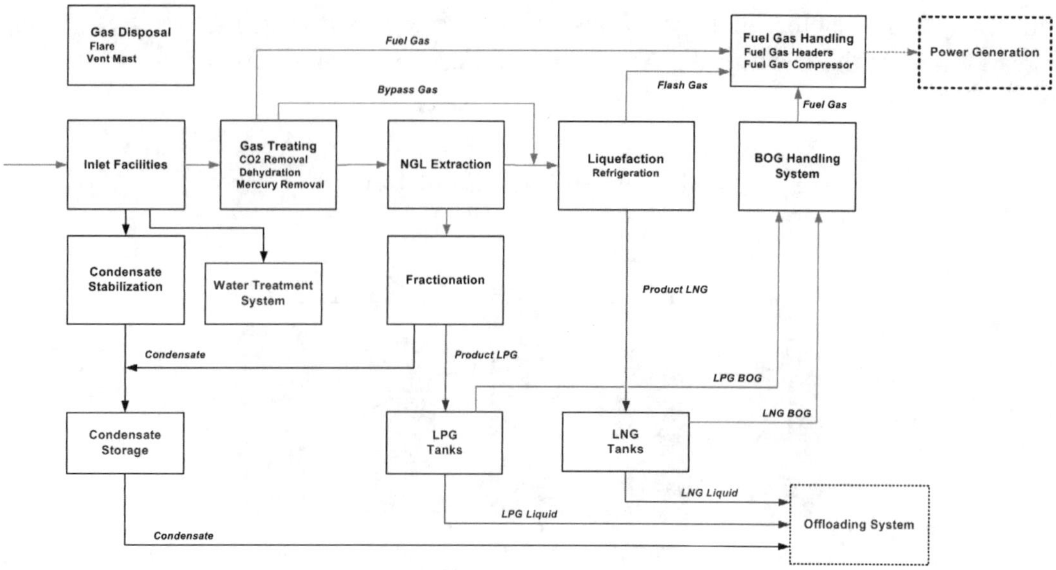

Fig. 3 Block Diagram of FLNG Topside Process

Liquefaction process flow

The liquefaction process applies a single mixed refrigerant (SMR) cycle which is separated into two refrigerant streams of different molecular weights by a two-stage refrigerant cycle compression and partial condensing downstream the compressor stages. The liquefaction section of the FLNG topside process applies LIMUM® process licensed by Linde AG to liquefy the treated gas. A schematic diagram of the process is shown in Fig. 4. The heavier hydrocarbon fraction (red line in Fig. 4) is used for pre-cooling of the natural gas, the lighter one (sky-blue line from separator downstream of HP MR compressor stage in Fig. 4) is first pre-cooled and then separated into two fractions from which the heavy one (yellow line in Fig. 4) is used for liquefaction and the light one for sub-cooling (sky-blue line in Fig. 4) of the natural gas.

In the liquefaction process, heavy hydrocarbons can solidify at LNG temperature (-162 ℃). Consequently, heavy hydrocarbons should be removed in the liquefaction unit to prevent the solidification. The natural gas is directed out of the main heat exchanger and into the heavy hydrocarbon recovery system, where the heavy liquids are removed. The vapor from the heavy hydrocarbon recovery system is returned to the main heat exchanger for liquefaction and sub-cooling.

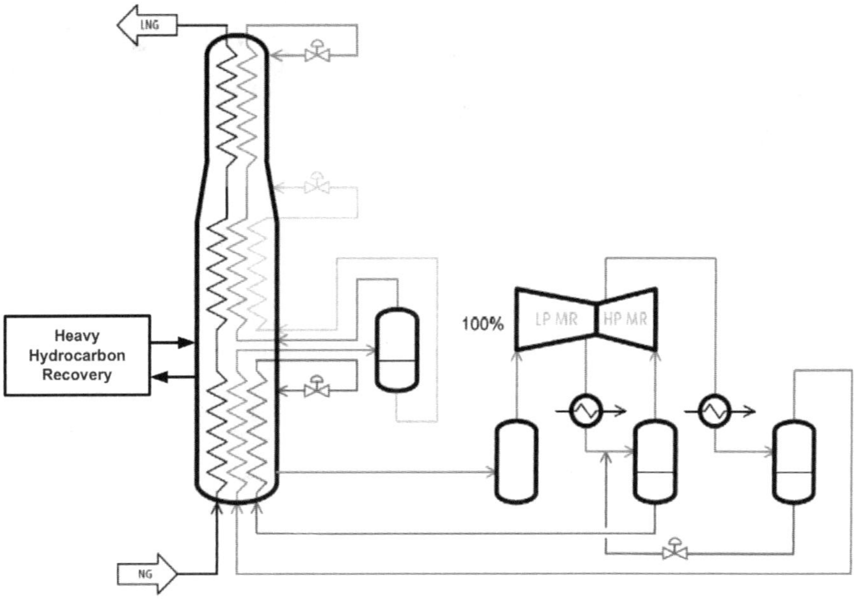

Fig. 4 Schematic Diagram of Liquefaction Unit (LIMUM® Process)

Refrigerant Loop

The liquefaction unit uses a closed loop refrigerant cycle system of compression, partial condensation, expansion, cooling and heating. The refrigeration loop contains two-stage compressors driven by gas turbines.

The low pressure refrigerant from the main heat exchanger is compressed to a midpoint pressure in the first stage of the refrigerant cycle compressor, and routed to the first inter-stage cooler. The first inter-stage drum separates the vapor and liquid. The liquid from the first inter-stage drum is the heavier hydrocarbon fraction used for pre-cooling of the natural gas. The vapor is sent to the second stage of the refrigerant cycle compressor. The refrigerant from the second stage of the refrigerant cycle compressor is cooled by the second inter-stage cooler. The liquid of the refrigerant is separated in the second inter-stage drum, exits the bottom of the drum and is routed after expansion to the first inter-stage drum. The vapor refrigerant enters the main heat exchanger where it is pre-cooled and partially condensed against the shell side refrigerant. Then it enters the refrigerant separator where it separates the light refrigerant as vapor and the heavy refrigerant as liquid. The heavy refrigerant is cooled in the main heat exchanger and is expanded by a Joules-Thompson valve. Thus the heavy refrigerant provides cooling duty for itself, the light refrigerant and liquefying the natural gas. The gaseous light refrigerant is routed to the main heat exchanger and is expanded by a Joules-Thompson valve into the shell side of the main heat exchanger. The low pressure refrigerant in the shell side migrates to the bottom of the main heat exchanger and vaporized. The refrigerant cycle is repeated.

Model test

The model test was performed to verify the sea keeping and offloading operability. The test was focused on the evaluation of the station keeping with the turret mooring arrangement and the offloading with the side-by-side arrangement.

The test demonstrated the safety for the turret mooring system and the side-by-side offloading system in the conditions of the wave, wind and current of 100 year return period and 1 year return period, respectively.

Fig. 5 Model Test of FLNG

Topside process layout

The primary consideration when laying out offshore facilities is their impact on people, the environment and asset integrity. The main criteria of the topside process layout are as follows;

- Compact, safe and cost efficient arrangement
- Use of optimized modules
- Fulfilling requirements for constructability and maintenance
- Integration of modules on main deck
- Minimize escalation of accidents between different units as far as possible
- High overall safety level
- Identify cost effective solutions, but keep conservative approach for the generic design
- Optimized regarding process and operation requirements

The plot plan was developed to identify the various areas and to obtain a safe layout and an optimum use of areas. Fig. 6 depicts the topside general arrangement of the FLNG.

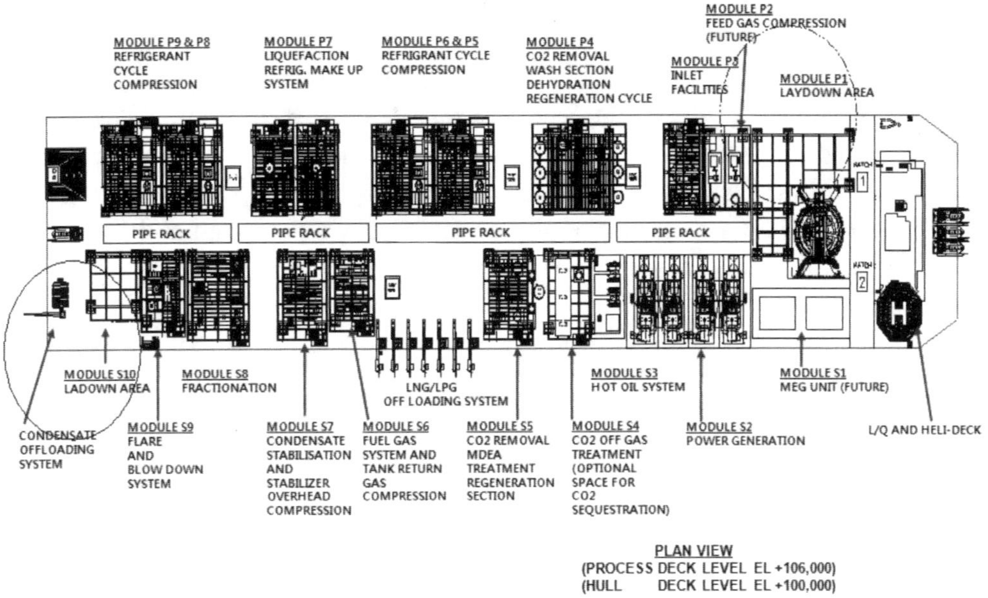

Fig. 6 Topside General Arrangement of Hyundai FLNG

The use of 3D model for each module ensures to maximize the degree of pre-fabrication and to minimize the amount on work in the integration yard. Fig. 7 shows the 3D Models for the refrigerant cycle and liquefaction unit, respectively. The refrigerant cycle module accommodates the compressors including inter-stage coolers, suction vessels and gas turbine. For the liquefaction unit module, the two modules are arranged side by side due to the weight limitation since for offshore facilities, the lift weight of the individual modules is critical.

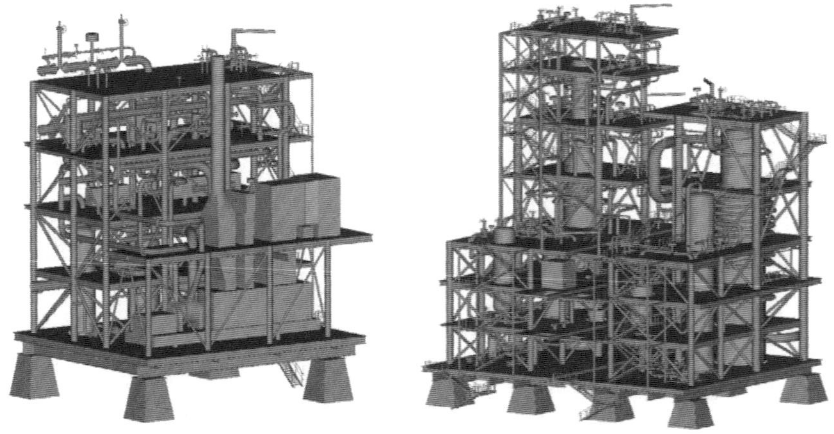

Fig. 7 3D Models for Refrigerant Cycle and Liquefaction Unit

Conclusions

The paper assessed the feasibility of the design of the FLNG topside process. The selection criteria of FLNG topside process focused on making two key decisions regarding the liquefaction units and the power generation systems.

For 2.5 MTPA LNG production, the SMR process offers more advantages of less equipment count, smaller footprint and higher efficiency than the nitrogen expander type.

The compressors of the refrigerant cycle of the liquefaction unit consume a huge amount of power, which requires a massive quantity of fuel gas. For the power generation of the compressors, the electrically driven compressors causes a loss of efficiency and an increase in cost due to converting the power into electricity. The application of the steam turbine is uneconomical since very large plants such as boilers, steam pipes and cooling systems are required with high seawater demand. Consequently, the mechanically driven type by gas turbines was applied.

The model test was performed to verify the sea keeping and offloading operability. The test demonstrated the safety for the turret mooring system and the side-by-side offloading system in the conditions of the wave, wind and current of 100 year return period and 1 year return period, respectively.

The plot plan was developed to identify the various areas and to obtain a safe layout and an optimum use of areas. The living accommodation was at the opposite end of the FLNG to the flare, upwind of any possible gas release.

Further investigation and development will be performed for the robustness of the design and the operability of the FLNG.

References

1. Kiil Nam, Daejun Chang, Kwangpil Chang, Taejin Rhee, In-Beum Lee, (2011), "Methodology of life cycle cost with risk expenditure for offshore process at conceptual design stage," *Energy*, 36, pp. 1554-1563.
2. Kiil Nam, Taejin Rhee, Kwangpil Chang, Daejun Chang, (2006), "An analysis of gas turbine propulsion system for LNG carriers based on reliability and economic evaluation." *The Society of Naval Architects and Marine Engineers: Maritime Technology Conference & Expo and Ship Production Symposium,* Paper No. B1, Fort Lauderdale, U.S.A.
3. OREDA Participants (2002), "Offshore Reliability (OREDA) Handbook." 4th ed. Høvik, Det Norske Veritas (DNV).
4. Ayyub BM (2003), "Risk analysis in engineering and economics." London, CRC Press Company

(66c) Application of the Anti-Sloshing Blanket to LNG Carriers and FLNG

Sangeon Chun, Samsung Heavy Industry Co. Ltd., Gyeongsangnam-Do, South Korea

Sloshing is long problematic issue on achieving safety and reliability of LNG carriers and FLNG if the liquid level is below at a certain point in the containment system. Sloshing motion can be significantly reduced by covering the free-surface using a patented floater type blanket which is formed by connecting multiple floating units.

This paper presents effects of the anti-sloshing blanket on minimization of dynamic energy and removal of impact loads, which results in an increase in structure safety.

(66d) Cost Effectiveness of 2-Row Membrane System for FLNG

Herbert Moon, Daewoo Shipbuilding & Marine Engineering Co. Ltd., Seoul, South Korea

One of the most critical issues in engineering and design of FLNG is to reduce top side weight along with its load distribution to the sub-structure. Concentrated top side loads will result in an increase in hull structure costs. The developed 2-row containment system allows flexibility of top side load distribution. This system allows flexible modularization of top side equipment and reduces in hull structure cots as well. This system also provides great benefits in minimization of sloshing effects of the membrane storage FLNG. FLNG storage capacity can be increased with the 2-row system without sloshing problem. This paper presents the cost effectiveness of the 2-row containment system adopted to FLNG.

(67a) Gas Separation Membrane Tutorial

Lubo Zhou, UOP LLC, a Honeywell Company, Des Plaines, IL

Due to its low energy cost, membrane has been applied to different gas separation processes such as acid gas removal from natural gas, hydrogen purification, nitrogen production from air, and hydrocarbon recovery from refinery off gas, etc. In most recent years, membranes have also been proposed for flue gas CO2 capture. In this tutorial, the author will start the introduction of fundamentals of membrane for gas separation, followed by the different types of membranes and their applications. Different membrane process configurations will also be introduced at the tutorial. At the end, large scale commercial membrane units in the field of natural gas purification will be presented as the successful membrane application examples.

(67b) Cost-Effective Catalytic Membranes for H2 Purification

Bhanu Vardhan Reddy Kuncharam, Artie McFerrin Department of Chemical Engineering, Texas A& M University, College Station, TX and **Benjamin A. Wilhite**, Artie McFerrin Dept. of Chemical Engineering, Texas A& M University, College Station, TX

Conventional membrane systems for high purity hydrogen production from hydrocarbons typically employ permselective Pd or Pd-alloy materials in conjunction with hydrocarbon reforming. Palladium is expensive, prone to poisoning by carbonaceous materials and suffers embrittlement over multiple thermal cycles. In the light of these challenges our research group is exploring the use of catalytic membranes to achieve hydrogen separation without relying upon costly and delicate permselective materials.

In this presentation, we will review our work in the development of catalytic membranes for hydrogen extraction from fermentation liquors[1]and demonstrate the extension of this technique to hydrogen purification from reformate mixtures (H_2, CO, CO_2, H_2O) using a catalytic membrane active for water-gas-shift reaction. Design simulations of this membrane system predict H_2:CO permselectivities that are competitive to palladium membranes at a fraction of the materials cost.

[1] B. Kuncharam and B.A. Wilhite. Design of a Composite Catalytic Membrane for Permselective Extraction of Hydrogen from Ethanol via Reaction and Diffusion.*Journal of Membrane Science, in review*(2011).

(67c) Offshore Gas Treatment Technology for Natural Gas

Saadet Ulas Acikgoz[1], Shain Doong[2], Pengfei Chen[2]and Lubo Zhou[2], (1)UOP LLC, a Honeywell Company, Des Plaines, IL, (2)UOP LLC, a Honeywell Company, Des Plaines, IL

Offshore liquefaction of natural gas is expected to be the next technological breakthrough for capitalizing stranded natural gas resources. There are several technology options that can be used for acid gas removal or pretreatment of FLNG (Floating Liquified Natural Gas) plants. The traditional amine absorption processes have been widely used for the treatment of feed gas for onshore LNG plants.Due to limited space available on the offshore facility, the gas treatment units on FLNG plants have to be designed with smaller weight and footprint.

A hybrid system is a combination of membranes for bulk CO_2removal and amine units as polishing systems. Membranes have been proven in multiple offshore applications and suitable for offshore environment due to their compactness, minimal weight, footprint and insensitivity to sea motion. These hybrid systems can reduce the size and weight of the total pretreatment unit, minimize the effect of sea motion on the amine column performance and offer a cost effective solution.

() A Novel Catalyst for Partial Oxidation of Natural Gas

Nicholas Xu, Research Institute, Dason Technology, Chicago, IL

A Novel Catalyst for Partial Oxidation of Natural Gas

Nicholas Xu* *Dason Technology Inc., Naperville, 60565, USA*

Yu Wu *Chongqing Chemical Industry Research Institute, Chongqing, 400023, China*

*nick.xu@dasontechnology.com

Introduction

Many people have worked on CPOX research. The main barrier in the area was catalyst lifespan being insufficient to meet industrial requirements. We developed a novel catalyst for partial oxidation of natural gas (CPOX) to synthesis gas. The experiment was carried out in the lab and pilot plant at 10-20 bar pressure. The feed consisted of a mixture of natural gas and oxygen at a ratio of C:$O_2$of about 2:1. The natural gas conversion and CO selectivity are at 95% and H_2selectivity is about 90%. The catalyst was continuously running for about a year without deactivation.

Experimental

The experiments were carried out in an adiabatic stainless steel reactor with a Rh and Ni catalyst system. The lab reactor size is 20 mm in diameter and the pilot plant capacity is 1000 ton/yr. There is inert material at the bottom and the top of the catalyst.

The feed was heated to a range from 150°C to 300°C in an electric heater. There is a thermocouple at the top of the catalyst bed and another thermocouple at the bottom of the inert material below the catalyst bed. The experiments were performed between 10-20 bar pressure. A gas sample was taken from the reactor inlet and outlet. A GC was used for the gas sample analysis.

Results and discussion

Our catalyst was running for about a year under 10 – 20 bar condition without deactivation. The pilot plant results are in the figure below.

CPOX is compared to steam reforming and ATR technology in the following table.

Technology	Advantages	Disadvantages
Steam Reforming (SMR)	• Mature, proven • Most commonly use for large scale operations	• Large reactor size, highest CAPEX and OPEX • Reaction requires heat (endothermic) and steam • Large CO2 emissions from furnace • High CO2 in syngas

Autothermal Reforming (ATR)	• Smaller reactor size, lower CAPEX and OPEX than SMR • Less CO2 in syngas product than SMR • Part of NG is burned in the reactor to supply heat for reforming, internal burning has higher efficiency than SMR furnace • Requires 20% more oxygen than reaction required (CH4:O2=1:0.6) • Must add steam to reduce soot (CH4:H2O=1:0.4-0.7)	
Catalytic Partial Oxidation (CPOX)	• Smallest reactor size, lowest CAPEX and OPEX • Highest carbon efficiency • Idea oxygen consumption (20% less than ATR) • Lowest CO2 in syngas product • Don't need steam • Low catalyst lifespan before Dason's invention • Need oxygen but less than ATR	

CPOX carbon efficiency is at least 15% higher than ATR as demonstrated in this chart.

	CH_4	CO	CO_2	H_2	CH_4Conv.	CO sel	H_2Sel
CPOX	1.68	33.0	1.23	62.4	95.3	96.4	89.0
ATR	1.0	24.5	7.62	66.9	96.98	76.28	100.2

CPOX data above is our pilot plant typical data. The ATR data is from V. Palma et al. (2010) and is a calculated results using GasEq software at the equilibrium condition.

The carbon efficiency for CPOX is calculated as

32.2/(1.68+33.0+1.23) = 91.9%,

and for ATR is

24.5/(1+24.5+7.62) = 74%.

For a large plant, the feed stock consumption difference at about 18% makes a huge difference in operating costs.

Reference

1. D. A. Hickman and L. D. Schmidt, "Syngas Formation by Direct Catalytic Oxidation of Methane", Science 259, 343-346 (1993).

2. Rostrup-Nielsen, "Syngas in perspective", *Catalysis Today* 71 (2002), pp. 243-247

(67d) An Innovative Mixed Zinc Oxide Based Sorbent Regenerable At Low Temperature for the Desulfurization of Syngas in IGCC or XTL Processes

Vincent Girard, Catalyse and Separation Division, IFP Energies Nouvelles, Solaize, France, Arnaud Baudot, Catalysis and Separation Division, IFP Lyon, 69360 Solaize, France, David Chiche, Catalyse and Separation division, IFP Energies Nouvelles, Solaize, France, Delphine Bazer-Bachi, GMD Division, IFP Energies Nouvelles, Solaize, France and Christophe Geantet, IRCELyon, Villeurbane, France

Integrated Gasification Combined Cycles (IGCC) and the second generation biofuels processes, and more generally Fischer-Tropsch based XTL processes, are attractive alternatives for future power production[1-3]. In order to prevent the corrosion of the industrial units, especially the combustion turbine blade used in IGCC processes or, in the others processes, to avoid the Fischer-Tropsch catalysts poisoning, deep desulfurization of syngas is achieved with metal oxides according to the following reaction :

$$MeO_{(s)} + H_2S_{(g)} \rightarrow MeS_{(s)} + H_2O_{(g)}$$

The large amounts of solid wastes produced are one of the major disadvantages of the in-situ desulfurization of syngas. The in-situ regeneration of the sulfided sorbent, back to the oxide, during the process could resolve this problem. From an industrial point of view, the oxidative regeneration is the simplest and most straightforward way to regenerate the sulfided sorbent, so it was mainly studied[4,5]. However, this reaction presents several drawbacks. The first one is its high exothermicity that can cause the sintering of the particles and the loss of reactivity due to the alteration of the textural properties[6,7]. Another drawback, which can be considered as the most important, is the formation of sulfates[5,8]. These refractory phases require high temperature to decompose, which causes important modification of textural properties and decrease of reactivity due to the sintering of particles. Moreover, the resulting temperature increase will impose the use of expensive alloys to make the reactors and will affect the thermal efficiency of the process. The formation of sulfate has to be avoided in order to maintain the sorbent durability and the temperature of the in-situ oxidative regeneration must be as low as possible, ideally the closest to the desulfurization step, to improve the global efficiency of this process.

The main goal of this work is to bring a new insight on the phenomena involved during the oxidative regeneration of the sulfided solids. The methodology developed is based on the confrontation between a global theoretical approach based on thermochemical data and an experimental study. This work is divided into two parts. The first one is the evaluation and the classification of the behaviors observed during both sulfidation and oxidative regeneration of simple oxides, in particular ZnO. The second one is the evaluation of the mixed oxides, selected thanks to the results of the first part and by the anticipation of potential synergetic effects. The synergetic effects are expected to avoid the formation of refractory sulfates and the resulting increase of temperature.

The first part led to a classification of the simple oxides based on the stability of their sulfide and sulfate phases. This classification led to a selection of different simple oxides whose mixture could bring synergetic effect. Finally, this methodology led to an innovative material which composition allows a fast regeneration, at a temperature 150°C lower than ZnO regeneration temperature and without sulfates formation.

Reference List

1. Daniel Vallentin, *Energy Policy*, 2008, **36**, 3198-3211.

2. Osamu Shinada, Akira Yamada, and Yoshinori Koyama,*Energy Conversion and Management*, 2006,**43**, 1221-1233.

3. Mark J. Prins, Krzysztof J. Ptasinski, and Frans J. J. G. Janssen,*Fuel Processing Technology*, 25-1-2005,**86**, 375-389.

4. R. V. Siriwardane and S. Woodruff,*Industrial & Engineering Chemistry Research*, 1995,**34**, 699-702.

5. W. F. Elseviers and H. Verelst,*Fuel*, 1999,**78**, 601-612.

6. T. Grindley and G. Steinfeld,*Proceedings of the Second Annuals Contractors Meeting on Contaminant Control in Hot Coal Derived Gas Stream.*, 1981.

7. J. Huang, J. Zhao, X. Wei, Y. Wang, and X. Bu,*Powder Technology*, 14-1-2008,**180**, 196-202.

8. R. V. Siriwardane and S. Woodruff,*Industrial & Engineering Chemistry Research*, 1-12-1997,**36**, 5277-5281.

() Musings on Management

Frank van Lier, The Lubrizol Corporation, Wickliffe, OH

It really is all about the people!

Hiring good people.

Retaining them.

Adjusting for different personal styles, cultures, encouraging diversity, embracing different perspectives.

Motivating people to work together in successful global teams.

Achieving top notch results that are better than any individual can achieve alone.

Maintaining the focus.

Getting the results.

Management is all about the excitement managers gets from having their people and teams grow and succeed.

(81a) Thermodynamic-Analysis-Based Design for Liquefied Natural Gas Receiving Terminal

Meiqian Wang, Jian Zhang and Qiang Xu, Dan F. Smith Department of Chemical Engineering, Lamar University, Beaumont, TX

Natural gas (NG) now has become the most promising alternate fuel energy. Recently, the estimation about large quantity of technically recoverable unconventional NG, such as shale gas, has been announced by the Energy Information Administration (EIA), which draws more attention on the development of natural gas. In the work, we developed a novel configuration for liquefied natural gas (LNG) receiving terminal, integrated with LNG regasification process, shale gas liquefaction and NGL recovery process. A great deal of cold energy from LNG regasification has been recovered through shale gas liquefaction and NGL fractionators' condensers.

The flexibility of the newly designed receiving terminal has also been addressed under different conditions, which include the changes of received LNG quantity and composition, transported shale gas quantity and composition, and heating value specified at local marketing place, etc. Moreover, heat duty integration has been involved in this design, where thermodynamic theory is employed to analyze and fundamentally evaluate the performance. All of the conditions together with thermodynamic and economic performance are integrated in a mixed-integer nonlinear programming (MINLP) model. This model is solved using optimization tool GAMS and validated via rigorous simulations in ASPEN Plus.

(81b) Design Approaches for Unconventional Gas Processing

John Harness, Technical Service, UOP, Des plaines, IL

Design Approaches for Unconventional Gas Processing

Membrane systems are well suited to remove CO2 and water to pipeline specifications although many people still consider membranes to be an "unconventional" technology. When applied to processing unconventional gas such as North American shale gas, membrane systems can offer distinct advantages.

This presentation will discuss the technology within membrane systems and present a case study from a gas plant that has been using membranes to process shale gas for the past 18 years. The presentation includes discussions on membrane theory and applications as well as discussion of a cost and benefit analysis comparing the use of membranes with Amine systems in unconventional shale gas applications.

UCSRP: A Flexible Sulfur Removal Process for Sweetening Natural Gas

Arun Basu, Howard Meyer, Dennis Leppin, Jim Zhou and Ajay Makkuni

Gas Technology Institute

ABSTRACT

With the recent activities in the area of unconventional natural gas (such as, Shale Gas) in the U.S., the need for a low-cost sulfur removal technology for processing relatively small as well as large CO_2-rich natural gas streams has increased significantly. In a conventional amine-based natural gas (NG) processing plant, the H_2S-rich acid gas stream from the stripper unit is treated in a Claus process for the production of elemental sulfur (the gas-phase reaction is $2H_2S + SO_2 = 3S + 2H_2O$). The Claus process has specific inherent limitations including (i) the need for a tail-gas clean-up unit (e.g., Shell Claus Off-gas Treatment: SCOT) to achieve the required H_2S recovery efficiency, (ii) inability to handle low sulfur production capacities (typically, for less than 20 tons/day sulfur), and (iii) the need for an additional CO_2-separation step prior to the Claus step if the gas stream from the Amine unit contains relatively high CO_2 (typically, if the CO_2 /H_2S ratio is greater than 2.3).

Gas Technology Institute (GTI) has been working with the University of California, Berkeley, for further development of their UCSRP (University of California Sulfur Recovery Process) technology. In this process, the Claus reaction is conducted in liquid phase at about 250-300 °F (and at any pressure) to produce liquid sulfur in the presence of an organic solvent and a homogeneous catalyst. A gas mixture consisting of a part of the H_2S-rich feed gas plus the unreacted H_2S stream from the UCSRP reactor is combusted with air in a furnace to form SO_2 (plus a small amount of sulfur) which is recovered in an absorber/stripper unit for recycle to the UCSRP reactor. As the overall sulfur recovery is 99.9+%, there is no need for a tail-gas clean-up unit. The UCSRP process can be scaled down to sulfur capacities significantly lower than 20 LT/day. For CO_2-rich feed streams from an amine plant, the gas can be processed directly in a UCSRP unit without a separate CO_2 removal unit. For direct processing of natural gas with relatively low CO_2 levels that already meet pipeline specifications, the UCSRP process will replace Amine/Claus-SCOT units; for higher CO_2 levels in feed natural gas, the exit gas from the UCSRP reactor will be treated to remove CO_2.without the complexity inherent when H_2S is also present in the stream.

For the processing of a H_2S-rich gas stream from an amine plant stripper with sulfur production at about 76 LT/day, Kellogg, Brown and Root Inc. had estimated the capital cost for an UCSRP plant to be about 60% and the operating costs to be about 68% of those for a three-stage Claus/SCOT plant.

INTRODUCTION

With the recent discoveries of relatively large quantities of unconventional natural gas (e.g., Shale Gas), the U.S. natural gas industry is interested in sulfur recovery processes that offer lower cost with high sulfur removal selectivity. Various amine-based technologies are commercially used to remove sulfur compounds from high-pressure natural gas feed as an acid gas which then requires a sulfur-recovery process. The modified Claus process coupled with a tail-gas treatment (TGT) process such as the SCOT (Shell Claus Off-gas Treating) process are typically used to recover elemental sulfur and produce a dischargeable plant tail-gas. Some plants produce sulfuric acid rather than elemental sulfur.

Gas Technology Institute (GTI) has been working with the University of California, Berkeley, for further development in a number of varying applications of their UCSRP (University of California Sulfur Recovery Process) technology for removal of sulfur compounds from various feedstock including natural gas and coal-derived synthesis gas.[1,2,3,4] In the UCSRP process, the Claus reaction ($2H_2S + SO_2 = 3S + 2H_2O$) is conducted in liquid phase (typically, using an organic solvent such as Diethylene Glycol Methyl Ether (DGM)) at about 250-300 °F (and at any pressure) to produce liquid sulfur. A homogeneous catalyst (e.g., 3- Pyridinemethanol), which does not dissolve in liquid sulfur, is added to the solvent in small quantities. At these conditions, as liquid sulfur is essentially immiscible in DGM solvent and over twice as dense, it is removed as a separate product from the reactor column. In this paper, we have (i) presented comparative capital and operating costs data (based on a prior study by Kellogg, Brown and Root Inc.[1]) for the UCSRP process to a conventional Claus/SCOT sulfur recovery system, and (ii) outlined key advantages of the UCSRP process for the sweetening of natural gas.

PROCESS OVERVIEW

Claus

The Claus process produces elemental sulfur by the partial oxidation of H_2S to SO_2 followed by the reaction of uncombusted H_2S with SO_2. The key reactions are:

$$2\,H_2S + 3O_2 = 2SO_2 + 2\,H_2O \qquad (1)$$
$$2\,H_2S + SO_2 = 3S + 2\,H_2O \qquad (2)$$

The overall reaction may be expressed as: $2\,H_2S + O_2 = 2S + 2\,H_2O$ (3)

In the original unmodified Claus process, the reaction was carried out in a single reactor (at about 400-650 °F) over a catalyst followed by a sulfur condenser.[2] With only one reactor, the sulfur recovery was limited to 75-90%. Most modern sulfur recovery plants use the modified Claus process, where the reaction is carried out in two steps. In the acid gas burner, 1/3 of the total H_2S in the feed gas is totally oxidized (using air) to SO_2. In the next step, the unburned H_2S and the SO_2 formed react in the reaction furnace without catalyst. This reaction proceeds rapidly in the high-temperature reaction furnace, and leads to a conversion of about 60% of the H_2S to sulfur vapor. The gas leaving the furnace is cooled to condense sulfur and then reheated for processing in a catalytic converter. Most modified Claus plants have two to four catalytic converters, each with its own condenser. Each succeeding converter operates at a lower temperature. The three-converter system typically provides a sulfur yield up to 98%. The gas leaving the final catalytic converter is typically sent to a Tail Gas Cleanup Unit (TGCU), such as a Shell SCOT unit. A schematic of the modified Claus process is shown in Figure 1.

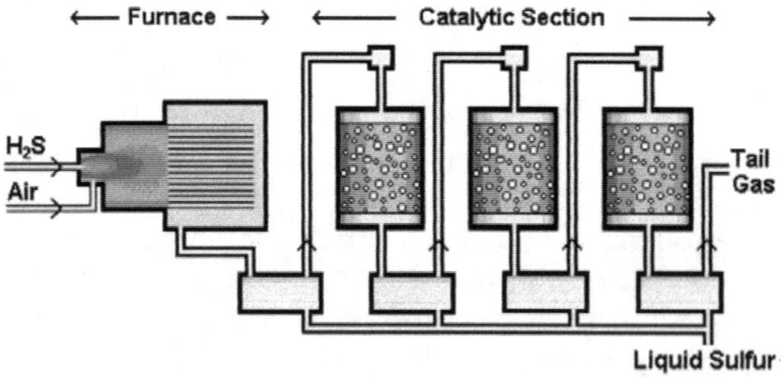

Figure 1: Schematic of the Modified Claus Process[5]

Several variations of the basic Claus process have been developed to handle acid gases containing different levels of H_2S. If the acid gas feed is relatively lean (less than 30 mol% H_2S on a dry basis), and if all of the acid gas is sent to the burner, the low level of H_2S would result in poor flame stability due to relatively low reaction furnace temperatures. Therefore, a fraction of the acid gas feed bypasses the burner to achieve suitable furnace temperatures necessary for flame stability. This configuration is referred to as the "Split-flow Claus". When the acid gas feed is sufficiently rich in H_2S (i.e., greater than 30-35 mol%), all of the acid gas is processed in the burner. This process configuration is known as the "Straight-through Claus".

A few key inherent problems of the modified Claus process are:

- As the furnace burns very fuel-rich, chances of soot formation increase significantly due to the conversion of hydrocarbons to COS and CS_2; this requires precise control of air flow to the furnace.
- Higher levels of CO_2 in the feed natural gas would significantly increase overall capital and operating costs of sulfur removal. As an example, if the CO_2/H_2S ratio is more than about 2.3, the off-gas from an amine stripper would have to be treated in an expensive CO_2 separation unit before the off-gas can be processed in a Claus plant.

SCOT

A schematic of a typical SCOT unit is shown in Figure 2.[2] The process was developed by Shell, and introduced in the early seventies as a process option for improving the efficiency of a Claus unit. By using the SCOT process, the sulfur recovery of a Claus plant can be increased to about 95-99.9%.

The SCOT process has two primary sections: (1) a hydrogenation/hydrolysis section, followed by water quench and (2) a selective amine gas treating section. In the first section, the tail gas from the Claus plant is heated to about 480-570 °F and then processed with H_2 (or a reducing gas) over a cobalt molybdate catalyst for converting all sulfur compounds (including SO_2,S, COS, CS_2) to H_2S. The effluent from the reactor is cooled to about 350 °F in a waste-heat exchanger (WHE) and the WHE effluent is further cooled by direct water quench. In the second section, the H_2S product stream is processed in an amine-based (e.g., MDEA or DIPA) absorber to recover most of the H_2S. The absorber overhead is usually incinerated (if the H_2S level is higher than 100 ppmv; using catalytic or thermal processes) to convert residual H_2S to SO_2 before discharge to the atmosphere. The H_2S-rich amine solution is processed in a regenerator to recover most of the H_2S which is recycled to the front end of the modified Claus unit. A schematic of the SCOT process is shown in Figure 2.

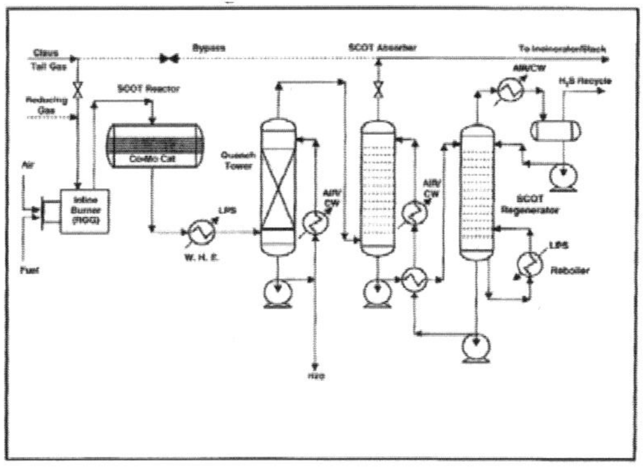

Figure 2: Schematic of the SCOT Process[2]

Although an integrated Claus/SCOT type process can meet current environmental regulations for sulfur emissions in the U.S., a SCOT-type tail-gas treatment process is quite expensive (typically costing as much as the modified Claus plant).[1]

UCSRP as a Replacement of the Claus/SCOT Processes

In the UCSRP concept (shown schematically in Figure 3), an organic solvent (e.g., DGM) containing a small quantity of a homogeneous catalyst is used to perform the Claus reaction in liquid phase at about 250-300 °F; this ensures that the reactor temperature is above the melting point (247 °F) of sulfur and below the polymerization temperature (310 °F) of elemental sulfur. The H_2S-rich acid gas is fed to the UCSRP reactor along with recirculated solvent plus SO_2 generated from combusting the reactor off-gases and part of the H_2S-rich feed with air in a furnace. As the liquid sulfur product is essentially immiscible and insoluble in DGM solvent and over twice the density as DGM, it is removed as a separate product from the reactor column thus shifting the chemical equilibrium toward sulfur formation.

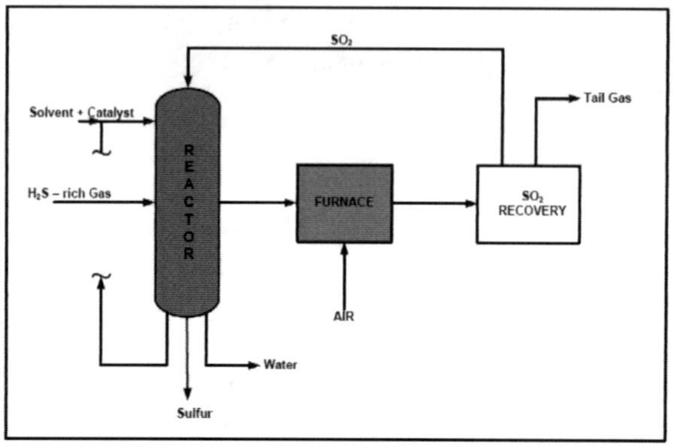

Figure 3: Schematic of the UCSRP Process

GTI is actively working with Prof. Scott Lynn (University of California, Berkeley), the inventor of the UCSRP process, to further improve the UCSRP technology, especially for high pressure (up to about 800 psig; the UCSRP-HP option) applications including sulfur removal from coal-derived syngas.[3,4] In the context of NG sweetening, GTI has demonstrated:

- The production of high-quality sulfur product (99.2% pure elemental sulfur) without any secondary treatment.
- Specific catalyst and solvent systems are stable under operating conditions based on a 1,000 hour test.
- Carbon steel is an acceptable material of construction for the UCSRP reactor

A few key advantages of the UCSRP process over the Claus process are: (i) the control of process stoichiometry is significantly simplified as the reactor is located before the furnace, (ii) slightly air-lean combustion precludes SO_3 and NO_x formation, (iii) hydrocarbons are burned primarily to CO_2 and water, (iv) relatively higher CO_2/H_2S ratios in the feed acid gas are tolerated, and (v) overall sulfur recovery is improved significantly.

- Cost Advantages of the UCSRP Process Over and Claus/SCOT: Scoping Economics Study by Kellogg Brown & Root, Inc.[1,2]

For sweetening of natural gas, KBR Inc. had conducted a scoping economics study to compare cost advantages of the UCSRP concept over the Claus/SCOT option in treating H_2S-rich gas from a conventional amine plant for the production of 76 long tons/day of sulfur.+1,2 A process flow diagram of the UCSRP process, as defined in the KBR study, is shown in Figure 4.

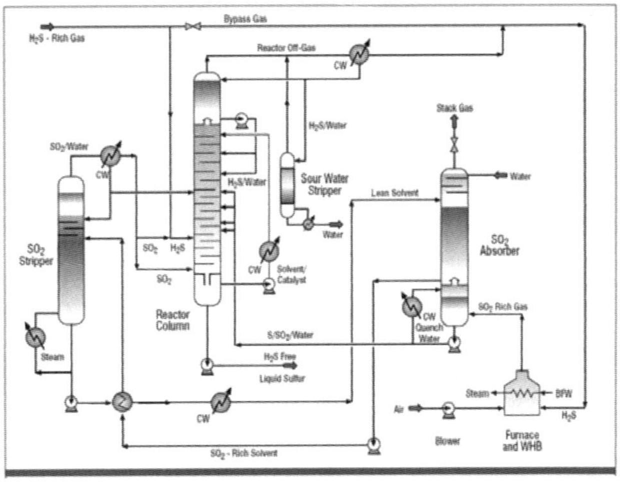

Figure 4: The KBR Study--Process Flow Diagram for the UCSRP Sulfur Recovery Process (for H_2S-rich Gas from an Amine Plant)[1]

As shown in Figure 4, the H_2S-rich gas stream enters the UCSRP unit at about 120 °F and 1.7 bar, typical of a Claus plant feed.[1] The reactor resembles a gas absorption column. A small fraction of the SO_2-rich gas stream enters the reactor column below the bottom bubble-cap tray. The remainder of the SO_2-rich gas together with about 75% of the H_2S-rich gas stream enters a tray or two higher and H_2S is in substantial excess at all points above. The remainder of the inlet H_2S bypasses the reactor and is sent to the furnace. The reactor is kept nearly isothermal by injecting water into the solvent flowing between trays to absorb the heat of reaction (from the Claus reaction) by vaporizing. The sulfur forms a second liquid phase. The solvent, which is recycled in a circulation loop from the bottom to the upper part of the reactor column, is saturated with elemental sulfur. The overhead vapor from the reactor is scrubbed with a water reflux to remove solvent vapor. A part of the water separated from the reactor off-gas is pumped back to the reactor as reflux. The net condensed water is sent to a sour-water stripper where dissolved H_2S and CO_2 are stripped overhead. The unreacted H_2S leaves the condenser and is mixed with the bypassed acid-gas feed and enters the furnace where it is burned with slightly sub-stoichiometric air to form SO_2. The hot combustion gas, containing SO_2, S, CO_2, H_2O and N_2 is cooled in a waste-heat boiler and is then fed to the SO_2 absorber where lean solvent from the SO_2 stripper is fed near the top. At the bottom of the absorber a water quench cools the combustion gas from the furnace to near ambient temperature, condensing a large fraction of the water vapors plus the sulfur vapor. The net dilute slurry of colloidal sulfur plus small amounts of higher sulfoxy acids is sent to the reactor column. The SO_2-rich solvent from the absorber is pumped to the SO_2 stripper to recover SO_2 for recycle to the UCSRP reactor.

As shown in Table 1, for the KBR study, two different H_2S/CO_2 feed gas ratios were evaluated:

Table 1: Feed Gas Composition (Design Basis: 76 LTPD Sulfur)[2]

	67/32 H_2S/CO_2	30/70 H_2S/CO_2
CH_4, mol%	0.2	0.2
H_2S, mol%	62.6	27.8
CO_2, mol%	30.2	64.9
H_2O, mol%	7	7
Feed Rate, kmol/hr	151.8	350.7
Claus Plant Configuration	Straight through flow	Split flow

Economic Results

- According to the KBR study[2], the relative capital cost for the UCSRP option is about 61% of a conventional Claus/SCOT unit (Figure 5) for a feed gas with 67% H_2S and 33% CO_2; for a feed gas with 30% H_2S and 70% CO_2, the estimate for the UCSRP option was about 60% of the Claus/SCOT plant.

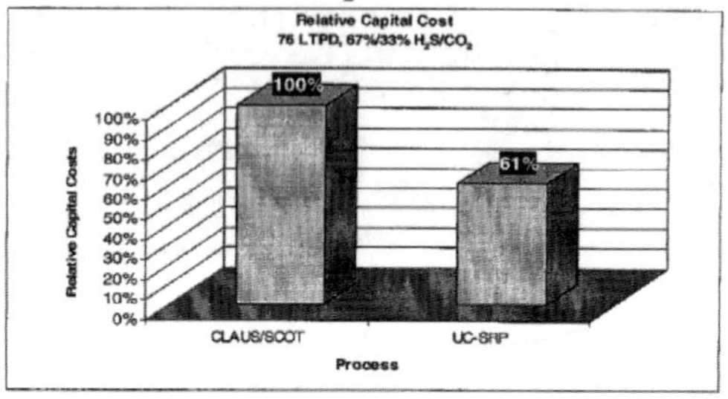

Figure 5: Relative Capital Cost – Claus/SCOT vs. UCSRP (Feed Gas with 67% H_2S, 33% CO_2)[2]

The relative Operating plus Maintenance costs for the UCSRP option were estimated at about (1) 68% of the Claus/SCOT design 67% H_2S/33% CO_2 Case, and (2) 67% of the Claus/SCOT design for the 30% H_2S/70% CO_2 Case.[2]

- Cost Comparison for a 935 LTD Sulfur Recovery plant[2]

As noted in Reference 2, another consulting firm performed an economic study for estimating relative costs for a 935 LTD sulfur recovery plant using the conventional Claus/SCOT and the UCSRP process. The relative capital cost for the UCSRP option was estimated to be about 62% of a Claus/SCOT system.

UCSRP as a Replacement of the Claus/SCOT Processes for a CO_2-rich Acid Gas Stream

If CO_2 level in a sour natural gas feedstock exceeds pipeline specifications (typically, less than 2 vol.%), an amine-type gas treating process is typically used to remove both H_2S and CO_2. In general, if CO_2/H_2S molar ratio is less than about 2.3, the off-gas from an amine stripper can be treated directly in a Claus/SCOT plant.[2] However, if this ratio is greater than 2.3, it may be necessary to carry out an expensive separation of the CO_2 prior to the Claus step. On the contrary, for an UCSRP unit, such a CO_2-rich off-gas from an "amine stripper" can be processed directly.

UCSRP as a Replacement of the Amine/Claus/SCOT Processes for a Natural Gas Stream

For direct sweetening of high-pressure natural gas feedstock, the high-pressure version of the UCSRP process would replace conventional amine plants integrated with a Claus/SCOT unit. A schematic of a two-stage UCSRP-HP concept, currently being evaluated at GTI for lab-scale demonstration, for high-pressure operations is shown in Figure 6; in this example, coal-derived syngas at about 800 psig is being processed to recover sulfur.[4] In this concept, the first UCSRP stage would be operated under "slightly excess SO_2" mode at the reactor outlet while the second stage would be operated under an "slightly excess H_2S mode". Typical sulfur balances for this example (in Figure 6) are shown in Table 2. In this example, the total SO_2 in the solvent plus syngas at the Reactor-1 outlet is about 9.7% of the SO_2 feed to this reactor; the H_2S content in the total solvent plus syngas flow at the Reactor-2 outlet is about 8.4% of H_2S in the syngas feed to the second reactor. For this example, the total syngas feed to the UCSRP reactors contains about 6,200 ppmv H_2S; about 1.34 vol.% of the feed bypasses the first reactor for processing in the second reactor. Based on GTI's design estimates, the exit syngas (sent to a CO_2 removal step) from the high-pressure separator (where Stream-11 is the inlet stream) would contain about 8 ppmv of H_2S plus COS.

Table 2: Stream Data for the Example Shown in Figure 6 (Basis: 1,000 lb-moles of total Syngas to the UCSRP Plant)[4]

Stream #	Stream Description	Lb-mole
1	Syngas to R-1[i]	986.6
2	Syngas to R-2[i]	13.4[ii]
3	SO_2 to R-1	3.1
4	Sulfur product	~ 6.2
5	Sulfur feed to Furnace	3.1
6	SO_2 from Furnace	3.1
7	SO_2 in solvent from R-1	0.26
8	SO_2 in syngas from R-1	0.04
9	H_2S in solvent from R-2	0.004
10	O_2 to Furnace	3.1
11	H_2S in syngas from R-2	0.003

i. Water vapor content: 1 mol%
ii. H_2S level: 0.083 lb-mole

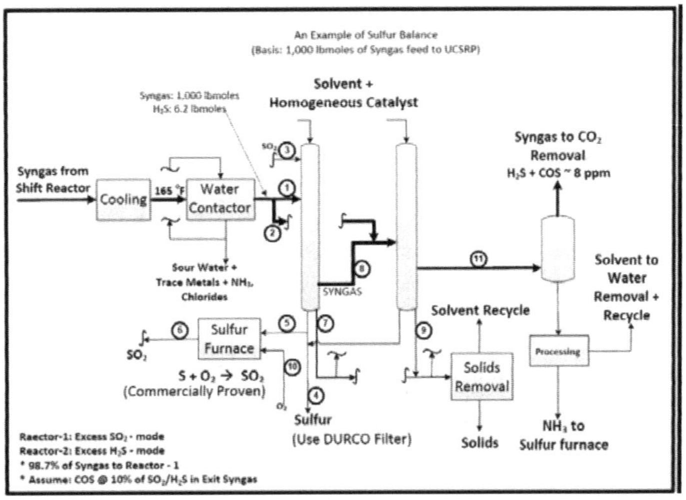

Figure 6: GTI's Two-stage UCSRP-HP Process Concept[4]

For direct sweetening of natural gas, an UCSRP-HP unit can replace an integrated Amine/Claus-SCOT plant. As shown conceptually in Figure 7, the USCRP-HP based plant designs would be relatively much simpler in handling various natural gas feedstocks with wide variations in (i) H_2S and CO_2 levels, and (ii) total gas flow rates.

As shown, if the CO_2/H_2S molar ratio is relatively high (~ greater than 2.3) in the H_2S-rich gas from an amine plant, we would need to first separate CO_2 before the gas can be processed in an integrated Claus/SCOT facility. For smaller plants (e.g., sulfur capacity less than 20 LT/days), we would need a "Liquid Redox" type process (and not a Claus/SCOT unit) to treat H_2S-rich gas (with CO_2/H_2S ratio of less than 2.3) from an amine plant. For the UCSRP-HP option, we can quite conveniently scale down to smaller sulfur capacities.[2]

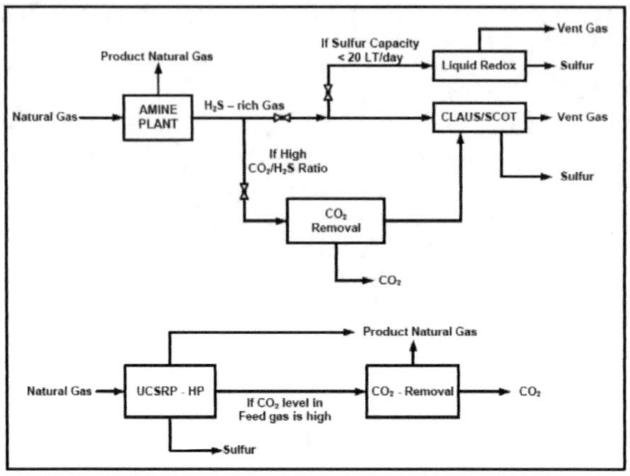

Figure 7: Relative Simplicity of an UCSRP-HP Process in Sweetening CO_2-rich Natural Gas

CONCLUSIONS

In summary, our key conclusions are:

- In comparison to the conventional Claus/SCOT process, UCSRP can reduce CAPEX by about 35-40% and OPEX by about 30-35%.
- In comparison to conventional Amine/Claus/SCOT for sweetening of natural gas feedstock, UCSRP-HP will reduce CAPEX/OPEX even more, especially for CO_2-rich NG streams.

References

1. Scot Lynn et al., "New Approach to Sulfur Removal Could Reduce Costs," GasTIPS, Winter 2002
2. Scot Lynn et al., "The University of California Berkeley's Sulfur Recovery Process: Claus Revisited," Paper presented at the 9th Gas Research Institute Sulfur Recovery Conference, October 24-27, 1999, San Antonio, Texas
3. S. J. Zhou, et al., "Economic Evaluation of the UCSRP-HP Process in IGCC Applications", Paper presented at the Clearwater Coal Conference, June 6, 2011.
4. Dennis Leppin, et al., "Economic Evaluation of the UCSRP-HP Process in IGCC Applications", SULPHUR 2011 International Conference & Exhibition, Houston, USA, November 6-10, 2011.
5. http://www.nelliott.demon.co.uk/company/claus.html

(89a) Increasing Propane Desuperheater and Condenser Reliability in PT Badak NGL

Mohammad Arief Setiawan, Technical, PT Badak NGL, Bontang, Indonesia

PT Badak NGL is the operator for natural gas liquefaction at Badak LNG Plant, Bontang, Indonesia. At present PT Badak NGL has 8 Trains. The mode for liquefaction is using the APCI process license. At the first step, Feed Gas is pre-cooled with propane refrigerant, followed by liquefaction process by using Multi Component Refrigerant (MCR) as cooler. The MCR is also using propane as its cooler. The propane refrigerant is using three level refrigerant pressures for different level of cooling.

The propane refrigeration system use sea water for condensing the propane after compression. To fulfill this, PT Badak NGL operates 2 parallel propane desuperheaters and 6 parallel propane condensers in each Train. The exchanger acts to condensed propane refrigerant prior being used as cooler media in refrigeration system. The units are all shell and tube exchangers, having sea water stream in the tube side.

The propane desuperheaters and condensers in Train G were frequently facing leakages problem. The last 2 years record showed at least 9 and 5 times repairs have been conducted in the propane desuperheaters and condensers respectively, while other Train's repair frequency only reached below 4 times. The problem has caused significant repair cost and reduction of propane inventory. Our analysis revealed that the sea water stream in the tube side has lower velocity than the minimum required specification of 6 ft/s. This has increased the tendency of fouling in the tube surface. Eventually the fouling effect will promote under deposit corrosion phenomenon which will lead to leakage.

Several options have been proposed for the trouble shooting actions, either by increasing the cooling water supply, installing special tube insert, or by modifying the service mode of the exchangers. PT Badak is planning on implementing the last option to solve the problem since it proposes the least cost of all options. To implement this program, the number of operating propane condensers will be reduced from 6 of 6 service mode to 5 of 6 service mode. This action is expected can increase the sea water velocity thus reducing under deposit corrosion and leakages phenomenon.

(89b) Absorption Refrigeration for Auxiliary Applications in LNG Plant

Jordan Belue, Energy Technology Company, Chevron, Houston, TX, Robert Havlik, Tengizchevroil, LLP, Farnborough, United Kingdom and Stanley Huang, Process Technology, Chevron Energy Technology Company, Houston, TX

ABSORPTION REFRIGERATION FOR AUXILIARY APPLICATIONS IN LNG PLANT

Jordan Belue[a], Robert Havlik[b], Stanley Huang[a],

a: Chevron Energy Technology Company, Houston, Texas

b: Tengizchevoil, LLP.

ABSTRACT

A Liquefied Natural Gas (LNG) plant demands a large amount of mechanical power for generating refrigeration. Over decades of development, the art of designing the prime mover trains utilizing steam or gas turbine drivers, combined with helper motors or synchronous generators, have been pushing the technological limits and approaching perfection in new constructions. This is witnessed by a rapid growth of global LNG capacity in the first decade of the 21stcentury. The largest gas turbines (GT) used are Frame 9 machines. Furthermore, various approaches to enhance the rotating equipment train and gas turbine output have been investigated and implemented to push the capacity boundaries. For example, inlet conditioning for power augementation utilizing fogging, inlet chilling and evaporative cooling of GT inlet air have been installed in various occasions.

In addition to gas turbine and prime mover capacity enhancements, the other frontier is improvement of thermal efficiency of the prime movers and power plant either to free up more gas towards LNG Production or utilize in a combined heat and power conceptual design to capture waste heat and integrate back into the LNG Process Plant. The most obvious approach is implementing waste heat recovery unit (WHRU) on the gas turbines. The recovered heat can be directed to various heat users in the plant which is known as Cogeneration designs. In most LNG plants, the recoverable waste heat is more than the demand of the process plant thermal users. Therefore, in some cases, the recovered heat is used for power generation, which is known as Combined Cycle (CC) designs. Relatively few cases were demonstrated to use recovered waste heat to enhance plant capacity, although a well-known steam-injection process is commonly practiced in the power industry.

In this presentation, comparison will be made between a selected air chilling and an alternative approach, called absorption chilling cycle, which directly converts heat to refrigeration. Absorption refrigeration has been used in refinery and other industries, but not LNG. Emphasis will be placed on dampening the impacts of seasonal temperature variations, instead of straightforward capacity enhancement, which requires enlargement of all supporting systems and underlying structures.

Work Output of Multicomponent LNG Mixtures in Two-Phase Expanders

- Dr. Linda St-Cyr, Ebara International Corporation, Sparks NV
Presentation ID# 244808

- Abstract: A two-phase expander recovers more work from a multicomponent LNG mixture than a single phase expander. The calculation of the recoverable work involves the complex thermodynamics of two-phase mixtures, and is best done utilizing property tables from the National Institute of Standards and Technology (NIST). The NIST tables are in turn based on equations developed by the European Gas Research Group (GERG-2008). Building on the work done by these distinguished groups, we are able to document the contributions to the enthalpy integral from the PV term versus the internal energy.

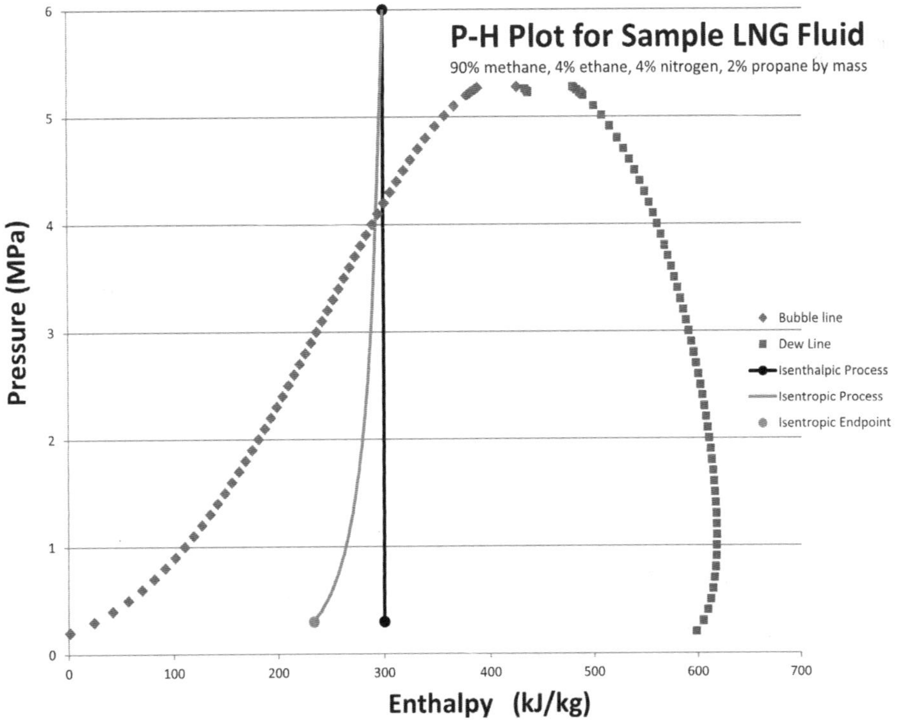

Expander Process vs Throttling Process

- The 2-phase region for a typical LNG fluid consisting of 90% methane, 4% ethane, 4% nitrogen and 2% propane is shown on this P-H diagram using values generated by the NIST REFPROP routine. The blue diamonds represent bubble points and the red squares represent dew points. Inside the 2-phase dome both liquid and vapor are present in equilibrium.
- An isentropic expansion process (green curve) and an isenthalpic expansion process (black line) show 2 possible fluid paths starting at h = 300 kJ/kg and P = 6 MPa and ending at P = 0.3 MPa.
- The isenthalpic line (black) represents the effect of using a throttling valve to reduce LNG pressure. This constant enthalpy process is known as a Joule-Thomson process.
- The isentropic curve (green) represents the most efficient expansion from the given starting position to an end state at 0.3 MPa. This trace follows a constant entropy curve and represents a 100% efficient (ideal) process. Expanders approximate this curve, usually at about 85% efficiency.
- For a given pressure, the farther to the left that a 2-phase point sits, the greater the percentage of liquid vs vapor. This Joule Thomson process ends at a liquid faction of 0.43 , while the isentropic process ends at a liquid faction of 0.57, which is 30% more liquid.
- Single-phase expanders operate only up to the beginning of the 2-phase region. After that, a Joule-Thomson valve is needed to continue the expansion to the desired ending pressure. For this example, the endpoint for the single-phase expander would lie at only 7% of the distance from the black line to the green curve.
- The endpoint for a 2-phase expander @ 85% efficiency would lie at 85% of the distance from the black line to the green curve. The expansion curve for an 85% efficient process can be approximated by a curve that lies at 85% of the distance from the black line to the green curve for each pressure from the beginning at 6MPa to the end at 0.3 MPa.

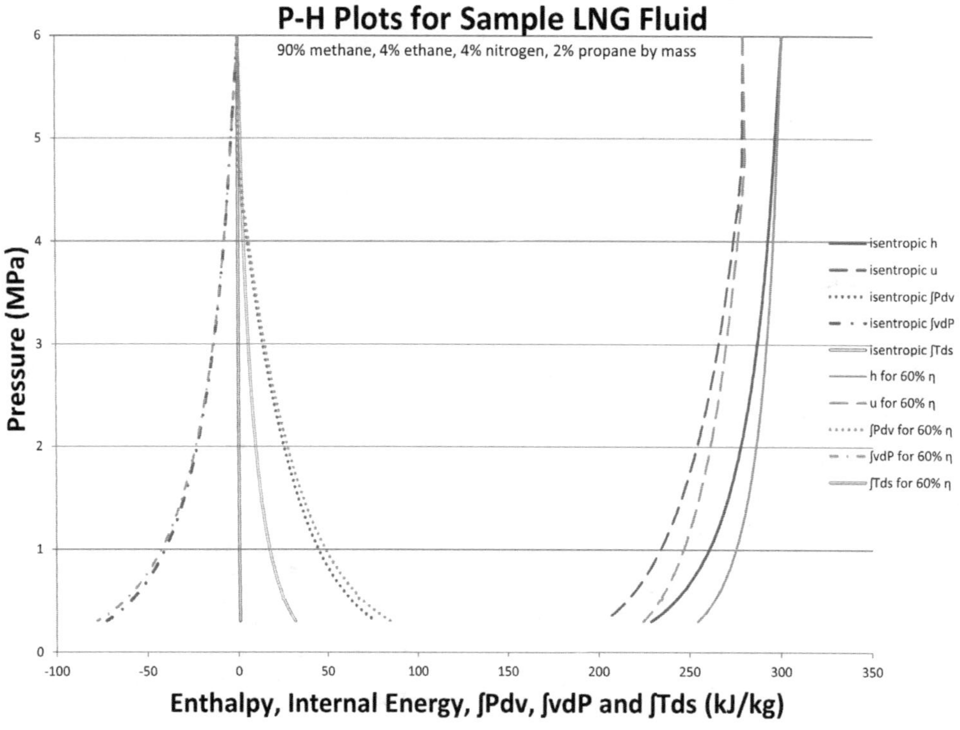

Enthalpy, Internal Energy, ∫Pdv, ∫vdP and ∫Tds

- The previous plot (made using NIST REFPROP) shows the process curves for enthalpy (h), internal energy (u), ∫Pdv, ∫vdP and ∫Tds for a 100% efficient process (blue) and for a 60% efficient process (orange). 60% was chosen so that there was enough of a difference to be clearly visible on the plot.
- Enthalpy per unit mass is defined as $h = Pv + u$, thus $\Delta h = \int Pdv + \int vdP + \Delta u$. From the plots and the tabulated data, it can be verified that at 100% efficiency, $\Delta h = \int vdP$ and $\Delta u = -\int Pdv$. An interesting thing to note is that the traces for Δu (long dashes) and ∫Pdv (dotted) both show a distinct kink at about P = 4.7 MPa that the other curves don't have. Indeed, for a closed volume, the reversible Pdv work done by a fluid decreases its internal energy u by exactly the amount ∫Pdv, so they are directly related. Similarly, Δh (solid line) and ∫vdP (dot-dash) are also directly related, since ∫vdP is the reversible work done by a steady-flow system.
- For a process that is not 100% efficient, we can write the Gibbs relation $\Delta h = \int vdP + \int Tds$. For this relation applied to an expansion, we note that $\Delta h < 0$, $\int vdP < 0$ and $\int Tds > 0$, the latter being another way to state that this process is *not* reversible. So the magnitude of ∫vdP must be greater than it was for an isentropic process by an amount that makes up for the irreversibilities in ∫Tds (the double-line plot). Thus we see that ∫vdP gives a measure of the available work only in the case of an isentropic process.

Numerical Experimentation with NIST REFROP

• By doing numerical experiments using properties from the NIST REFPROP tables, it is possible to clarify the thermodynamics of mixtures and 2-phase systems.

• The NIST tables for LNG components are based on equations developed by the European Gas Research Group (GERG-2008).

NIST
National Institute of
Standards and Technology

Dr. Eric W. Lemmon
Thermophsical Properties Division
Theory and Modeling of Fluids Group
325 Broadway
Boulder, CO 80305, U.S.A.
Phone: +1-303-497-7939
Web: http://www.nist.gov/cstl/properties/fluids_modeling/index.cfm

Dr. Mark O. McLinden
Thermophsical Properties Division
Experimental Properties of Fluids Group
325 Broadway
Boulder, CO 80305, U.S.A.
Phone: +1-303-497-3580
Web: http://FluidProperties.NIST.Gov

Dr. Marcia L. Huber
Thermophsical Properties Division
Theory and Modeling of Fluids Group
325 Broadway
Boulder, CO 80305, U.S.A.
Phone: +1-303-497-5252
Web: http://www.nist.gov/cstl/properties/fluids_modeling/index.cfm

(89d) Two-Phase Pump Expanders for Liquefied Propane Gas

Hans E. Kimmel, Ebara International Corporation, Sparks, NV

Thermodynamic power cycles at cryogenic temperatures require new and different designs and materials than there are in conventional power plants operating at high temperature.

The working fluids in cryogenic power cycles are liquefied gases like ammonia, nitrogen and hydrocarbon gases. Most of these cryogenic working fluids are flammable or corrosive or both.

The development of a particular design of a two phase pump expander for liquefied propane gas was necessary to meet the requirements for a reliable and safe operation of the rotating machinery within an efficient cryogenic power cycle. The applications and the design of two-phase pump expanders for liquefied propane gas as working fluid are presented and the performance characteristics discussed.

(89f) Cryogenic Expander Application in Propane Liquefaction Process

Michael Cords, Ebara International Corporation, Sparks, NV

The use of cryogenic liquid expanders in the propane liquefaction process can increase the amount of liquefied product by reducing the boil-off losses. Current practice is to use a Joule-Thomson valve to reduce the high pressure condensate. Losses associated by throttling across a J-T valve result in additional heat causing unwanted boil-off. A cryogenic expander reduces the enthalpy of the liquid during expansion which sub-cools the liquid, significantly reducing the boil-off losses, and can export this extracted energy as electrical energy.

The thermodynamic properties of propane are well suited to exploit the benefits of a cryogenic expander. The technology for propane cryogenic expanders exists and has been proven in LNG liquefaction, both in the LNG stream and the Mixed Refrigerant cycle. Mixed Refrigerant is comprised of mostly propane, making the technological transition from MR expander to propane expander negligible and low risk. The benefit of increased production already proven in LNG liquefaction can be applied to propane liquefaction resulting in overall process efficiency and reduced costs.

(93a) A Photo Safari Though a Petroleum Refinery

Matthew J. Reisdorf, Process, Fluor Corporation, Sugar Land, TX

Cut a crude assay, dehead a coke drum, and stalk wild naphtha in a 30-minute whirlwind tour of a petroleum refinery. This tour of refining products and processes is aimed at chemical engineers who want a better understanding of how and why the different parts of a refinery work together as an integrated system.

(93e) Planning for a Career in Refining - What Every Chemical Engineer Must Know

Shankar Vaidyanathanand Kevin Turini, Fluor Corporation, Sugar Land, TX

The paper will present an overview of how a chemical engineering graduate could put together a self managed development program to prepare for a career in the refining industry. It will cover the basic skills that are required and how to pursue them. It will suggest various resources that are available in the public domain and a suggested reading list. The paper will present an outline of a proposed action plan as well. The attempt is to enhance the knowledge and training of the individual chemical engineer by a structured learning program that can be adjusted based on individual needs.

(93b) From World Scale Refineries to State of the Art Nano Labs and Super Computers. From Coker to Clean Rooms, From KBPD to LPH. How to Translate Your Past Experience Into a Competitive Advantage?

Ahmed Khogeer, KAUST & Thuwal development Projects, Saudi Aramco, Thuwal, Saudi Arabia

In this paper we will cover some case studies on how a chemical engineer with a specialization in one area can find himself in a position where he is required to work on the design and construction in other chemical engineering discipline that he never practiced in the past. Some challenges, lesson learned and tips to translate what may be thought irrelevant experience into a competitive advantage.

Refinery-Wide Topics for Profitability, Reliability and Safety

Mike Siravo, Fluor Corporation, Sugar Land, TX

Introduction

The objective of this presentation is to raise your awareness of the multitude of problems lurking among the many activities performed in a complicated facility like an oil refinery. These problems take many forms and have varied causes. If not identified they can and do result in:

- Serious safety and environmental incidents
- Community relations breakdown, loss of reputation
- Unnecessary capital expenditure
- Damage to equipment and expense increases
- Loss of production and profitability

We will review several case studies of design and operational incidents to identify learnings. We will discuss what went wrong and why. It is beneficial to learn from other's mistakes, but hopefully the negative impact of these case studies on people, businesses and, possibly, careers will do more than provide specific learnings. Hopefully these negative impacts will instill in you the proper level of anxiety and accountability to provide the motivation to remain vigilant and alert as you perform your work.

Identifying and mitigating potential pitfalls in your work is not easy, especially when you may be encountering new assignments and the unfamiliarity that brings. Time constraints overlay everything we do and provide possibly the largest impediment to quality results. Although in these case studies you will encounter the bad news of things gone wrong, the good news is that the skills to root out problems before they occur can be developed. As we go through these case studies, look for the common themes of poor communication, lack of attention to detail and violation of procedures.

Case Study 1: Know Procedures and Follow Procedures

A refinery Crude Unit was nearing the end of its four week shutdown/turnaround. The lack of crude processing over the turnaround period had significantly reduced the rates of those refinery units that stayed on-line; units like the FCCU, Alky and Hydrocracker. These units and others were able to operate due to inventories of feedstock that were generated before the Crude Unit was shutdown. Gasoline blending was also partially reliant on inventories of certain blend components. These inventories were getting low as planned, but it was imperative that the Crude Unit start up on time.

An operating group can become extremely unpopular if they are unable to maintain schedule and achieve "oil-in" on the expected date. If delay were to occur, there would be some heated conversations and the plant manager would be making telephone calls to his boss, the President of Refining, at headquarters. Calls that he desperately wants to avoid, if possible. In summary the refinery staff was feeling a very strong sense of urgency for this startup.

One of the activities during the turnaround was to perform a light oil wash of some of the heavy oil equipment. Once the wash was completed, removal and disposal of this oil was necessary to proceed with commissioning of the unit for startup. Around that time, in the evening, it was discovered that the original plan for disposing of the contaminated wash oil was no longer an option. A scramble for an alternative ensued. Some ingenious soul discovered that a series of lines existed that could be used to connect the Crude Unit vessel holding the oil to a heavy oil tank in the Offsites area of the refinery. As often happens in an older refinery, sometime in the past, the need for flexibility had added these connections for a reason no one could now remember. This appeared to be very fortuitous.

Telephone conversations were initiated between the Crude Unit and the operations foreman in the tank farm area. The Offsites tank in question normally held a heavy oil used in blending asphalt. It was a fixed roof tank with a steam coil to keep the heavy oil fluid and a roof vent to the atmosphere. As luck would have it, the tank was essentially empty. The tank farm foreman was, correctly, not interested in receiving transfer of the wash oil. The foreman resisted the transfer because he knew that his operating procedures limited use of the tank to heavy oils with a high flash point. He did not know exactly what this oil was and the Crude Unit was not particularly forthcoming with information.

The standoff remained for the next few hours until the tank farm foreman relented. Much of his thinking was rooted in the culture of the refinery which held that the oil processing units was where the money was made. Any operations that were part of the refinery offsites and utilities were support functions that were never to be the cause of a problem in a place like the Crude Unit. This was a frequent source of friction between the two areas.

About three hours after the transfer was made, the manager of the refinery offsites area who lived about ten miles from the refinery was awakened by a loud, rumbling sound at about four AM. Not long after falling back to sleep, the manager's pager began buzzing. A return phone call to the refinery provided the news that one of the tanks in the manager's area had exploded. Miraculously, no one was injured. The roof had been launched some distance damaging another tank. The heat from the fire damaged the adjacent tanks. The ignition source was determined to be either pyrophoric iron in the tank or a thunderstorm that passed overhead around that time.

Putting everything else aside, the Offsites manager was going to spend the foreseeable future leading an incident investigation. The Plant Manager got to call his boss, the President of Refining, at headquarters, which is what he wanted to avoid in the first place.

Learnings:

- Have proper and clear procedures
- Train to the procedures
- Audit the understanding of the procedures
- Use weekly safety meetings to review and discuss procedures
- As often as necessary, using all reasonable means, management needs to reinforce to every person in the facility that they have the authority to shutdown any operation that poses an immediate threat to safety. Everyone needs to know that they will be lauded for such an action. Conversely, there should be no perception of reward for anyone that violates a procedure or takes risks to benefit production.

Case Study 2: Poor Unit Monitoring

During a shut down, a 30 MBPD diesel hydrotreater reactor was loaded with the latest generation of fresh catalyst. Startup proceeded smoothly and without issue. When catalyst is new the reactor temperature needed to meet the product sulfur specification is at its lowest point. As time goes on and the catalyst ages or deactivates, the reactor temperature is increased to overcome the lower activity of the catalyst to continue to meet spec. The predicted catalyst life was 2 years, meaning that at that time the reactor temperature will have been increased slowly over time until the maximum operating temperature is reached. Generally the reactor temperature is increased linearly for the first 75% of the 2 year run and the at an increasingly faster rate until end of run (EOR) temperature is reached.

About 5 months into the run, the operations supervisor noticed that to meet the product sulfur specification he needed to increase reactor temperature more often and in larger increments than was his experience with prior runs. Discussions were held with the process engineering department which agreed with the assessment. More intense monitoring of unit performance was initiated. Investigation showed that the catalyst deactivation rate (rate of required reactor temperature increase) was approximately three times higher than expected. At this deactivation rate the catalyst would reach EOR after only 8 months, 16 months earlier than planned.

Process engineering resources and support staff from the catalyst supplier began a concerted effort to understand the cause of the rapid deactivation. After several weeks three causes were identified:

- With this current run, the feed to the unit included a 15% increase in cracked stock. Cracked stock in this case included diesel boiling range material from the FCCU and Coker unit. Cracked stock has a higher hydrogen consumption than straight run diesel from the Crude unit.
- The higher activity catalyst increased hydrogen consumption compared to the catalyst used in prior runs due to an increase in aromatics saturation.
- The hydrogen content of the recycle gas in the diesel hydrotreater unit was designed to be 80 mole%. The centrifugal compressor, therefore, was designed for a recycle gas of 80% hydrogen. For some time, even in the prior run, the recycle gas purge rate was set too high. Make up hydrogen at 99 mole% hydrogen entered the unit on pressure control. So since the purge gas rate was high, the make up hydrogen rate was higher than it needed to be to maintain recycle gas purity at 80%. In fact the make up rate was so high that the recycle gas purity was running 90%. Generally, higher recycle gas hydrogen purity should be good for hydrotreater performance. However, as the hydrogen purity increased the molecular weight decreased significantly. Since the recycle gas compressor was a centrifugal design, this drop in molecular weight decreased discharge pressure and led to a reduction in recycle gas rate. The net effect was a reduced rate of hydrogen available for reaction.

Unfortunately, all three of the above combined to reduce the hydrogen partial pressure in the reactor by 35%. This greatly escalated coke formation on the catalyst causing rapid deactivation.

Indeed, as predicted, the catalyst lasted only 8 months although feed rate needed to be reduced to even achieve this. A fresh load of catalyst needed to be procured on short notice at a premium price. The unplanned shut down caused disruption to the refinery operating plan just prior to entering the profitable gasoline season. The total cost of the incident was several million dollars. The process engineering department, responsible for daily monitoring of the unit, had failed to do their job. And it was obvious to everyone.

Learnings:

- Specific and up to date operating guidelines should have been in place providing targets for recycle gas rate and hydrogen purity. In fact operating guidelines should be in place across the entire unit.
- The unit process engineer should have trended unit performance daily to identify abnormalities as early as possible. With proper monitoring the change in feed quality, low recycle gas rate and high hydrogen consumption all could have been

identified in the first month of operation. Corrective actions could have taken place to ensure the 2 year run length.

- Proper unit monitoring requires the process engineer to understand the reasoning behind all process conditions on a unit. In this case an understanding of the importance of hydrogen partial pressure and the variables that affect it were necessary. If you do not feel you have this level of understanding, discuss this with your supervisor and ask him or her to help you develop a plan to acquire the required competency.

Case Study 3: Installation Failure

An FCCU debutanizer was being revamped to increase capacity to allow for increased vapor and liquid traffic. This was necessary to improve separation between the bottoms product gasoline and the overhead product C3/C4 stream. The objective was to recover additional C4's from the bottoms product to reduce its blending vapor pressure and reduce the C5 content of the overhead product since C5's have a negative impact on the Alkylation Unit which is the ultimate destination for the C4's.

It was determined that the reboiler, condenser and reflux pumps all had the necessary spare capacity to allow for the increase in stripping required to improve the separation. An evaluation of the trays showed that a vapor-liquid bottleneck existed but only in the trays above the feed point or in the rectification section of the column.

A redesign and replacement of the trays was found to fall short of adding the required capacity as did the possibility of replacing the trays with random packing. The only possibility for gaining the needed capacity was replacing the trays with structured packing. Simulation and packing vendor data showed that structured packing would work. At the time of this revamp structured packing had been successfully used extensively in the refining industry but not in services handling C4's and lighter materials. In fact some industry literature discussed concerns with this application. Additional research including discussion with fractionation experts was successful alleviating enough of the perceived risk for the structured packing revamp to proceed.

In the final revamp design the eighteen trays above the feed point were to be replaced with two beds of structured packing. Above the top bed was a perforated tray that would take the reflux stream and drop it evenly across the top of the top packed bed through several hundred small (3/8" diameter) weep holes. Even distribution of liquid across the packing was essential, and this design would serve that purpose. Vapor leaving the top of the packing rose through this perforated tray through eight rectangular risers. Each riser was six inches wide and several feet in length. The risers extended one foot above the tray to ensure that liquid could only reach the packing below through the weep holes. Upon exiting the risers the vapor would leave the top of the column as normal on its way to the condenser.

Since there were two packed beds, there was a similar liquid distribution tray between the beds. This tray was to collect the liquid exiting the bottom of the upper packed bed and redistribute it evenly through the several hundred weep holes onto the lower packed bed. Vapor from the lower packed bed rises through eight vapor risers that were almost identical to those just described for the upper bed distributor. To eliminate the possibility that liquid falling from the upper bed could fall into the riser, each of the risers for the redistributor tray between the packed beds was equipped with a cover. The cover took the form of a gutter sitting five inches above the top of the riser and extending the full length of the riser. The gutter has side walls so that any liquid that lands on it can be routed to and run off its ends. This collected liquid then falls harmlessly onto the perforated tray deck. Since there are eight risers there are eight gutters. The five inch gap between the top edge of the riser and the bottom of the gutter provided the area for vapor to exit the riser.

Despite the concern of the new application of structured packing to such a revamp, the early phase of the column start up proceeded smoothly as the FCCU was brought on line. A problem developed, however, as the FCCU and column rate approached 60% of the design. The newly installed differential pressure instrument for the upper packed bed alarmed on high pressure drop. Overhead and bottoms product sampling confirmed the problem as the degree of separation was very poor.

For several days everything imaginable was done to determine what was causing the column flooding so it could be corrected. It was no small effort to start up the FCCU, and there was no desire to shut it back down so the column could be entered for inspection. On the other hand the economic impact of limiting the FCCU capacity to 60% of plan for much longer was also brutally painful to consider. A significant part of the thought process working against shutting down and entering the column was the fear that no problem would be found. And then what? It was unthinkable to go through the time, money, effort and embarrassment to enter the recolumn only to find later that it was not necessary to correct the problem.

Once all of the diagnostics and theories were exhausted, there was no other choice but to shutdown the FCCU, prep and enter the column for inspection. This took another day and a half, all the time the refinery was hemorrhaging money at the rate of about one half million dollars a day.

Once the column manways were opened and the air mover was in place. The FCCU unit process engineer and the process engineer from the project group responsible for the revamp design began inspecting what they could from the manways in the hope that something obvious could be found. As often is the case there was some good news and some bad news.

The good news was that the cause of the problem was painfully obvious. From one of the manways it was very easy to see the liquid redistributor tray that was located between the two packed beds. At first it seemed fine but something was not quite right. No part of it was loose or damaged, but it became clear that the space provided in the design between the vapor risers and the liquid collection gutters was somehow blocked. The gap between the two was slightly less than one inch instead of the design gap of five inches. How was all of the column vapor supposed to flow up through the risers when the cross sectional area for flow was restricted by more than 80%. How was this possible?

The explanation lay in a mistake made by the maintenance craft during the tray installation. The tray deck sections came with the risers already in place. During installation it was necessary for the crafts to mount the gutters above the risers using the supplied support strips and bolts. The crafts did not properly read the tray drawing. Thinking that the gutters were like a long hat sitting on top of the risers, they installed them upside down. This meant that instead of extending upward to collect liquid, the side walls of the gutters were extending downward, blocking off most of the area allotted for vapor flow.

As a result the pressure drop for vapor flow through the tray was high, much higher than design. With this high pressure drop the liquid level on the tray (designed to be only a few inches high) increased until it was above the top of the vapor risers. This further restricted vapor flow increasing pressure drop even more. Ultimately the liquid level rose into the top packed bed increasing bed pressure drop (thus the high pressure drop alarm) and destroying the degree of separation.

In a relatively short time the position of the gutters was reversed, the column was closed and the FCCU was restarted, reaching design rate in the typical amount of time.

The bad news was that the installation error could have been easily avoided by proper supervision of the crafts during installation. Certainly it is a reasonable expectation that the drawings be followed. But the responsibility for the final inspection prior to the column being closed lay with the process engineers who are generally tasked with this inspection since they are the most familiar with the details of the design intent.

Learnings:

- Since there were two process engineers involved, accountability for the final inspection became uncertain. The engineer involved in the revamp design had been reassigned to other project work before the column shut down and packing installation. The unit engineer was less experienced and was not directly informed of the responsibility to inspect all such changes to equipment in his area.
- In a perfect world one or both of the process engineers would have felt the need to perform the final inspection. One could argue that management had a responsibility to assign the task. Since none of this happened and luck didn't play out either, they all got to share the substantial amount of blame involved.

Modern Loop Tuning Tutorial

by

Tim Olsen
Emerson Process Management
Tim.Olsen@emerson.com

and

Jim Dunbar
Emerson Process Management
Jim.Dunbar@emerson.com

Prepared for Presentation at the

AIChE 2012 Spring Meeting
Houston, TX
April 1-5, 2012

Session 93, Young Professional Tutorial - Topics in Refining
Paper 93d

Abstract

The 2012 AIChE Spring Meeting presentation 93d is a tutorial on tuning PID process controllers. The course will explain the operation of the PID controller, typical variations of the PID controller, common process responses and how to identify the dynamics of the process. The instructor will demonstrate how to calculate PID tuning parameters from the process dynamics and the desired controller response.

Introduction

This paper is intended to encourage young professional engineers to understand the dynamics of their processes and use this tuning tutorial to improve process control performance. Modern loop tuning is a simple and inexpensive tool that any engineer can utilize for better operation.

Step 1 – Understand the Process and Operating Objectives

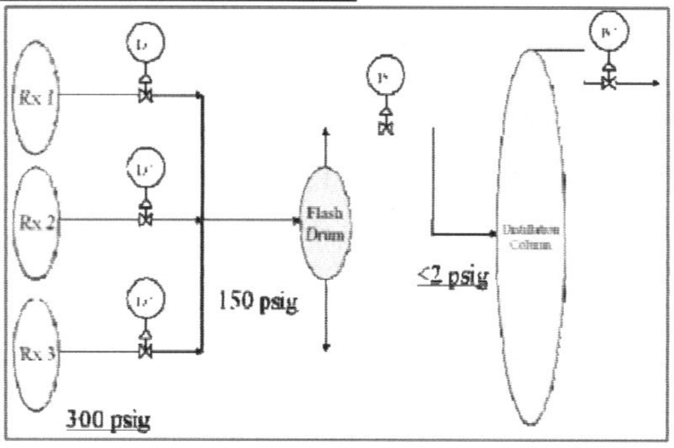

One of the first steps before actually starting to tune a loop is to understand the process and the objectives of the process. The process on the right, Figure 1, has several reactors on individual level control operating at 300 psig flowing into a flash drum at 150 psig and then vapors into a distillation column less than 2 psig. Each control loop was tuned individually, but the overall process was unstable and difficult to control the distillation column including product quality. Changes in the reactor to maintain level sent a disturbance to the flash drum, but the flash drum was tuned very aggressively (standard deviation of 0.1 psi from 150 psig setpoint) and the flash drum vapors significantly disrupted the performance to separate in the distillation column.

The objective of this process is not to have very accurate pressure control of the flash drum, but to maximize yield of desired products from the reactors and then effectively separate impurities from the product in the downstream distillation column.

Step 2 – Prioritize the Control Loops

Based on information gathered in step 1, you are now ready to prioritize the control loops to determine which should execute fast and which should execute slower. Process control loops interacting with each other will fight each other if given equal priority in tuning. This is easiest to understand in a cascade controller where steam flow is the inner loop and temperature is the outer loop (process temperature setpoint). The inner control loop is tuned 5-10 times faster than the outer loop to quickly compensate for changes in steam pressure. Since the setpoint of the inner loop is a function of the outer loop, you will always interact; if the inner and outer loop tuning is equal priority, this will cause instability or cycling because the inner loop cannot track the setpoint from the outer loop.

For the reactor-flash drum-column example in Figure 1 above, the primary objective is to maintain proper levels in the reactors to ensure maximum product yield. The downstream separation process is next in priority to ensure product quality at the lowest energy usage required for separation. The flash drum is just a vessel in-between the reactors and distillation column for a quick, basic separation. The 150 psig is not a critical value since the separation of flashing is simple separation based on pressure drop from the 300 psig reactors. The flash drum can be used as a variability buffer to absorb disturbances from the reactors maintaining level, so to minimize disturbances to the distillation column at less than 2 psig.

As good practice, one should utilize levels as variability buffers to absorb disturbances from upstream processes, thereby minimizing the impact to downstream processes. This can be a culture change for many facilities where console operators like seeing levels at 50% with a flat line process variable matching the setpoint. Surge drums, distillation column bottoms and overhead drums are good examples where the level can fluctuate from 40 and 60% (around the 50% setpoint) to maintain steady flow downstream with gradual changes as needed to rebalance the process flow. Note, some levels need to have minimal variability such as the example with the reactors in Step 1, so not all levels can be variability buffers, but many can.

Step 3 – Step Test each Loop and Identify Hardware issues

A step test is when you place a controller in manual and change the output a small percentage; the size of the output step should be large enough to observe the response above any process noise that may be present, yet small enough to minimize process upsets.

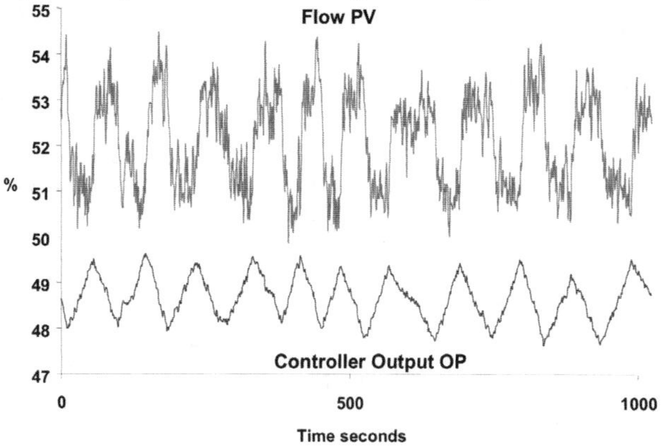

Figure 2: Limit cycle in a flow loop due to a sticky valve

Ensure you understand the process before executing step tests to prevent unwanted process upsets (for example, do not reduce hydrogen quench flow to cause a hydrocracker reactor temperature runaway; start with increasing hydrogen and then changing direction back to normal hydrogen use). Start with small steps such as 0.25%, 0.50% or 1% depending on sensitivity of the process. If you do not see a change in flow / temperature / pressure, continue making step changes until you do see a change. It is not uncommon to see control loops requiring an output change in excess of 2% to see changes from the final control element movement (Figure 2 above shows a limit cycle due to a sticky valve). Under-performing field devices add process constraints that cannot be tuned out or compensated with advanced process control.

Step 4 – Fix Hardware Issues

As you identify hardware issues like sticky control valves, hopefully you have the ability to make fixes by isolating the control valve to make repairs (bypass), or plan the repairs at the next upcoming turnaround. Smart positioners have made significant improvements over the past couple decades and help control valve performance for better response and positioning.

Step 5 – Re-Step Test Loops that had Hardware Fixes

Now that control valves move appropriately when the controller asks it to move, you can place a controller in manual and change the output a small percentage (same as step 3).

As shown on the right in Figure 3, the response is the most common type, a first-order plus dead time.. There are many types of responses that could be observed such as integrating, second order response, inverse response and others. This paper will address the first-order plus dead time response.

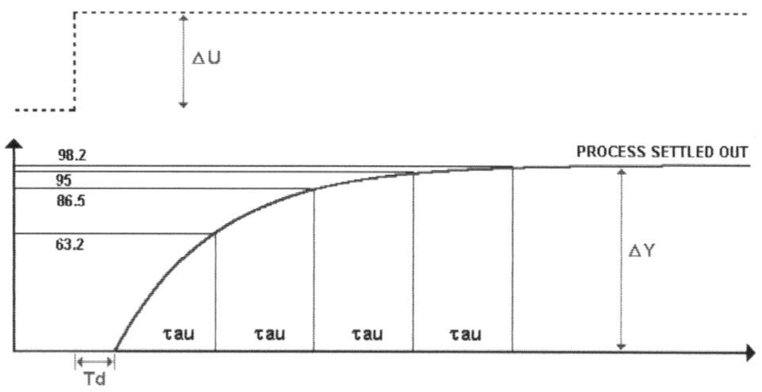

Figure 3: Typical first order step response

The primary process dynamic information attained from a step test is steady-state process gain, dead time, and the time constant. Typically these parameters are an average of several step tests around the normal operating range.

The steady-state gain can be calculated by taking the measured process value before and after the step test and then dividing by the output change (normally the gain is converted to units of %PV/%Output because most controllers calculate their output in dimensionless units). The time constant Tau is the time required to reach 63.2% of the final process change. The dead time is the amount of time after the step test is made until the process actually begins to move.

Step 6 – Calculate Tuning parameters

Once an engineer understands the process dynamics and control loop priorities, model-based tuning such as Lambda Tuning can be applied to set the speed of response for each controller based on operating objectives and priorities. "Lambda Tuning" is a tuning method where the control loop speed of response is a selectable tuning parameter; the closed loop time constant is referred to as "Lambda" (λ). Lambda Tuning originated with Dahlin [1] in 1968; it is based on the same IMC theory as MPC [2, 3]. It is model-based and uses pole-zero cancellation to achieve the desired closed loop performance. A recommended Lambda starting point to ensure robustness is three times the larger of Tau or

Deadtime. The time for the loop to reach setpoint is approximately four times the selected Lambda value.

Self-Regulating Process (First-order with deadtime model, Classical PID) [3, 4]

τ = time constant (from step test)

T_d = dead time (from step test)

K_p = steady state process gain (from step test)

λ = closed loop time constant (Lambda – user defined speed of response)

Controller Settings

$T_r = \tau$

$K_c = (1 / K_p) * T_r / (\lambda + T_d)$

Note, tuning parameters setup for normal operation may not work as effectively during turndown rates. If turndown rates continue for a long period of time, one should consider re-evaluating the required tuning parameters for the new operating conditions. Write down the old tuning parameters for when you return to normal operation.

Step 7 – Input New Tuning Parameters and Observe

There is an old saying that if you change the tuning parameters on a controller, you own it. If you make changes to the tuning, you need to observe the controller ability to minimize disturbances while maintaining the overall operating objectives effectively. A common way to test the speed of response is to install the new tuning parameters and then perform a small setpoint to confirm that the response is as expected.

Keep a log book for tuning parameters, changes including time and date, along with who made the changes. Some modern systems will electronically log changes and the user that made the tuning changes.

Step 8 – As needed, Follow-up on Tuning to Ensure Desired Performance

Control valves have moving parts and degrade over time, and the performance of these final control elements will impact the performance of tuning. A control loop that worked just fine in the past may not be working today due to hardware issues or a change in the operating parameters. If operating conditions change and will continue for awhile, a re-evaluation of the process and operating objectives along with new tuning parameters may be in order.

Conclusions

Young professional engineers have the ability to understand more about the dynamics of their processes and then use this information to improve process control performance. Modern loop tuning is a simple and inexpensive tool that any engineer can utilize. These steps listed above can be used to identify control valve deficiencies, understand overall process dynamics, and assist with tuning control loops. Additional examples and details will be shown in the presentation at the meeting.

REFERENCES

1. Dahlin E.B., *Designing and Tuning Digital Controllers,* Instr and Cont Syst, 41 (6), 77, 1968.
2. Morari M. and Zafiriou E., *Robust Process Control,* Prentice Hall, 1989.
3. Chien I-Lung and Fruehauf P.S., *Consider IMC Tuning to Improve Controller Performance*, Hydrocarbon Processing, Oct. 1990.
4. J. Martin, A.B. Corripio and C.L. Smith, *How to Select Controller Modes and Tuning Parameters from Simple Process Models*, ISA Transactions, Vol 15, April 1976, pp 314-319.

(95a) Switching Between Hydrogenolysis and Bifunctional Hydrocracking on An Unsulfided Co/MoO3/SiO2-Al2O3 Catalyst

Heinz J. Robota, University of Dayton Research Institute, Dayton, OH and Jhoanna Alger, Energy and Environmental Engineering, University of Dayton Research Institute, Dayton, OH

In previous work using hexadecane as a feed molecule, we found that under typical catalytic hydrocracking conditions, an unsulfided $Co/MoO_3/SiO_2$-Al_2O_3catalyst exhibited primarily hydrogenolysis. Two types of hydrogenolysis were observed. One type involved terminal cleavage of methyl groups, in which the feed n-C_{16}underwent progressive cleavage to first n C_{15}, then, to a lesser extent to n-$C_{14}$and so on. This resulted in high quantities of product methane. The second type of hydrogenolysis reaction resulted in the cleavage of interior C-C bonds with near equal probability. This second pathway yielded ethane in molar quantities similar to other cracked products in the $C_3$to C_{13}range. Further, the overall yield of isomerized cracked products was quite low, typically less than 10 mole percent. However, when feeding a mixture of n-C_{26}/n-$C_{28}$to the same catalyst under similar conditions, methane yields were low, ethane was barely detectable, and isomer content of the cracked products was in excess of 80 mole percent. These latter observations are consistent with a bifunctional hydrocracking mechanism. Further, a comparison of the product selectivities as a function of conversion in these two cases are consistent with hydrogenolysis in the n-C_{16}case and bifunctional hydrocracking in the n-C_{26}/n-C_{28}case. In an effort to understand the reasons for this feed dependent reaction behavior, we have investigated the cracking of n-C_{18}under similar conditions. We have found that we are able to switch between hydrogenolysis and bifunctional hydrocracking by altering reaction conditions so as to either favor or suppress the relative olefin content of the reacting C_{18}. While holding the feed rate of hydrocarbon constant, hydrogenolysis is nearly exclusively observed at 300° C when $H_2$is fed at a molar ratio of 10:1 H_2:n-C_{18}. However, at the same absolute $H_2$and n-C_{18}feed rate, but with a supplementary flow of Ar at twice the H_2feed rate, nearly exclusively bifunctional hydrocracking behavior is observed. In contrast, when only $H_2$is fed at a 30:1 H_2:n C_{18}rate, an even higher rate of exclusively hydrogenolysis is observed. Similarly, we found that when n-$C_{16}$is fed with the supplementary Ar flow, it also exhibits predominantly bifunctional hydrocracking behavior. We believe that when the olefin content of the hydrocarbon reactant is sufficiently high, bifunctional hydrocracking predominates. Hence, since the equilibrium olefin content of a C_{26}/C_{28}feed is considerably higher than in a C_{16}feed, at equivalent molar flux, bifunctional behavior dominates in the C_{26}/C_{28}case. For this reason, when seeking to process a typical Fischer-Tropsch wax, where the dominant molecular sizes are in the $C_{26}$and larger range, quite typical bifunctional hydrocracking is exhibited by this catalyst.

(95b) The Effect of Chain Length of Feed on Hydrocracking Over Pt/SiO_2-Al_2O_3Catalyst

Jungshik Kang, Wenping Ma, Wilson D. Shafer, Keogh Robert, Gary Jacobs and Burtron H. Davis, University of Kentucky, Center for Applied Energy Research, Lexington, KY

The hydrocracking of hydrocarbons with carbon number $C_{16}$and $C_{28}$and their mixture was studied over Pt promoted amorphous silica-alumina catalyst. At various reaction conditions we determined the conversion, product selectivity, and the degrees of isomerization on the catalyst of each carbon number. We found out that gas phase hydrocarbons (at the reaction conditions) passed through the catalyst bed much faster than liquid phase hydrocarbons so that essentially only the liquid hydrocarbons were converted. We found that at a low hydrocracking conversion range of wax, all the liquid phase reactants show about the same hydrocracking reactivity. So that we can conclude that the hydrocracking reaction depends not only on the carbon chain length but the phase of reactant at the reaction conditions if predominantly in the gas phase it is not or only slightly converted. Therefore the dependence of conversion on carbon number is directly related to the liquid/gas phase equilibrium.

(95c) Middle Distillates Production Via Fischer Tropsch Synthesis with Product Upgrading Under Supercritical Phase Conditions

Sihe Zhang, Rui Xu, J. Ed Durham and Christopher B. Roberts, Department of Chemical Engineering, Auburn University, Auburn University, AL

Fischer-Tropsch (FT) synthesis involves surface catalyzed polymerization reactions that inherently yield a high selectivity towards methane and a broad product distribution of gaseous light products, middle distillate hydrocarbons (olefins + paraffins), and heavy wax, as well as $CO_2$and water. Subsequent product upgrading via catalytic oligomerization and hydrocracking/isomerization can be employed to convert light olefin FT hydrocarbons into middle distillate range liquids, while also converting the long-chain hydrocarbons of heavy wax into fuel range products. This results in a reduction in olefin selectivity, generation of aromatics, suppression of heavy hydrocarbons selectivity and an enhancement of cold-flow properties.

In an effort to further direct the FT product distribution towards middle distillate transportation fuel range products, we have undertaken a series of experiments where we have sequentially incorporated catalytic oligomerization and hydrocracking/isomerization stages subsequent to FT in single pass operation. A multiple fixed bed reactor system was designed and utilized in this investigation. A precipitated iron-based FT catalyst was applied due to its low price and high tendency to yield olefins, which can be upgraded in the oligomerization, hydrogenation and isomerization reactions within the sequential reaction beds. The reaction temperatures were maintained differently in each bed to provide optimal reaction performance, while the system pressure was kept uniform in all three reaction beds.

Moreover, we have utilized the enhanced heat transfer properties of a supercritical fluid reaction medium in the FT and subsequent upgrading stages in order to reduce the selectivity towards $CH_4$and CO_2. The use of the supercritical solvent provides prolonged catalyst life and activity maintenance. Distinguishable enhancement of propagation probability was observed under the supercritical phase operation. In particular, this study involves a systematic investigation into the influence of the supercritical fluid reaction medium on the overall product distribution and selectivity towards fuel range products.

(95d) Oxygenate/Hydrocarbon Distribution Compared Across Different Group I Promoters for Iron Catalysts

Wilson D. Shafer[1], Robert A. Keogh[2]and Burtron H. Davis[1], (1)University of Kentucky, Center for Applied Energy Research, Lexington, KY, (2)Center for Applied Energy Research, University of Kentucky, Lexington, KY

Oxygenate and Hydrocarbon Distribution Compared Across Different Group I Promoters for Iron Catalysts.

Wilson D. Shafer, Robert Keogh, Burtron H. Davis, Center for Applied Energy Research, University of Kentucky, Lexington, Ky

Abstract

A main catalyst used in FT currently is iron; this is driven by cost and the ability of Fe to play a role in water-gas-shift alongside the production of liquid fuels. Iron catalysts can be dispersed amongst a support and promoted with potassium or copper. The activity of iron catalysts promoted with potassium display an increase in the FTS and WGS activity [1]. Specific testing was done on different K promoter loading to find that greater than 5%molar K/Fe loading did not increase the CO conversion [2]. Low atomic loadings of potassium have been shown to increase the FTS conversion by playing a role in increasing the CO dissociation but slightly decreasing the H_2dissociation [3]. If this is the case for potassium; the other alkali metals, containing the same chemical characteristics should play a similar role.

A factor in describing the effects seen is the basicity of the promoters described by Roper et al [4], using a fluidized bed reactor, where the alkaline basicity was inversely correlated to the amount methane produced. In addition, 2002 specific work was done with slurry phase operations investigating the role in other row I alkali metals in comparison to potassium in iron catalysts [5]. This specific work was done looking into the WGS and FT activity for a slurry reactor system with similar alkaline promoter loadings and specific reactor conditions. The results based on the previous work, display varying results for each promoter across the same Fe catalyst. It was found that of the entire row I alkali metals, potassium promotes both FTS and WGS the best, in fact Li, Cs, and Rb promoted catalysts seem to show lower activity when compared to the unpromoted Fe catalyst.

This specific analysis work was to investigate the role I metal promoters by looking at the specific hydrocarbon/oxygenate distribution. The addition of each row I metal exclusively adds to the mole per carbon amount of olefin and oxygenated material over the unpromoted Fe catalyst. If, as previously mentioned, the addition of K as a promoter assists in the dissociation of CO and in turn suppression of H_2dissociation, this then could be seen in the overall product distribution for each row I promoter. The amounts of oxygenates and olefins increase linearly with each promoter added, where Li and the unpromoted produced the smallest amounts and Cs, Rb promoted Fe catalysts displayed the most. This distribution work can then compared to the alkali promoted Fe catalysts work done in 2002, which described basicity as a factor, others include atomic size and the ability to suppress $H_2$and assist in CO dissociation, to better explain the metals activity and why K seems to be the best row I promoter.

References

1.D.B. Bukur, D. Mukesh, S.A. Patel, Ind. Eng. Chem. Res. 29 (1990) 194.

2. M. Luo, R. J. O'Brien, S. Bao, B. H. Davis; Appl. Catal. A: General 239 (2003) 111-120

3.Gaube, J.; Klien, H.-F.; J. of Mol. Catal. A 283 (**2008**) 60

4. M. R"per, in: W. Keim, (Ed.), Catalysis in C1 Chemistry, D. Reidel Publishing Company, Dordrecht, 1983, pp. 41"88.

5.W. Ngantsoue-Hoc, Y. Zhang, R.J. O'Brien, M. Luo, B. H. Davis; Appl. Catal. A: General 236 (2002) 77-89

(106a) Synthesis of PVA Based Membranes Modified by Benzimidazole Group for Direct Methanol Fuel Cells

Pratima Gajbhiye, Chemical Engineering, Indian Institute of Technology, Kanpur, State of Uttar Pradesh, India, Jayant K. Singh, Chemical Engineering, Indian Institute of Technology, Kanpur, Ka, India and Anil Kumar, Department of Chemical Engineering, Indian Institute of Technology Kanpur, Kanpur, India

Abstract

In recent years, there has been a great effort to look for alternate membrane material and Polybenzimidazole (PBI) membranes have been synthesized where the conducting moiety of this polymer is benzimidazolium groups. The monomers used to synthesize these polymers are bis conjugated amines and diacids are expensive. In this work, we have synthesized grafted polymers of PVA as membrane materials where the grafts are either Poly Acrylic Acid, Poly maleic anhydride (PMA) or Poly Methacrylic acid (PMAA) and these materials were denoted by PVA-g-PAA, PVA-g-PMA, and PVA-g-PMAA. We then synthesized C1, C3 & C4 compounds (formula given at the bottom of Table S1) having bis conjugated amines whereas compound C2 having only one conjugated amine group. One of the steps in synthesizing C1 compound is nitration of phenyl ring, where there is a possibility of formation of two isomers out of which only one of them has strictly bis conjugated amines and can be polymerized with Isophthalic acid giving the formation of PBI. The other byproduct formed is an impurity and is minimized by carrying out the nitration step at low temperature but this cannot be completely removed from the product which is a mixture. Table S1 presents the summary of our experimental results on characterization of membrane, In the C1 to C4 compounds and in some cases (membranes 9, 10, 11), Sulfocuccinic acid were added to the casting polymer solution as additional crosslinking agent in order to achieve higher conductivity. We added the polymer, thus formed, to the casting solution, we first compared the results of the Fuel cell formed by adding C1 (C3 and C4) compound alone and heating at 180°C for some time. Table S1 has two sets of results, one on characterization (columns 1, 2, 3,4,5,6, and 7) of membranes while the other on Fuel cell (columns 8 to 11). For the latter columns 10 and 11 gave results on characterization on Fuel cell while columns 8 and 9 gave results when the Fuel cell was run on control mode for higher efficiency.

Experiments on membrane pore sizes are given in column 1 which lies between 8.88 to 9.52nm and they all are ultra filtration type. The values of the pore size are mostly governed by the nature of the graft chains, e.g. membranes 1,3,4,7 and 12 consist of PAA grafts and were found to have similar pore sizes. Similarly membranes 2,5,8,9 were of the similar pore size because the PVA have the same grafts of PMA. The next important thing is to determine the wall charges of these membranes and they were estimated by experimentally determining the rejection of Chromic (VI) acid solution through these membranes. Subsequently, a space charge model was proposed for the flow of the solution through their pores and a semi analytical solution was derived for the overall problem of separation of chromic acid solution. We observed that the high surface charge was because of the benzimidazolium groups present on the surface and they ranged from minimum of 465 to maximum of 525mV. The thermal degradation (TGA) experiments were carried out which showed that, through the modification of the polymer by benzimidazole groups, lead to stable membrane and the degradation temperature varies from 150°C for 10% weight loss and for approximately 80% weight reduction, a minimum of 300°C to 700°C.

The twelve membranes reported in Table S1 were first characterized completely in terms of pore size, wall charge, conductivity, IEC and thermal degradation. In fuel cell applications, higher temperatures give large current density and it is observed that some of the membranes (rows 1, 2, 5, 6, 7, 8 and 12) could easily be operated upto 150°C without the membrane degrading. However in our case, we insisted that the membrane of direct methanol fuel cell were operated at 90°C for which most of the fuel cell results for Nafion 117 are available. The catalysts used on the anode and cathode side were Pt-Ru and Pt loaded on carbon respectively, which were synthesized by following procedure given in literature. We find that in each case, the power density and the current density were higher and for some of our results (membranes 1,6,12) were atleast two to three times higher than that of Nafion 117 as seen in column 10 and 11.

It was recognized that the membrane material has hydrophilic PVA matrix and hydrophobic benzimidazolium groups at the surface formed by reaction of the C1 to C4 compounds with grafts (see Fig.S-1). When the reaction occurs (confirmed by IR and NMR), the hydrophobic benzimidazolium are formed which tends to separate as confirmed by the SEM results. In view of this we assumed that the pores of the membrane would have a thin hydrophobic layer with positive charge walls. This is confirmed by the contact angle measurements where the minimum contact angle is 51° and varied upto 89°. Even though most of the polymer matrix is PVA with its grafts, when the ions move through the pores (giving rise to current in Fuel cell), the pore walls get highly charged (as measured by column 2 of Table S1). Some of the membranes have Sulfo succinic acid SSA as part of the matrix (e.g. the membrane 9 to 11) which would tend to give sulfonate groups inside the PVA matrix as well as at the phase boundaries, this way giving rise to bipolar nature of the membrane (benzimidazolium group being strictly cationic in nature). In addition to this, some of the compounds such as C1 and C2 compounds have bonded sulfonate groups which would give the positive and negative charge separated by the molecular distance, once again producing a bipolar nature to the membrane. In remainder of the cases the charges would be induced on the phase boundaries as seen in Fig.S-1. The conductivity observed in our membranes is very high and could be attributed to the special bipolar structure of our polymers.

Table S1. Fuel cell and Membrane Characterization

Compound	Membrane	1	2	3	4	5	6	7	8	9	10	11

			MWCO (pore size, nm)	Wall Charge	Conductivity (S/cm)	IEC	Thermal Degradation temp (°C) 10% - 80%		Water Uptake	Contact Angle	Average Voltage Gain (V)	Average time cycle (min)	Power Density (mW/cm)	Current Density (mA/cm)
C1	1	PVA-g-PAA-C1	9.1	488	0.04	2.7	199.732	268.70	7.5	72	0.523	55.75	50.47	74.88
	2	PVA-g-PMA-C1	9.35	465	1.23×10^{-2}	1.8	142.659	447.6	10.691	74.917	0.568	42.896	0.305	3.81
	3	PVA-g-PAA-modified	9.52	488	1.51×10^{-2}	2.12	115.123	706.53 (28.93%)	13.665	75.581	0.778	69.85	37.37	64.44
C2	4	PVA-g-PAA-C2	9.52	525	6.59×10^{-2}	2.38	119.558	691.93	13.4	51.2	0.704	39.83	23.0020	50.55
	5	PVA-g-PMA-C2	9.36	519	1.04×10^{-2}	2.17	181.59	562.348	14.9	50.24	0.77	161.93	34.72	62.11
	6	PVA-g-PMAA-C2	8.99	482	1.16×10^{-2}	2.18	201.055	472.568	13.8	51.8	0.6066	147.81	76.72	92.33
C3	7	PVA-g-PAA-C3	9.0	512	2.51×10^{-2}	2.23	204.908	453.03	8.624	70.368	0.6883	82.55	36.397	75
	8	PVA-g-PMA-C3	9.5	510	3.51×10^{-2}	2.27	166.819	505.293	13.513	62.944	0.726	37.557	36.11	64.381
	9	PVA-g-PAA-SSA-C3	9.36	497	1.71×10^{-2}	2.51	154.914	813.769 (24.32%)	10.5	54.8	0.8096	99.423	23.45	57.778
	10	PVA-g-PMA-SSA-C3	9.83	480	2.28×10^{-2}	2.46	115.12	428.713	8.5	76.2	0.663	224.22	28.561	58.358
	11	PVA-g-PMAA-SSA-C3	8.88	525	3.66×10^{-2}	1.85	139.772	681.715 (30.78%)	7.2	89.2	0.77	18.36	20.832	48.11
C4	12	PVA-g-PAA-C4	9.5	495	2.56×10^{-2}	1.62	189.87	451.86	8.927	59.132	0.704	74.19	41.1	71
	13	Nafion 117			0.1								21.01	48.33

Compound C1, [Bis-3-Amino-4-{3-(triethylammoniumsulfonato) phenylamino}-phenyl sulfone] hydrochloride

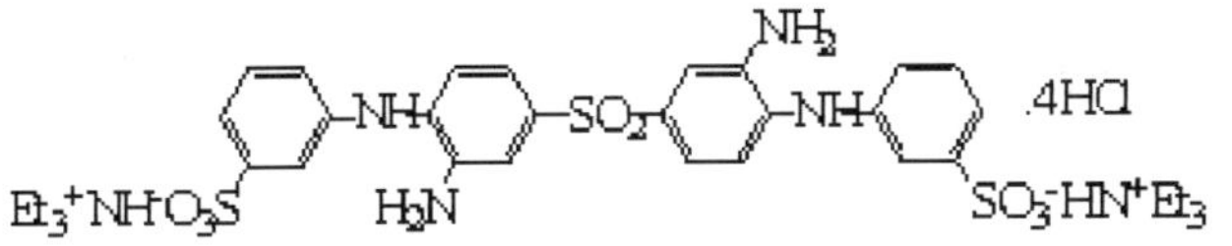

Compound C2, 3-Amino -4-[3-(triethyl ammonium sulfonato) phenyl amino] phenylene hydrochloride

THE GRAPHIC ON THIS PAGE WAS NOT AVAILABLE ON THE E-MEDIA VERSION OF THIS PAPER.

Compound C3, Bis -4 [1, 2 diamino phenyl]-phenyl sulfone) hydrochloride

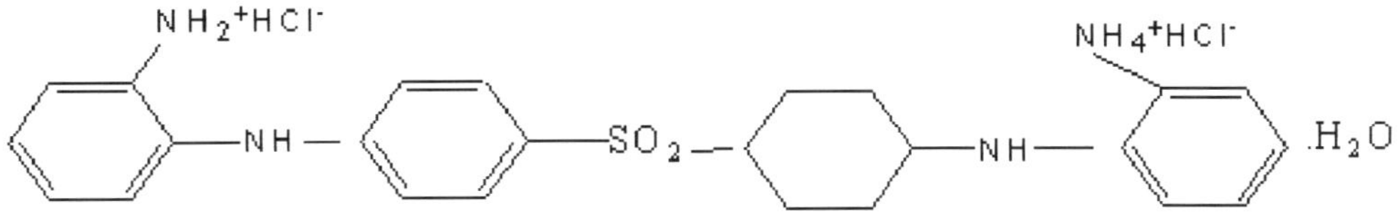

Compound C4, 1, 4 [1, 2 diamino phenyl] Phenylene hydrochloride

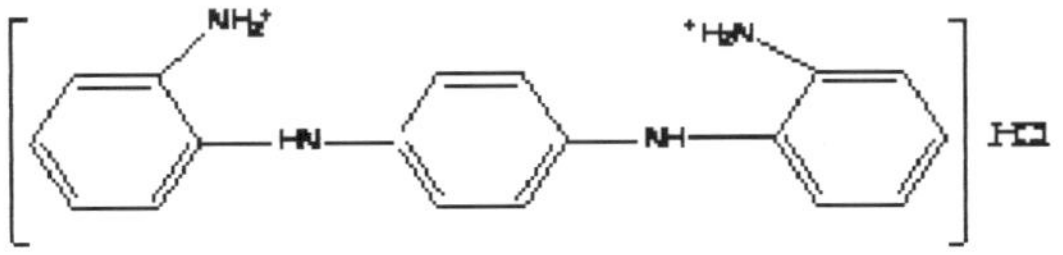

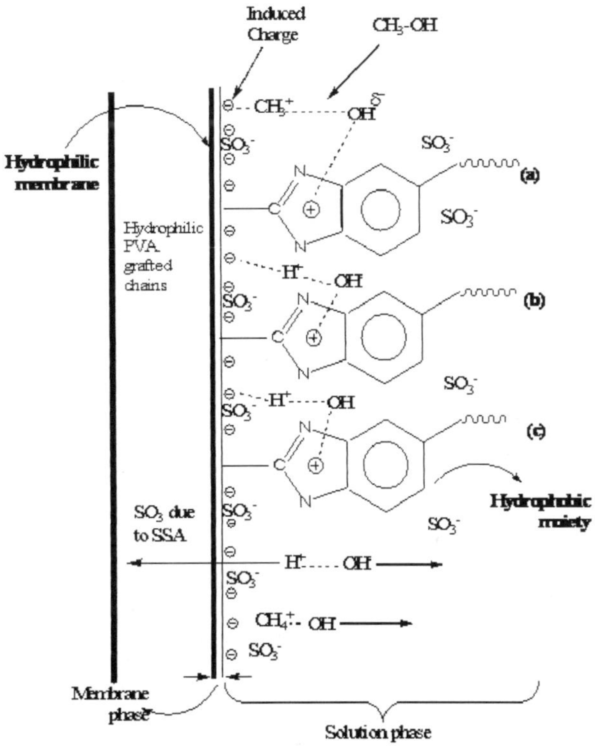

Figure S-1. The Bipolar nature of the membrane surface

(106b) Operation and Controls of Operating Direct Methanol Fuel Cells with PVA Based Membranes Modified by Benzimidazolium Groups

Pratima Gajbhiye, Chemical Engineering, Indian Institute of Technology, Kanpur, State of Uttar Pradesh, India, Pranjal Joshi, Chemical Engg, IIT Kanpur, Kanpur, India, Jayant K. Singh, Chemical Engineering, Indian Institute of Technology, Kanpur, Ka, India and A. Kumar, Department of Chemical Engineering, Indian Institute of Technology Kanpur, Kanpur, India

Operation and Control of Direct Methanol Fuel cell with PVA based membranes modified with benzimidazolium groups

Pratima Gajbhiye, Pranjal Joshi, Jayant K.Singh, Anil Kumar

Department of Chemical Engineering, Indian Institute of Technology, Kanpur 208016, India

Abstract

In recent years, in an effort to find alternate membrane material (other than Nafion 117), Polybenzimidazole (PBI) membranes have been synthesized where the conducting moiety of this polymer is benzimidazolium groups. The monomers used to synthesize these polymers are bis conjugated amines and diacids are expensive. In this work, we have synthesized grafted polymers of PVA as membrane materials where the grafts are either Poly Acrylic Acid, Poly maleic anhydride (PMA) or Poly Methacrylic acid (PMAA) and these materials were denoted by PVA-g-PAA, PVA-g-PMA, and PVA-g-PMAA. We then synthesized C1, C3 & C4 compounds (formula given at the bottom of Table S1) having bis conjugated amines whereas compound C2 having only one conjugated amine group. One of the steps in synthesizing C1 compound is nitration of phenyl ring, where there is a possibility of formation of two isomers out of which only one of them has strictly bis conjugated amines and can be polymerized with Isophthalic acid giving the formation of PBI. The other byproduct formed is an impurity and is minimized by carrying out the nitration step at low temperature but this cannot be completely removed from the product which is a mixture. Table S1 presents the summary of our experimental results on characterization of membrane, In the C1 to C4 compounds and in some cases (membranes 9, 10, 11), Sulfocuccinic acid were added to the casting polymer solution as additional crosslinking agent in order to achieve higher conductivity. We added the polymer, thus formed, to the casting solution, we first compared the results of the Fuel cell formed by adding C1 (C3 and C4) compound alone and heating at 180°C for some time. Table S1 has two sets of results, one on characterization (columns 1, 2, 3,4,5,6, and 7) of membranes while the other on Fuel cell (columns 8 to 11). For the latter columns 10 and 11 gave results on characterization on Fuel cell while columns 8 and 9 gave results when the Fuel cell was run on control mode for higher efficiency.

Experiments on membrane pore sizes are given in column 1 which lies between 8.88 to 9.52nm and they all are ultra filtration type. The values of the pore size are mostly governed by the nature of the graft chains, e.g. membranes 1,3,4,7 and 12 consist of PAA grafts and were found to have similar pore sizes. Similarly membranes 2,5,8,9 were of the similar pore size because the PVA have the same grafts of PMA. The next important thing is to determine the wall charges of these membranes and they were estimated by experimentally determining the rejection of Chromic (VI) acid solution through these membranes. Subsequently, a space charge model was proposed for the flow of the solution through their pores and a semi analytical solution was derived for the overall problem of separation of chromic acid solution. We observed that the high surface charge was because of the benzimidazolium groups present on the surface and they ranged from minimum of 465 to maximum of 525mV. The thermal degradation (TGA) experiments were carried out which showed that, through the modification of the polymer by benzimidazole groups, lead to stable membrane and the degradation temperature varies from 150°C for 10% weight loss and for approximately 80% weight reduction, a minimum of 300°C to 700°C.

The twelve membranes reported in Table S1 were first characterized completely in terms of pore size, wall charge, conductivity, IEC and thermal degradation. In fuel cell applications, higher temperatures give large current density and it is observed that some of the membranes (rows 1, 2, 5, 6, 7, 8 and 12) could easily be operated upto 150°C without the membrane degrading. However in our case, we insisted that the membrane of direct methanol fuel cell were operated at 90°C for which most of the fuel cell results for Nafion 117 are available. The catalysts used on the anode and cathode side were Pt-Ru and Pt loaded on carbon respectively, which were synthesized by following procedure given in literature. We find that in each case, the power density and the current density were higher and for some of our results (membranes 1,6,12) were atleast two to three times higher than that of Nafion 117 as seen in column 10 and 11. In normal Fuel cell operations, there is decay in the cathode catalyst which is further increased by the methanol crossover. The voltage of the Fuel cell increases, the average volatage gained is reported in column 8 and it is found that the voltage gain depends on the nature of the membrane which is not the same for all of them. In column 9, we have measured the average time cycle and we found that some of the membranes (membranes 6, 7, 9, 10) gave extremely good results.

It was recognized that the membrane material has hydrophilic PVA matrix and hydrophobic benzimidazolium groups at the surface formed by reaction of the C1 to C4 compounds with grafts (see Fig.S-1). When the reaction occurs (confirmed by IR and NMR), the hydrophobic benzimidazolium are formed which tends to separate as confirmed by the SEM results. In view of this we assumed that the pores of the membrane would have a thin hydrophobic layer with positive charge walls. This is confirmed by the contact angle measurements where the minimum contact angle is 51° and varied upto 89°. Even though most of the polymer matrix is PVA with its grafts, when the ions move through the pores (giving rise to current in Fuel cell), the pore walls get highly charged (as measured by column 2 of Table S1). Some of the membranes have Sulfo succinic acid SSA as part of the matrix (e.g. the membrane 9 to 11) which would tend to give sulfonate groups inside the PVA matrix as well as at the phase boundaries, this way giving rise to bipolar nature of the membrane (benzimidazolium group being strictly cationic in nature). In addition to this, some of the compounds such as C1 and C2 compounds have bonded sulfonate groups which would give the positive and negative charge separated by the molecular distance, once again producing a bipolar nature to the membrane. In remainder of the cases the charges would be induced on the phase boundaries as seen in Fig.S-1. The conductivity observed in our membranes is very high and could be attributed to the special bipolar structure of our polymers.

Table S1. Fuel cell and Membrane Characterization

Compound	**Membrane**		**1**	**2**	**3**	**4**	**5**		**6**	**7**	**8**	**9**	**10**	**11**
			MWCO (pore size, nm)	**Wall Charge**	**Conductivity (S/cm)**	**IEC**	**Thermal Degradation temp (°C) 10% - 80%**		**Water Uptake**	**Contact Angle**	**Average Voltage Gain (V)**	**Average time cycle (min)**	**Power Density (mW/cm)**	**Current Density (mA/cm)**
C1	1	PVA-g-PAA-C1	9.1	488	0.04	2.7	199.732	268.70	7.5	72	0.523	55.75	50.47	74.88
	2	PVA-g-PMA-C1	9.35	465	1.23 x10^{-2}	1.8	142.659	447.6	10.691	74.917	0.568	42.896	0.305	3.81
	3	PVA-g-PAA-modified	9.52	488	1.51 x10^{-2}	2.12	115.123	706.53 (28.93%)	13.665	75.581	0.778	69.85	37.37	64.44
C2	4	PVA-g-PAA-C2	9.52	525	6.59 x10^{-2}	2.38	119.558	691.93	13.4	51.2	0.704	39.83	23.0020	50.55
	5	PVA-g-PMA-C2	9.36	519	1.04 x10^{-2}	2.17	181.59	562.348	14.9	50.24	0.77	161.93	34.72	62.11
	6	PVA-g-PMAA-C2	8.99	482	1.16 x10^{-2}	2.18	201.055	472.568	13.8	51.8	0.6066	147.81	76.72	92.33
C3	7	PVA-g-PAA-C3	9.0	512	2.51 x10^{-2}	2.23	204.908	453.03	8.624	70.368	0.6883	82.55	36.397	75
	8	PVA-g-PMA-C3	9.5	510	3.51 x10^{-2}	2.27	166.819	505.293	13.513	62.944	0.726	37.557	36.11	64.381
	9	PVA-g-PAA-SSA-C3	9.36	497	1.71 x10^{-2}	2.51	154.914	813.769 (24.32%)	10.5	54.8	0.8096	99.423	23.45	57.778
	10	PVA-g-PMA-SSA-C3	9.83	480	2.28 x10^{-2}	2.46	115.12	428.713	8.5	76.2	0.663	224.22	28.561	58.358
	11	PVA-g-PMAA-	8.88	525	3.66 x10^{-2}	1.85	139.772	681.715	7.2	89.2	0.77	18.36	20.832	48.11

		SSA-C3						(30.78%)						
C4	12	PVA-g-PAA-C4	9.5	495	2.56×10^{-2}	1.62	189.87	451.86	8.927	59.132	0.704	74.19	41.1	71
	13	Nafion 117			0.1								21.01	48.33

Compound C1, [Bis-3-Amino-4-{3-(triethylammoniumsulfonato) phenylamino}-phenyl sulfone] hydrochloride

Compound C2, 3-Amino -4-[3-(triethyl ammonium sulfonato) phenyl amino] phenylene hydrochloride

Compound C3, Bis -4 [1, 2 diamino phenyl]-phenyl sulfone) hydrochloride

Compound C4, 1, 4 [1, 2 diamino phenyl] Phenylene hydrochloride

THE GRAPHIC ON THIS PAGE WAS NOT AVAILABLE ON THE E-MEDIA VERSION OF THIS PAPER.

Figure S-1. The bipolar nature of PVA-g-PAA-BASPAPS membrane which reduces the methanol crossover due to the molecular dissociation. In (a), the imidazolium group is shown to separate out from the PVA moiety of the chain into a separate phase. In (b), the water is shown to complex with the SO and the imidazolium group. As a result of this, interaction in (d) and (e), the water and the methanol is shown to split into anion and cation and the anion permeates through the membrane giving bipolar effect

(106e) Numerical Investigation in Heat Conduction Effect on Heat Exchange for Mini Cross-Flow Heat Exchanger (MCHE)

Hongyue Duan[1], Wei Mu[1], **Xigang Yuan**[1] and Lingai Luo[2], (1)State Key Laboratory of Chemical Engineering, Tianjin University, Tianjin, China, (2)LOCIE Chambery, Savoie University, Savoie, France

In a traditional heat exchanger, the effect of heat transfer within the wall material driven by the temperature gradient in axial direction of the fluid flow is usually very small because the area of the cross section of the tube wall is much smaller than that of the tube for heat exchange. However, in a mini cross-flow heat exchanger(MCHE), where the heat conducting medium is typically a cubic-shaped block of metal and the hot and cold streams are put into heat exchange via two sets of channels in the block, the heat conduction in the wall material in the direction parallel to the fluid flow may have a significant effect on the performance of the heat exchanger.

In this paper, a 2-channel-layer MCHE was simulated and a periodic boundary condition was applied so that the results could be applicable to entire MCHE. The proposed model and the mesh grid size were proved by the good agreement of the simulated overall heat transfer coefficient with that found in the experimental reported in the literature[1]. Then the effect of heat conduction on heat transfer performance in a MCHE was studied numerically. The simulation results suggest that the performance of the MCHE can be improved by increasing the wall thickness between the adjoining channels in the same layer or by decreasing the wall thickness between the adjacent layers. The heat conduction effect can also be weakened by optimizing the material conductivity or by increasing the diameter of the channels. New correlations were developed based on the simulated results to predict the value of the overall heat transfer coefficient as a function of the operating parameter such as Re in the channel and the design parameters of the MCHE such as the diameter of the channel, wall thickness and thermal conductivity.

[1]Luo L A, Fan Y L, Zhang W W*et al.*Integration of constructal distributors to a mini crossflow heat exchanger and their assembly configuration optimization[J]. Chemical Engineering Science, 2007, 62 (13): 3605-3619.

(106g) The Effects of Out Gassing Through Cylindrical Walls on the Transport Coefficients

Luma. H. Yaseen[1], **Nada B. Nakkash**[2] and M. N. Latif[1], (1)Al-Nahrain University, Baghdad, Iraq, (2)Chemical Engineering, Al-Nahrain University, Baghdad, Iraq

Abstract

This work concerns with studying the influence of blowing on friction factor and mass transfer coefficient when the flow is normal to a tube bank with outgassing porous cylinders. The friction factor and the mass transfer coefficients are measured for a range of Reynold's number and blowing rate and expressed in terms of various dimensionless groups.

The test rig consists of a rectangular duct (20.3 × 11.3 cm), five rows of three porous tubes in staggered arrangements fixed across the duct with 1.5pitches (both longitudinal and transversal). The outside and the inside diameters of tubes were 25.4 and 21.4 mm, respectively.

The friction factor measurements was studied for Reynold's number range of

$3000 < Re < 17{,}000$ and blowing rate up to 4×10^{-3}. The results show that the blowing intothe boundary layer for the cylinders increases the overall friction factor from 10%to 50% depending on the blowing rate. The mass transfer coefficient was measured for Reynold's number range of $5000 < Re < 16{,}000$ and blowing rate of $10^{-5} < m/G_\infty < 10^{-3}$.

A new definition of the blowing rate was considered and expressed by the mass transfer driving force. Two successful formulas were obtained to account the outgassing effect on the mass transfer coefficient. The mass transfer coefficient for the outgassing boundary layer was obtained first by determining the non-blowing coefficient and multiplying it with the correction factor .

The first model was modified to account the density variation associated with the boundary layer change:

$0.51\ Re^{-0.5} \rho_s$

$St_m Sc^{2/3} =$ -------------- (------)$^{0.4}$

$(1 + \beta_m)^{4.7} \rho_\infty$

With the coefficient of determination R2 = 0.93.

The second model depends on the flat plate correction factor:

$Ln(1+ \beta_m)$

$St_m Sc^{2/3} = 0.5\ Re^{-0.5}$(---------------)$^{9.94}$

β_m

With the coefficient of determination R2 = 0.92.

These two models are found to be in agreement with the previous literature and a reduction of about 80% in the mass transfer coefficient has been obtained due to the outgassing.

(106h) Down the Hubbert Curve

Ian Sutton, Sutton Technical Books, Houston, TX

Down the Hubbert Curve

In March 1956 M. King Hubbert, who was at that time Chief Geologist with the Shell Development Company published a seminal paper with the title "Nuclear Energy and the Fossil Fuels" (Hubbert 1956). In this lengthy and authoritative paper, which was presented before the American Petroleum Institute in San Antonio, Texas, Dr. Hubbert showed how key fossil fuel resources in the United States, particularly coal and oil, were of finite size and that their output would reach a maximum, and then decline. Dr. King's forecasts for the production of oil in the United States were accurate; he predicted that oil production would peak in the early 1970s, and that is what has happened. (The widely-used term "Peak Oil" is frequently accredited to Dr. Colin Campbell (2001)).

In the fifty years since the publication of Dr. King's paper great advances have been made in finding and extracting oil from what were considered to be marginal fields. But Dr. King's overall conclusion is still valid. Production of oil in the United States remains below the 1970's peak. Although Dr. Hubbert's analysis was confined to the United States, subsequent analysts applied his insights to other parts of the world. For example, the book *Twilight in the Desert: The Coming Saudi Oil Shock and the World Economy* (Simmons 2005) by the late Matthew Simmons achieved wide recognition by showing that the concept of peak oil most likely applies to reserves in Saudi Arabia, and other nations such as Iran and Kuwait.

The purpose of this paper is to examine how oil production and consumption have changed in the last five years, and to examine whether or not the concept of Peak Oil remains valid. Key to this discussion is an understanding of the concept of Energy Returned on Energy Invested (EROEI). It is concluded that the concept of Peak Oil, as first postulated by Dr. King, remains valid and current, and the production of oil peaked in the period 2005-2011 and is now set to move down the right hand side of the Hubbert Curve.

The title of this session is *Challenges in Instituting Energy Savings Project in the Present Market Conditions.* If the production of oil and has indeed already passed a peak then the challenge is not to do with "instituting" energy savings; those savings are going to occur whether we like it or not. The challenge will be to determine how the process industries can restructure themselves given that the energy resources available (both as direct energy and as petrochemical feedstock) will be in decline and that there will be intense demand for those same resources from other areas of the economy.

(106j) Statistical Decision Making for In-Service Inspection of Underground Waste Storage Tanks

Emily Mitchell, Emory University, Atlanta, GA and **Stephen P. Harris**, Savannah River National Laboratory, Aiken, SC

Pitting is a localized form of corrosion that occurs on passivated metal surfaces exposed to aggressive chemical environments. The pitting process involves the breakdown of the passive oxide film and subsequent anodic dissolution of the metal at a local site, while the remainder of the surface remains protected by the film.

Deep pits may result in lowering the maximum level of waste that made be stored in the tank and eventually lead to loss of confinement. Pit depths, established by ultrasonic testing, are subject to variables that cause uncertainty, including the repeatability of metric location, time between measurements, different transducers, and different inspectors.

The generalized extreme value (GEV) distribution is fit to the data gathered from ultrasonic testing and statistical projections are made for the full tank wall interior. A comparison is made with results from other well known statistical distributions. The estimated parameters are used in statistical tolerance limits for bounding pit depth in a waste tank.

The tolerance limits give an indication of extreme values of pit depth for the current condition of the tank walls. The upper tolerance limit for pit depth will then be compared to action levels for pitting. If the upper tolerance limit for pit depth exceeds the action level, more frequent examinations or increased tank surface area inspections are conducted.

(106k) Thermal Inbreathing Requirements of Low Pressure Storage Tanks At Elevated Temperatures

Donald E. Brooks, Pressure Relief Analysis, Lloyd's Register Energy Americas, Inc., Houston, TX

ISO 28300 (API 2000) "Petroleum, petrochemical and natural gas industries - Venting of atmospheric and low pressure storage tanks" provides guidance on determining the required thermal inbreathing flow rates to protect atmospheric and low pressure storage tanks against vacuum due to weather changes (e.g. a decrease in atmospheric temperature)." For heated, un-insulated tanks with vapor space temperatures above 58.9 °C (120 "F), ISO 28300 recommends an engineering review of thermal inbreathing requirements.

This paper describes one approach to evaluating thermal inbreathing requirements for un-insulated storage tanks with vapor space temperatures exceeding 58.8 °C (120 "F)." The tank is assumed to be exposed to the cooling effects of a sudden rain. A simplified unsteady state heat transfer analysis is performed to determine the tank wall and vapor space temperatures versus time. Initial gas mass and pressure are assumed to remain constant in the tank. Gas volume in the tank is calculated as a function of vapor space temperature using the ideal gas law. The rate of volume change is calculated for each time increment. The rate of volume change goes through a maximum value. This maximum value is selected as the thermal inbreathing requirement of the tank.

The calculation methodology described in this paper can be programmed in a spreadsheet for ease of use.

Dynamic Operational Risk Assessment with Bayesian Networks

Shubharthi Barua, Xiaodan Gao, M. Sam Mannan
Mary Kay O'Connor Process Safety Center
Texas A&M University, Jack E. Brown Engineering Bldg., 3122 TAMU, Room 244, College Station, TX 77843
shubharthi.barua@tamu.edu, xiaodan_gao08@mail.che.tamu.edu, mannan@tamu.edu

Extended Abstract Prepared for the Poster Presentation at the
8th Global Congress on Process Safety
Houston, TX
April 1 – 4, 2012

UNPUBLISHED

8TH GLOBAL CONGRESS ON PROCESS SAFETY

Dynamic Operational Risk Assessment with Bayesian Networks

Shubharthi Barua, Xiaodan Gao, M. Sam Mannan

Mary Kay O'Connor Process Safety Center
Texas A&M University, Jack E. Brown Engineering Bldg., 3122 TAMU Room 244, College Station, TX-77843
shubharthi.barua@tamu.edu, xiaodan_gao08@mail.che.tamu.edu, mannan@tamu.edu

Keywords: risk assessment, Bayesian network, dynamic Bayesian network, fault tree, event tree

Abstract

Oil/gas and petrochemical plants are complicated and dynamic in nature. Dynamic characteristics include ageing of equipment/components, season changes, stochastic processes, operator response times, inspection and testing time intervals, and timing of safety system operations, all of which are time dependent criteria that can influence dynamic processes. Conventional risk assessment methodologies can quantify dynamic changes in processes with limited capacity. Several methods are proposed for dynamic risk assessment such as C-DET, D-DET, DORA, etc., but all of them have deficiencies of some kind. Bayesian networks, a graphical inference technique, have some advantageous characteristics for risk analysis over other techniques. This research focuses on providing a framework for dynamic risk analysis based on Bayesian networks. A case study will be provided to illustrate this model.

1.0 Introduction

Oil/gas, chemical, and petrochemical process plants consist of numerous equipment and unit operations, thousands of control loops, and exhibit dynamic behavior. Different hazards and several types of risk are associated with these process facilities. For example, large process areas have the potential to stock a large amount of hazardous chemicals, which can result in the process area being highly congested. It is reported that process risk, i.e., risk in process facilities due to fires and explosions, greatly contributes to the total risk of offshore installations. Also, reactivity, toxic, and mechanical hazards have a significant effect on the total risk of these facilities. Therefore, it is very important to identify hazards, perform risk assessments, and take proper initiatives to minimize/remove hazards and risks; else a catastrophic accident may result.

From case histories, it has been observed that catastrophic accidents have a significant effect on people, environment, and society. Catastrophic accidents such as the Flixborough disaster, the Bhopal incident, and the Piper Alpha disaster caused fatalities and unbearable economic loss. In the past few years, several disastrous accidents in refineries, power plants, and offshore platforms have significantly affected people's perception, and consequently raised concern to emphasize process safety. A series of explosions in 2005 resulting from insufficient process safety and lack of risk reduction measures at the BP Texas City refinery accident resulted in 15 fatalities, 180 injuries, and $1.5 billion in losses [1]. The Kleen Energy power plant explosion in

Connecticut in 2010 is considered to be one of the worst industrial explosions in the USA [2]. In 2010, a fire and explosion resulting from a blowout at the Macondo well resulted in 11 deaths and 17 injuries. Also the continuous spill from the wellhead for 87 days had disastrous effects on the environment and wildlife surrounding the Gulf of Mexico. It has been reported that most of the accidents could be prevented or their severity could have been minimized by effective risk assessment and process safety management. But existing risk assessment methodologies cannot properly quantify various time-dependent effects. Therefore, the objective of this research is to develop a risk assessment methodology with the capability to capture variations due to different time-dependent effects.

2.0 Literature Review

N. Siu [3] summarized available methodologies for risk assessment. For example, the fault tree and event tree methods are two most common methodologies for risk assessment. But both techniques have limited ability to address different time-dependent effects such as equipment or component ageing, stochastic processes, operator response times, and inspection and testing time intervals. N. Siu [3] also described other methodology limitations while applying dynamic system analyses. Therefore, for dynamic systems, conventional techniques are not able to provide accurate results because of their inadequate ability to describe the variation of operational risks as time-dependent deviations, or the changes occurring in the process. Continuous event trees [4], discrete event trees [5], and Markov theory [6] methods have been proposed for dynamic reliability and safety analysis. But all of the proposed techniques have some deficiencies. For example, the Markov theory based model does not consider the effect of inspection on system-state transitions. X. Yang et al. [1] proposed the Dynamic Operational Risk Assessment (DORA) technique, which is able to take inspection and testing time intervals and repair time into account for evaluating operational failure.

Recently, Bayesian networks are receiving wide acceptance in reliability assessment, fault diagnosis, safety, risk, and maintenance analysis. Several authors have [7,8] compared some commonly used methods for accident analysis with Bayesian networks, and described several advantages of Bayesian networks over these methods. N. Khakzad et al. [2] illustrated parallelism between Bayesian networks and fault trees, and discussed the potential of Bayesian networks as a tool for process safety and accident analysis. H. Pasman and W. Rogers [9] described an approach to incorporate Bayesian networks into layer of protection analysis to make it more effective. These studies demonstrate the potential advantages of Bayesian networks application in risk assessment. This research provides a methodology for dynamic risk assessment based on Bayesian networks.

3.0 Methodology and Expected Result

3.01 Bayesian Networks

A Bayesian network, which is a probabilistic graphical model, describes causal influence relations among variables. A set of random variables and their conditional dependencies are represented by Bayesian networks via a directed acyclic graph (DAG). It has the ability to represent dependency among events clearly, accommodate multi mode and continuous random variables, and incorporate information i.e. generic, system specific and expert judgment to support optimum decision making.

A simple Bayesian network is shown in Figure 1. In a binary network, variables are represented by nodes and causal relationships among different nodes are shown by an arc. Conditional probability tables or defined probabilistic relations among nodes represent how one variable is linked with one or many variables. The nodes that influence another variable and have unconditional probability are called parent or root nodes. Nodes that are conditionally dependent on their direct parents are called intermediate nodes. The top node is defined as a leaf node, which is similar to the top event of a fault tree and initiating event for an event tree.

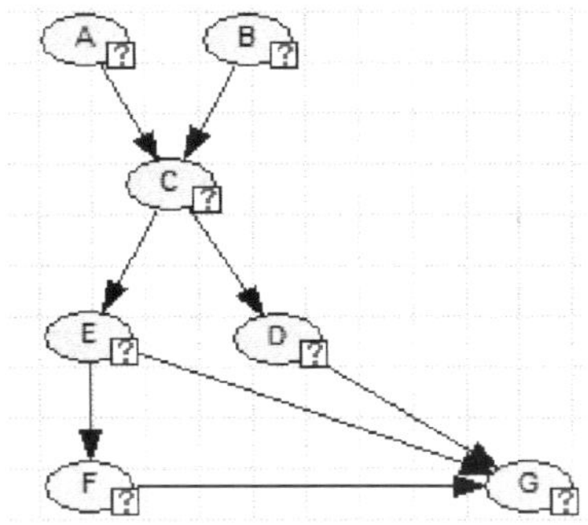

Figure 1: A simple Bayesian network

Let $N = (G, P)$ be a Bayesian network, where

$G = (V, E)$ is a directed acyclic graph; V (random variables) represents nodes; and E represents edges between pairs of nodes of DAG.

P represents probability distribution over V and $V = \{X_1, X_2, \ldots, X_n\}$ is either discrete or continuous random variables [10]. These random variables are assigned to the nodes and the edges.

Bayesian networks can be represented by the joint probability distribution P(V);

$$P(V) = \sum_{X \in V} P\{X|pa(X)\} = P(X_1, X_2, \ldots, X_N) = \sum_{i=1}^{n} P\{X_i|pa(X_i)\}$$

Here $pa(X_i)$ = parent nodes of X_i.

A main advantage of the application of Bayesian networks in risk analysis is the ability to update prior data using Bayes' theorem by incorporating new information as demonstrated by N. Khakzad et al. [2]. Also, Bayesian networks have an advantage of handling different types of uncertainty.

3.02 Dynamic Bayesian Networks

A general Bayesian network is a static model, i.e., the joint probability distribution is usually a representation of a fixed point or an interval of time [11]. A dynamic Bayesian network is a Bayesian network that has the ability to describe the evolution of joint probability distribution over time. Dean and Kanazawa [12] proposed modeling time in a discrete manner to represent the progression of time in a dynamic Bayesian network. In a dynamic Bayesian network, arcs links node/nodes from previous time slice to the node/nodes of the next time slice to represent temporal dependencies among them.

3.03 Methodology

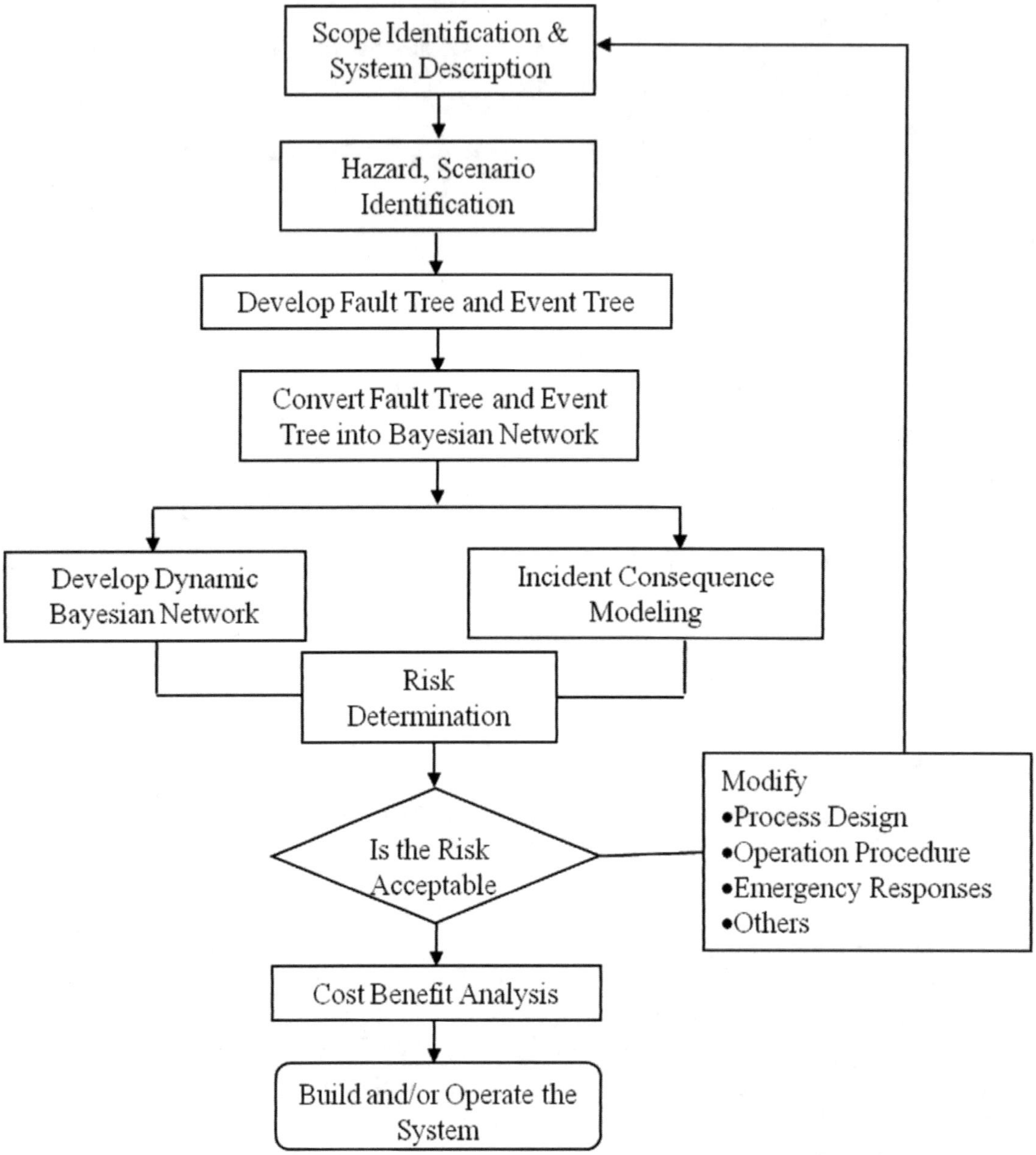

Figure 2: Framework for the dynamic Bayesian network based risk assessment

The objective of this research is to propose a dynamic risk assessment method based on dynamic Bayesian networks. According to the framework, a fault tree and event tree are developed for a system. The objective for following step is to convert the developed fault and event tree into a Bayesian network. Bobbio et al. [13] provided a methodology for conversion of fault trees into Bayesian networks. N. Khakzad et al. [2] showed equivalency between fault trees and Bayesian networks, and described the advantages of Bayesian networks for process safety analysis. Bearfield and March [14] proposed another methodology for generalizing event trees into Bayesian networks. The method has the ability to generalize simple event trees into Bayesian networks in a relatively easy manner, but the method becomes complicated for large event trees. Therefore, in the first phase, a methodology is proposed to convert fault trees and event trees into a static Bayesian network. In the second phase, a dynamic Bayesian network is developed from the static Bayesian network. The dynamic Bayesian network provides probability of a system at different time-slices for different states of the system. The probabilities are then combined with consequence modeling levels to calculate risk. GeNIe, which was developed at the Decision Systems Laboratory at the University of Pittsburgh, is used to create the model [15].

4.0 Conclusions

This article briefly describes a framework based on Bayesian network for dynamic risk assessment for the oil/gas, chemical, and petrochemical industries. This research illustrates the capability of Bayesian networks to represent all component interactions in complex systems, and to update all prior system probabilities in the presence of new information. The advantages of modeling flexibility and efficient computational mechanisms using dynamic Bayesian networks are also discussed.

4. References

[1] Yang, X. and Mannan, M.S., "The development and application of dynamic operational risk assessment in oil/gas and chemical process industry", Reliability Engineering and System Safety 95 (2010) 806-815

[2] Khakzad, N., Khan, F. and Amyotte,P., "Safety analysis in process facilities: Comparison of fault tree and Bayesian network approaches", Reliability Engineering and System Safety 96(2011) 925-932

[3] Siu,N. "Risk assessment for dynamic systems: an overview", Reliability Engineering and System Safety 43 (1994) 43-73

[4] Devooght, J. and Smidts,C. "Probabilistic reactor dynamics-I: the theory of continuous event trees", Nuclear Science and Engineering 1992; 111:229-40

[5] Amendola, A. "Accident sequence dynamic simulation versus event trees", Reliability Engineering and System Safety 1988; 22:3-25

[6] Howard, R. "Dynamic probability systems", Vols. I and II, New York, 1971

[7] Sklet, S. "Comparison of some selected methods for accident investigation", Journal of Hazardous Materials 2004; 111:29-37

[8] Nivolianitou, Z.S., Leopoulos, V.N. and Konstantinidou, M., "Comparison of techniques for accident scenario analysis in hazardous systems", Journal of Loss Prevention in the Process Industries 2004; 17"467-75

[9] Pasman, H. and Rogers, W.; "BBN, a tool to make LOPA more effective, QRA more transparent and flexible, and therefore to make safety more definable", 14th Annual Sysposium, Mary Kay O'Connor Process Safety Center, College Station, Texas, October 2011

[10] Donohue, S.K. and Dugan, J.B.; "Modeling the "Good enough to release" decision usin V&V preference structure and Bayesian belief networks", Reliability and Maintainability Symposium, 2003 Annual, 2003

[11] McNaught, K.R. and Zagorecki, A. "Using Dynamic Bayesian networks for prognostic modeling to inform maintenance decision making", IEEM 2009, Hond Kong, January 2010

[12] Dean, T. and Kanazawa, K., "A model for reasoning about persistence and causation", Computational Intelligence 1989; 5:142-150

[13] Bobbio A, Portinale L, Minichino M and Ciancamerla E., "Improving the analysis of dependable systems by mapping fault trees into Bayesian networks", Reliability Engineering and System Safety 71(3): 249-260

[14] Bearfield G and Marsh W, "Generalising event trees using Bayesian networks with a case study of train derailment", Available online at http://www.agenarisk.com/resources/white_papers/Generalising_Event_Trees.pdf , last modified: 23 February 2010, accessed: 7 January 2012

[15] Decision Systems Laboratory, University of Pittsburgh, GeNIe (Graphical Network Interface and SMILE (Structural Modeling, Inference, and Learning Enginer), Version 2.0 software, 2010. Available online at http://genie.sis.pitt.edu

(137b) Comparing the Performance of Compact Heat Exchange Reactor to CSTR for Fischer Tropsch Synthesis

Hani Gadalla[1], Zhijun Jia[1], Steven J. Vallee[1], Burtron H. Davis[2]and Dennis Sparks[2], (1)Chart Energy & Chemicals, La Crosse, WI, (2)Center for Applied Energy Research, University of Kentucky, Lexington, KY

A commercial Fischer Tropsch catalyst was studied in two Compact Heat Exchange Reactors (CHER) with different channel sizes, and in a continuous stir tank reactor (CSTR). The CHERs were constructed by Chart Energy and Chemicals with one reactor having a 5mm channels and the other having a 6mm channels, while maintaining the same catalyst volumeThe catalyst was crushed and sieved retaining particles ranging in size from 75-125µm. Testing was performed at the University of Kentucky Center for Applied Energy Research in Lexington, KY. The CO conversion, selectivity and hydrocarbon productivity were measured for each reactor over a wide range of operating conditions, as a function of temperature, pressure, H2:CO feed ratio, and flow rate. Both reactor types showed excellent temperature control, but the performances varied at different operating conditions. The larger channel CHER had a higher conversion than the 5mm channel, potentially due to slightly higher reactor bed temperature. The two CHERs approached and even exceeded the productivity of the CSTR for liquid hydrocarbons per gram of catalyst. At higher flow rates the methane selectivity increases at a lower rate for either of the CHERs than for the CSTR.

(114a) Advances in FCC Propylene Production and Recovery

Andrew Mezera, UOP, Des Plaines, IL

World propylene demand is projected to increase dramatically in the coming years. Conventional steam cracking and emerging technologies dedicated to propylene production will only be able to provide a portion of the demand. FCC Units are expected to contribute a significant portion of the shortfall. This talk will focus on two main aspects of propylene production from FCC Units: 1) increasing FCC propylene yields through operations, catalyst and equipment changes, and 2) improving propylene recovery in the Gas Concentration Unit.

(114c) An Experimental Study of Sulfur Release in Gas Products of Coal Steam Hydrogasification

Qian Luo, CEE, UCR, Riverside, CA

An Experimental Study of Sulfur Release in Gas Products of Coal Steam Hydrogasification

Qian Luo, Chan S. Park, Joseph M. Norbeck

Bourns College of Engineering

Center for Environmental Research and Technology (CE-CERT)

University of California, Riverside, California 92521-0425.

Steam hydrogasification reaction (SHR) is the thermochemical conversion of carbonaceous materials into synthetic gas in steam and hydrogen environment. It has been developed by the Bourns College of Engineering - CE-CERT at the University of California, Riverside (UCR). Like other thermochemical process, the formation of gaseous sulfur species from the solid sulfur in the feedstock becomes inevitable which necessitates the cleaning of gas to make it end-use possible. Therefore, a better understanding of the sulfur release in the gas phase in SHR could help to find the suitable method to remove sulfur containing species for the protection of downstream catalysts.

The temperature, steam and H_2amount have an effect on the distribution of the various gaseous sulfur species. Experiment results showed that sulfur in the feedstock is mainly converted to H_2S in the SHR process but COS and $CS_2$are undetectable. The $H_2$and steam environment in the SHR process are favorable for the formation of H_2S and suppress the COS and CS_2formation. Furthermore, increase of H_2S concentration was observed with the rise in temperature from 700°C to 800°C. However, the increasing the partial pressure of H_2 decreased the H_2S concentration released in the gas phase.

(114d) An Innovative Mixed Zinc Oxide Based Sorbent Regenerable At Low Temperature Used for the Desulfurization of Syngas in IGCC or XTL Processes

Vincent Girard, Catalyse and Separation Division, IFP Energies Nouvelles, Solaize, France, Arnaud Baudot, Catalysis and Separation Division, IFP Lyon, 69360 Solaize, France, David Chiche, Catalyse and Separation division, IFP Energies Nouvelles, Solaize, France, Delphine Bazer-Bachi, GMD Division, IFP Energies Nouvelles, Solaize, France and Christophe Geantet, IRCELyon, Villeurbane, France

Integrated Gasification Combined Cycles (IGCC) and the second generation biofuels processes, and more generally Fischer-Tropsch based XTL processes, are attractive alternatives for future power production[1-3]. In order to prevent the corrosion of the industrial units, especially the combustion turbine blade used in IGCC processes or, in the others processes, to avoid the Fischer-Tropsch catalysts poisoning, deep desulfurization of syngas is achieved with metal oxides according to the following reaction :

$$MeO_{(s)} + H_2S_{(g)} \rightarrow MeS_{(s)} + H_2O_{(g)}$$

The large amounts of solid wastes produced are one of the major disadvantages of the in-situ desulfurization of syngas. The in-situ regeneration of the sulfided sorbent, back to the oxide, during the process could resolve this problem. From an industrial point of view, the oxidative regeneration is the simplest and most straightforward way to regenerate the sulfided sorbent, so it was mainly studied[4,5]. However, this reaction presents several drawbacks. The first one is its high exothermicity that can cause the sintering of the particles and the loss of reactivity due to the alteration of the textural properties[6,7]. Another drawback, which can be considered as the most important, is the formation of sulfates[5,8]. These refractory phases require high temperature to decompose, which causes important modification of textural properties and decrease of reactivity due to the sintering of particles. Moreover, the resulting temperature increase will impose the use of expensive alloys to make the reactors and will affect the thermal efficiency of the process. The formation of sulfate has to be avoided in order to maintain the sorbent durability and the temperature of the in-situ oxidative regeneration must be as low as possible, ideally the closest to the desulfurization step, to improve the global efficiency of this process.

The main goal of this work is to bring a new insight on the phenomena involved during the oxidative regeneration of the sulfided solids. The methodology developed is based on the confrontation between a global theoretical approach based on thermochemical data and an experimental study. This work is divided into two parts. The first one is the evaluation and the classification of the behaviors observed during both sulfidation and oxidative regeneration of simple oxides, in particular ZnO. The second one is the evaluation of the mixed oxides, selected thanks to the results of the first part and by the anticipation of potential synergetic effects. The synergetic effects are expected to avoid the formation of refractory sulfates and the resulting increase of temperature.

The first part led to a classification of the simple oxides based on the stability of their sulfide and sulfate phases. This classification led to a selection of different simple oxides whose mixture could bring synergetic effect. Finally, this methodology led to an innovative material which composition allows a fast regeneration, at a temperature 150°C lower than ZnO regeneration temperature and without sulfates formation.

Reference List

1. Daniel Vallentin, *Energy Policy*, 2008, **36**, 3198-3211.

2. Osamu Shinada, Akira Yamada, and Yoshinori Koyama,*Energy Conversion and Management*, 2006,**43**, 1221-1233.

3. Mark J. Prins, Krzysztof J. Ptasinski, and Frans J. J. G. Janssen,*Fuel Processing Technology*, 25-1-2005,**86**, 375-389.

4. R. V. Siriwardane and S. Woodruff,*Industrial & Engineering Chemistry Research*, 1995,**34**, 699-702.

5. W. F. Elseviers and H. Verelst,*Fuel*, 1999,**78**, 601-612.

6. T. Grindley and G. Steinfeld,*Proceedings of the Second Annuals Contractors Meeting on Contaminant Control in Hot Coal Derived Gas Stream.*, 1981.

7. J. Huang, J. Zhao, X. Wei, Y. Wang, and X. Bu,*Powder Technology*, 14-1-2008,**180**, 196-202.

8. R. V. Siriwardane and S. Woodruff,*Industrial & Engineering Chemistry Research*, 1-12-1997,**36**, 5277-5281.

(122a) Tuning Equation-of-State to Match Experimental Data for near Critical Fluid: A Case Study of NKJ

Adel Elsharkawy, Petroleum Engineering, Kuwait University, Kuwait, Kuwait and Bashayer . Al-Enize, Kuwait Oil Co, Kuwait, Kuwait

Tuning equation-of-State to Match Experimental Data for Near Critical Fluid: A Case Study of NKJ

Bashayer h. Al-Enize and Adel Elsharkawy

College of Engineering & Petroleum, Kuwait University

Po Box 5969 Safat 13060 Kuwait

Abstract

Equations-of-state (EOS) are widely used in compositional simulation and surface facility design. Volatile oils and gas condensates require special treatment due to their near-critical nature; the classical material balance cannot be applied to these near-critical fluids because their molar compositions change as the reservoir is depleted. Thus, compositional simulation is necessary to determine the best economical and technical exploitation. However, the performance of cubic equations of state is questionable in their predictive mode when they applied to petroleum mixtures. Tuning of the EOS is, generally, necessary for improving the predictions of compositional reservoir simulators, since the plus fraction of reservoir fluids has some uncertainty in its molecular weight and critical properties, equation-of-state, EOS, are generally not predictive without tuning its parameters to match experimental data.

We describe the optimum characterization and regression techniques that could provide tuned Equation-of-state for the near-critical volatile oil and gas condensate from the North Kuwait Jurassic reservoirs (KNJ). The PVT simulator PVTsim is used for the purpose. The results show that the implemented methods provide a reliable and efficient tool for representing PVT properties and phase behavior of the complex reservoir fluids. In this paper detailed of two samples is provided: one gas condensate and a volatile oil sample. We discuss the optimum sampling technique in this region and the quality control needed to perform the PVT analyses in order to ensure the validity of data to be used in EOS tuning.

Equations of state EOS parameters such as critical temperatures, critical pressures and acentric factors of the pseudo-components are matching parameters used to EOS tuning. In this study A characterization procedure is described to simulate the plus fractions of reservoir fluids and optimal selection of the regression parameters in determining the quality of the tuned fluid model.

(122b) Dynamic Simulation: Value During LNG Process Design and Beyond

Kenneth V. Allsford, Honeywell Process Solutions, Honeywell, Houston, TX, Vijay KumarPatel, Honeywell Process Solutions, Honeywell International, Houston, TX and Robin K Conwell, Honeywell Process Solutions, Honeywell International, Lakewood, CO

A contributor to complex/large project management difficulty is uncertainty gaps – gaps between the information required and that readily available within the project organization. Moreover, for large projects, information is generally spread among many professional personnel across multiple organizations.

In the process industries, LNG engineering teams are finding that – both in front-end engineering design (FEED) and during detailed design – dynamic simulation is a cost effective tool to reduce uncertainty gaps and facilitate information communication. This is all the more so as both inherent functionality and exploitation capability of commercially available dynamic process simulation software expands, enabling the scope of rigorous dynamic modeling to cover from wells to export jetty within a single software application.

But where should engineering effort be focused? What are the expectations from dynamic simulation during the design phase? And what happens after the design phase completes? To address these questions, the authors will reference the literature and discuss some of their generalized experience. Technical focus items will include:

- Representing wells, pipelines and more

- Assurance evaluation of high integrity pressure protection system and/or pressure relief valve scenarios

- Defining and evaluating control strategies associated with tight thermal integration, including for anticipated (even if rare) off-normal operations

- Checking the integrity of available operating procedures and/or providing logistic information that can feed into sequencing start-up, shutdown and other operating procedures for the fully integrated facility

- Example of exchanger tube rupture unanticipated potential consequences

Evaluation of Reverse Osmosis - Electrode-ionization (RO – EDI) Pilot Project in PT Badak NGL

Fany Arfianto, PT Badak Natural Gas Liquefaction, Kalimantan Timur, Indonesia

Background

PT Badak NGL utilizes groundwater for Plants and Community water consumption. To produce ultrapure water for Boiler Feed Water (BFW), PT Badak NGL services demineralization process. Figure 1 shows the simplified water treatment process diagram of PT Badak NGL. The Demineralization unit uses ion exchange process. For regeneration, it utilizes sulphuric acid and caustic, which has high safety risk. The sulphuric acid sometimes leaks out from the pipe and connections that can endanger the worker and plant facilities around the leakages due to its aggressiveness.

Figure 1. Simplified Process Diagram of Water Treatment Facilities

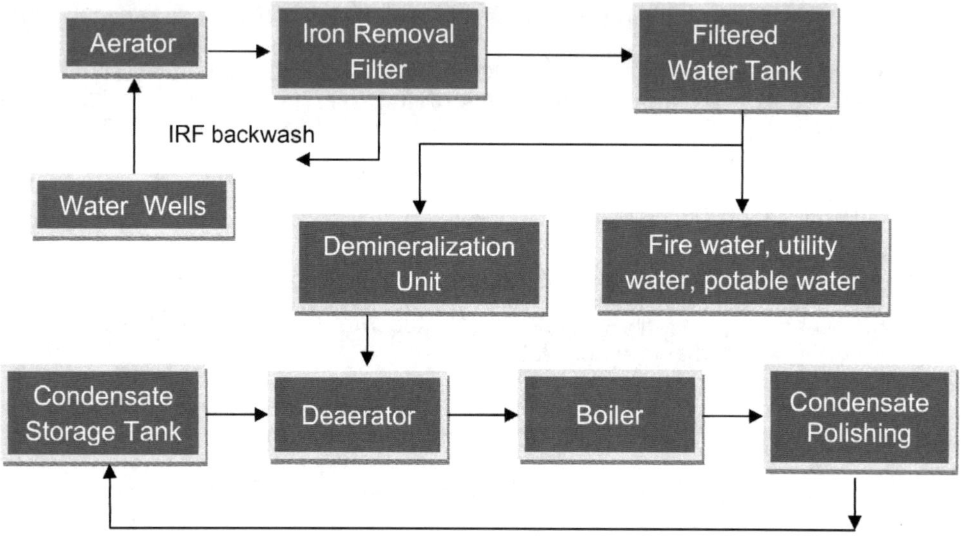

PT Badak NGL intends to reduce the risk of sulfuric acid usage by implementing a more safe process that use less chemicals such as RO-EDI process. The RO-EDI is a new water treatment technology that utilizes less hazardous chemicals.

General Process of RO – EDI

The implementation of new technology is high risk and high capital investment, therefore PT Badak NGL conducted a literature study and a pilot project of RO – EDI in order to ensure that the parameter of safety, product quality, operability and feasibility are acceptable. The previous RO – EDI study literature was conducted in 2009. This study was intended to develop the methodology of membrane technology utilization in PT Badak NGL waste water and demineralization unit. The study recommended implementing pilot project of RO – EDI process in Demineralization unit.

RO – EDI is a process to produce pure water by combining a membrane separation process and an Ion Exchanger. The membrane separations are used in a wide range of industrial processes to separate biological macromolecules, colloids, ions, solvents and gases. RO is a part of membrane processes which utilize porous polymeric or inorganic films. EDI process is removing ionized species from water using a combination of ion exchange membranes, ion exchange resins and a DC electric potential.

Figure 2. Simplified Process of RO – EDI

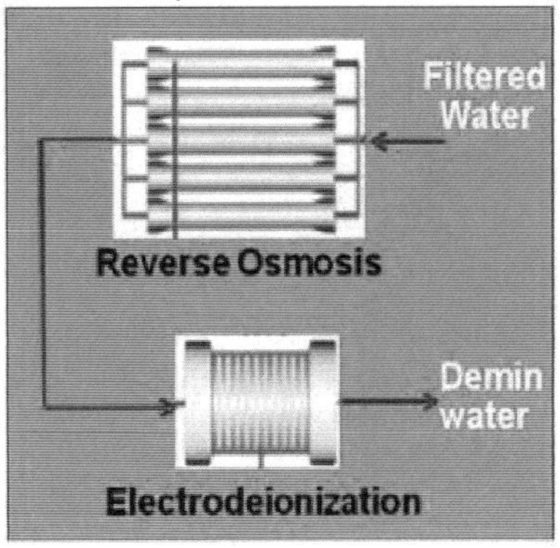

To fulfill the PT Badak demineralization water requirement, RO – EDI pilot plant PT Badak NGL Water Treatment Plant consists of RO module and EDI unit. Single-stage of RO module still need a support from polishing unit to ensure that the produced water meet the specification. The polishing unit is Electrodeionization (EDI). According to previous study, RO water product quality does not meet with PT Badak NGL demineralization water specification (see Table 1), specifically in conductivity and Silica. Therefore combination of RO and EDI is required.

Table 1. Demineralization Water Quality

Water Product	Conductivity (μmhos/cm)	pH	Silica (ppm)	Na (ppm)
PT Badak Spec.*	**< 5.0**	**6.5 – 8.5**	**< 0.12**	**< 0.3**
Ion Exchange**	**0.87**	**6.1 – 7.1**	**0.03**	**0.2**
RO	2.50	5.54	0.29	0.18
EDI	0.30	6.20	0.01	0.00

Note: *) PT Badak NGL demin water specification
**) actual Demineralization unit quality

Pilot Project of RO – EDI

The simplified process diagram of RO – EDI pilot plant can be seen in Figure 3 and the related mass balance is shown in Table 2.

Table 2. Mass Balance of Pilot Project

Stream	1	2	3	4	5	6	7	8	9	10
Flow, m3/h	3.7	2.8	1.1	0.7	1.1	1.0	1.0	0.1	1.0	1.0

Figure 3. Schematic Process Flow Diagram of Pilot Project

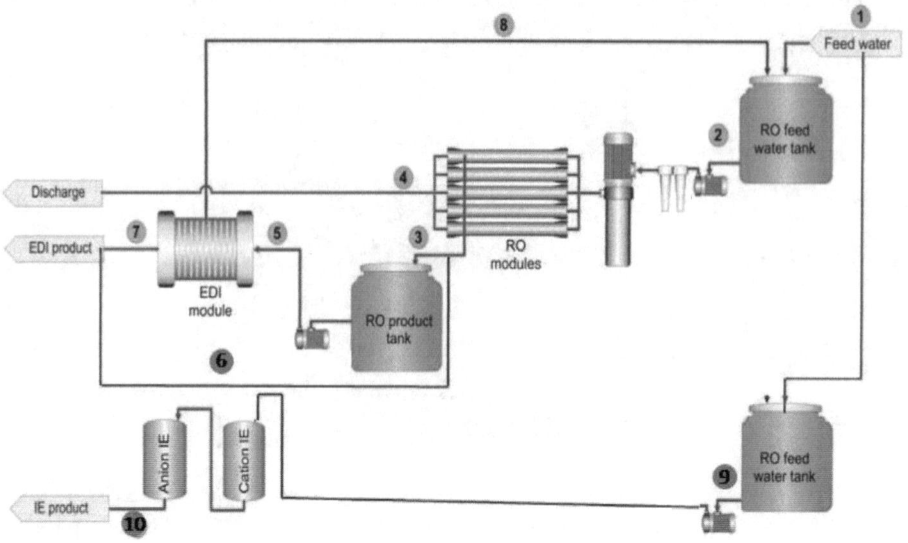

Key Performance Indicators

The water product quality was observed during pilot project field test observation, such as a product water recovery, pH, conductivity, silica and sodium contents, as well as product water recovery. The product water recovery is the ratio (percentage) of product to feed. The recovery in RO-EDI is proportional to the process efficiency. The target water product recovery of RO-EDI pilot project was between 80-90% and it was achieved 82.5% during the pilot project test.

The RO-EDI field test result can be seen in Table 3. The achievements of pH and conductivity were out of the target specification. However, the conductivity of RO – EDI product is still much lower than existing Ion Exchange unit product. The sodium target specification of the pilot project was 0.001 ppm, whereas the laboratory device detection limit is only 0.003 ppm. See detail field test result in Figures 6 – 9.

Table 3. Key Performance Indicators & Achievement of RO- EDI Field Test

Performance Items	PTB spec.	Pilot Project Results		
		Ion Exchange	RO-EDI	
			Target	Achievement
1. Product Recovery, %	N/A	94	80 – 90	82.5
2. Electrical Consumption, kwh/m^3	N/A	0.6	1.97	2.64
3. Specific Water Production Cost, $/m3	N/A	0.66	0.35	0.4
4. Demin water Quality				
- Conductivity, μmosh/cm	< 5.0	0.87	0.1	0.3 – 0.4
- pH	6.5 - 8.5	6.1 - 7.1	8 - 8.5	6.2 – 6.3
- SiO_2, ppm	< 0.12	0.03	0.005	0.00 – 0.01
- Na, ppm	< 0.3	0.2	0.001	<0.003*)

Note: *) the lowest analyzer detection level

The phenomenon of low pH of low conducitivity water are depicted in Figures 4 and 5, where conductivity level is correlated with pH of water. As shown in those figures, water with conductivity level of 0.3 – 0.4 , μmosh/cm will probably has pH of 6.2 or 8.2 (Figure 4); or even lower than 6 or higher than 8.5 (Figure 5).

Figure 5 shows the theoretical resistivity values of water following the addition of hydrochloric acid (HCl) or sodium hydroxide (NaOH). The resistivity of ultrapure water is 18.18 MΩ.cm and the pH is 6.998 (***Estelle Riché, et al. 2006***).

However, one cannot measure the pH of high-purity water simply by dipping the electrodes of a pH meter into a beaker full of water. This measurement is challenging due to the quasi-absence of ions that would enable electron transport between the measuring and reference sides of the pH electrode. It often results in erratic and meaningless pH readings. Pure water readily absorbs carbon dioxide (CO_2) when exposed to the atmosphere, which forms carbonic acid. ***(Estelle Riché, et al. 2006).***

Dissolved CO_2 in water (CO_2) is suspected substance which present in RO – EDI product water that influence pH. Ultrapure water such as EDI product readily absorbs CO_2 when exposed to the air in Aerator or during measurement which forms Carbonic acid. However, the lower pH can be solved by adjusting amine injection in water effluent of RO – EDI to meet the Boiler Feed Water pH specification of 6.5 – 8.5.

Figure 4. Relationship between pH and Conductivity

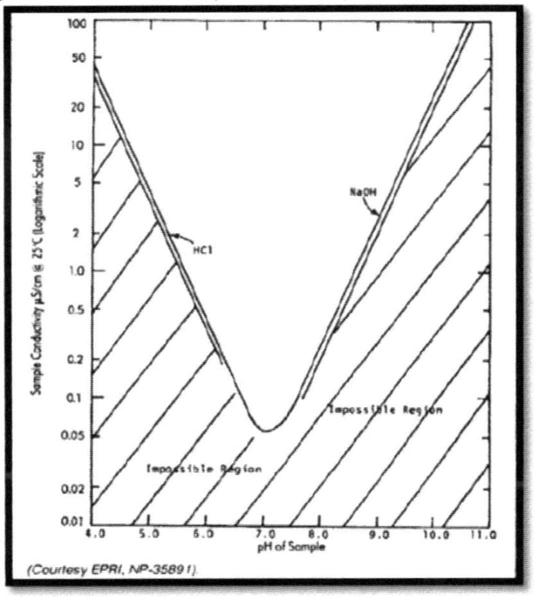

Figure 5. pH vs Resistivity of Ultrapure Water

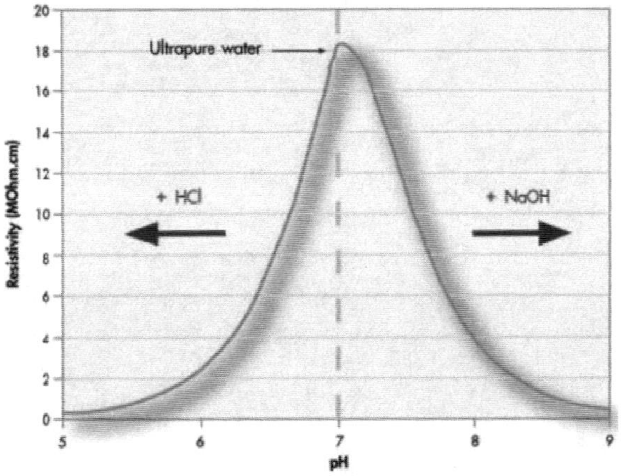

Meanwhile, conductivity, sodium and silica content of effluent RO – EDI is lower than water effluent Ion Exchanger. It means the product of RO – EDI is more pure. This condition is potential to improve boiler water quality. Based on the above comparison, RO – EDI produces water effluent quality better than Ion Exchanger.

The electric consumption was 2.64 kwh/m^3 which is 34% higher than the target, therefore it causes the specific water production increases up to 0.4 US$/m^3 than the target of 0.35 US$/m^3. However the RO-EDI specific water production cost of pilot project is remain lower than PT Badak specific water production cost (using Demineralization unit) of 0.66 US$/m^3.

The water product recovery of Ion Exchanger is higher (94%) than RO – EDI (83%). This item indicates that Ion Exchanger is more efficient than RO - EDI for producing demineralization water with similar water flow rate and quality. Meanwhile, the product recovery of RO – EDI is still in range of target 80% – 90%.

EDI is continuous electro-chemical process of water purification. Therefore, electrical power contributes significant portion on RO – EDI operational. RO – EDI consumes electrical power of 2.64 kwh/m^3; higher than Ion Exchanger Unit (0.6 kwh/m^3). The higher electrical consumes power continuously for auto regenerating in cation and anion resin inside.

The demineralization water operating cost components are electricity, chemicals and maintenance. The higher operating cost of demineralization water produced by Ion Exchanger unit is contributed by chemical and maintenance. Meanwhile, significant operating cost of RO – EDI unit is contributed by electrical power. During field test, pilot project of RO – EDI has not consumed chemical for supporting its operational. However, for further operation, chemical cleaning and injection of antiscale agent should be performed. It shows that RO – EDI was not chemical free technology because it will require chemical consumption even though the chemical are less hazardous than Sulphuric acid.

Based on field test result, in overall the quality of RO – EDI effluent is good and from safety point of view, RO – EDI is potential to be implemented for replacing existing Ion Exchange unit regarding the utilization of hazardous chemical.

Sensitivity Analysis

The NPV at 10th year and IRR are simulated in various EDI capacities (see Figure 6). It concludes that EDI capacity significantly contributes to RO – EDI investment.

Figure 6. Sensitivity Analysis at Various EDI Capacities

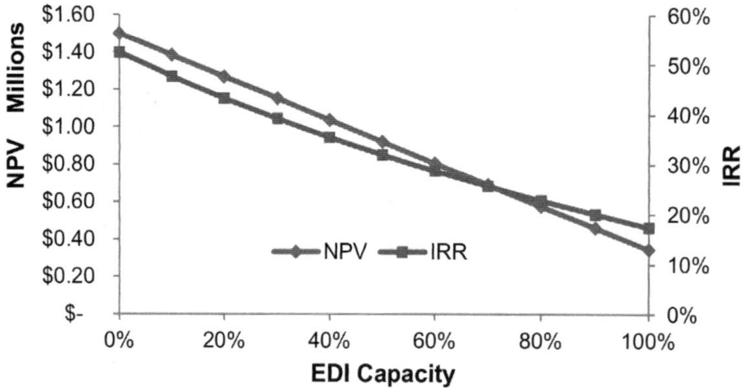

Variation of demin water quality and variation of EDI unit capacity for blending of RO product and EDI product also investigated. All of the variation resumed in Table 4. The minimum EDI unit capacity is 60 % to meet PT Badak NGL demineralization water specification.

Table 4. Variation of EDI Capacity and its Effect to Demin Water Quality

	Conductivity, mS/cm	pH	Silica, ppm	Na, ppm
RO	**2.5**	**5.54**	**0.29**	**0.18**
EDI	**0.3**	**6.2**	**0.01**	**0**
PTB Spec.	**<5**	**6.5 – 8.5**	**<0.12**	**<0.3**
EDI Capacity				
0	2.50	5.54	0.29	0.18
10	2.28	5.61	0.26	0.16
20	2.06	5.67	0.23	0.14
30	1.84	5.74	0.21	0.13
40	1.62	5.80	0.18	0.11
50	1.40	5.87	0.15	0.09
60	1.18	5.94	**0.12**	0.07
70	0.96	6.00	0.09	0.05
80	0.74	6.07	0.07	0.04
90	0.52	6.13	0.04	0.02
100	0.30	6.20	0.01	0.00

Cost & Benefit Analysis

Economical analysis was conducted to measure the feasibility of the project. In order to get the optimum capacity of RO – EDI, the analysis was performed for various RO – EDI design capacities.

RO – EDI application will reduce demineralization water specific operating cost from US$ 0.66 per m^3 to be US$ 0.40 per m^3 (the specific cost of RO – EDI included estimated chemical consumption cost and replacement of RO and EDI module cost). The margin of the specific water cost is gained from chemical cost and maintenance cost reduction. Meanwhile, total operating cost is calculated from sum of RO – EDI and Ion Exchange operating costs. Project feasibility is determined by the parameters of Payback Period (PBP), Net Present Value (NPV) and Interest Rate of Return (IRR), as shown in Table 5.

Table 5. Economic Analysis

RO – EDI Capacity, m^3/hr	CAPEX, USD	PBP, years	NPV at 10th years, USD	IRR, %
42	471,635.55	> 10	-2,465,953.84	-29.14%
77	667,407.31	> 10	-1,249,352.82	-10.92%
108	840,805.16	> 10	-398,558.96	5.79%
135	**991,829.08**	**5.82**	**294,096.08**	**15.89%**
165	1,159,633.45	5.93	278,047.18	15.43%

Based on the above analysis, RO – EDI with design capacity of 135 m^3/h is the optimum option compared to the others with higher Net Present Value (NPV) & Internal Rate of Return (IRR) and lower period of payback period.

The benefit analysis is developed to determine relative advantage and disadvantage of producing demineralization water, as determines in Table. 6. RO - EDI offers a variety of benefits compared with Ion Exchanger unit. The most obvious is the elimination of the regeneration process and its associated hazardous regeneration chemicals (acid and caustic). Since RO - EDI operates through a combination of ion-transfer across the resins and membranes, as well as electrochemical regeneration of a portion of the bed, the resins and membranes are always functional as long as the DC voltage is applied. The resin in an Ion Exchanger unit only purifies water when in its active (regenerated) form. It means RO – EDI does not require conducting the regeneration in separate process while Ion Exchanger does.

If RO – EDI is designed to supply make up demineralization water for supporting 4 Process Trains service with design capacity of 135 m^3/hr, it will replace the existing Ion Exchanger unit to supply make up demineralization water. It will leads to the elimination of the regeneration process and its associated hazardous regeneration chemicals (acid and caustic). So, the risk to personnel regarding the usage of hazardous chemicals can be reduced.

Meanwhile, RO – EDI implementation has a reliability concern due to short time of field test that cannot cover some maintenance program such as antiscale injection, maintenance activity (membrane and EDI unit replacement) and chemical cleaning activity. Beside reliability concern, limited information about operating experiences in big capacity of industrial applications should be considered. Detail comparison of Ion Exchanger and RO – EDI unit can be seen in Table 6.

Table 6. Benefit Analysis of Existing Ion Exchanger Unit and RO – EDI Application

Item	Ion Exchange	RO – EDI
Chemicals for regeneration, anti scaling and organic fouling	Hazardous chemicals (NaOH & H_2SO_4)	Less hazardous chemicals (Citric acid)
Safety and Environment Aspect	High risk of hazardous chemicals impact	Low risk of hazardous chemicals impact
Acid regeneration infrastructure	Should be maintained	Not required
Temperature	High temp. (60°C)	Ambient temp.
Pressure	Low pressure (3-5 kg/cm^2)	High pressure (12 kg/cm^2)
Feed quality	More stable against feed fluctuation	Sensitive against feed quality fluctuation
Unit capacity	Fixed	Expandable

Conclusions

1. The implementation of RO – EDI potentially improves boiler water quality.
2. Safety and environment issues can be eliminated since no hazardous chemical handling.
3. The optimum RO – EDI capacity to be installed at PT Badak NGL for supporting 4 Process Trains is 135 m^3/h.

(122d) Reducing Black Smoke and CO2 Emissions in PT Badak's Ground Flare

Ferry A. Perdana, Technical, PT Badak NGL, Bontang, Indonesia

Reducing Black Smoke and CO_2 Emissions

in PT Badak's Ground Flare

Ferry Adhi Perdana

Process Engineer

PT Badak Natural Gas Liquefaction

Bontang, Kalimantan Timur, Indonesia

E-mail :ferry@badaklng.co.id

During plant upset (such as start up or shutdown), the waste liquid hydrocarbon is burned at ground flare and results in black smoke and CO_2 emission. Approximately 35 kilotons / year of CO_2 emission is generated only from sweep gas burned to maintain ground flares in service. To support PT Badak's commitment to the environment issues and government regulation, the study is required with the intention to evaluate the possibility eliminating the black smoke and reduce CO_2 emission at ground flare.

PT Badak has five ground flares to burn the waste liquid hydrocarbon from all plant area such as process trains, LNG & LPG storages and loading dock. Once plant upset, firstly the liquid hydrocarbon is sent to Hydrocarbon Condensate Stabilizer unit for its recovery by delta pressure as a driving force. After the delta pressure is low, the remaining liquid hydrocarbon is sent to ground flare by gravity and burned. At ground flare, combustion reaction is occurred between hydrocarbon and air with carbon dioxide and water as product. When air supply for combustion is lower than requirement, incomplete combustion is occurred and it causes black smoke. There are three options to prevent black smoke namely Steam Educator Assisted, Liquid Atomizing Burner and Liquid Hydrocarbon Recovery. From the evaluation, Steam Educator Assisted and Liquid Atomizing Burner will still produce CO_2 emissions and it will not offer an additional benefit. Meanwhile, for Liquid Hydrocarbon Recovery (LHCR), it will provide advantages in eliminating black smoke and CO_2 emissions, saving sweep gas and also recovering liquid hydrocarbon. So, based on the evaluation, LHCR is the preferred option to reduce black smoke and CO_2 emissions in PT Badak's ground flare. In LHCR, the waste liquid hydrocarbon from plant upset will be collected in the surge drum and processed in Condensate Stabilizer unit and Fuel Gas system. LHCR will reduce CO_2 emissions around 35 kilotons/year and save sweep gas approximately 1,818 kNm^3/hr.

Keywords: waste liquid hydrocarbon, black smoke, CO_2 emissions

Modeling of Ground Flare Pits

Nicolas Ponchaut, Ph.D., PE, CFEI [1,2]
Harri Kytömaa, Ph.D., PE, CFEI, CFI [2]
Christopher DesAutels [2]
Shoichi Kaganoi [3]
Olivier Becu [4]
Gareth Cardiff [3]

Abstract

Some new LNG liquefaction facilities have been configured to include large ground flares to allow for the safe combustion of large volumes of combustible gasses in the event of extremely rare, short duration, facility upset conditions. A recent example is the planned INPEX LNG development near Darwin, Australia. Ground flares can exceed the size of football fields, measuring greater than one hundred yards in length. A typical ground flare consists of several hundred burners, surrounded by porous (slotted) radiation protection fences. Questions have been raised as to whether the plumes of hot combustion gasses may impact the facility, the work force, or nearby roadways in the event that the wind happens to be blowing in the wrong direction at the wrong speed.

This study examines the use of Computational Fluid Dynamics (CFD) to model plumes from ground flares and the downwind temperature distribution. The CFD model is first validated by comparing its results with commonly used dispersion models such as PHAST (Process Hazard Analysis Software Tool) and CALPUFF. The flow features of a ground flare plume rising in a crosswind are then discussed. In particular the plume acts as what is commonly referred to as a "jet in cross flow." In this type of flow, counter rotating vortices are known to form. These vortices are not represented in commonly used Gaussian plume models.

The use of two adjacent flare pits is then investigated. The interaction of the two pairs of counteracting vortices is discussed, as it influences the downwind temperature distribution. The results show that the distance between adjacent flare pits is of importance when planning multi-flare pit operations.

Introduction

Some new LNG liquefaction facilities have been configured to include large ground flares to allow for the safe combustion of large volumes of combustible gasses in the extremely rare, short

[1] Corresponding author: nponchaut@exponent.com
[2] Exponent Inc. Natick, MA, USA
[3] INPEX Operations Australia Pty Ltd Yokohama, Japan
[4] TOTAL Paris, France

duration event of facility upset conditions. These ground flares, an example of which is shown in Figure 1, typically consist of several hundreds of burners surrounded by porous (slotted) radiation protection fences.

Figure 1: Typical ground flare pit. The flare pit is composed of many individual burners. It is surrounded by radiation fences that are not impermeable.[5]

The side length of these flare pits can reach several hundred feet. During upset conditions, these large pits are able to burn thousands of tons of gas every hour, generating tens of thousands of MW of heat.[6] As an example, the planned LNG development near Darwin, Australia is expected to include 4 large flare pits (see Figure 2), designed to be able to burn more than 2,500 tons of gasses per hour.

Given the very large amount of heat that can be released by these flare pits; concerns have been raised that these upset conditions may result in elevated downwind temperatures near the ground. In particular, the downwind temperatures are of a concern for human safety. This publication quantifies the downwind ground and elevated temperature for worst-case flare pit operating conditions.[7] The flow features of the plume resulting from the operation of flare pits are discussed, and in particular the effect of simultaneously operating two adjacent flare pits is investigated.

[5] Multipoint Ground Flares. Hamworthy Combustion. http://www.hamworthy-combustion.com

[6] Note that these rare upset conditions only last up to 15 minutes and this high combustion rate would remain short lived.

[7] Elevated temperatures refer to temperatures above 2 meter (7 feet) elevations.

The results were obtained using CFD.[8] The CFD model is first validated by comparing its results to the results obtained by the dispersion models PHAST and CALPUFF, as well as to an analytical model published in literature.

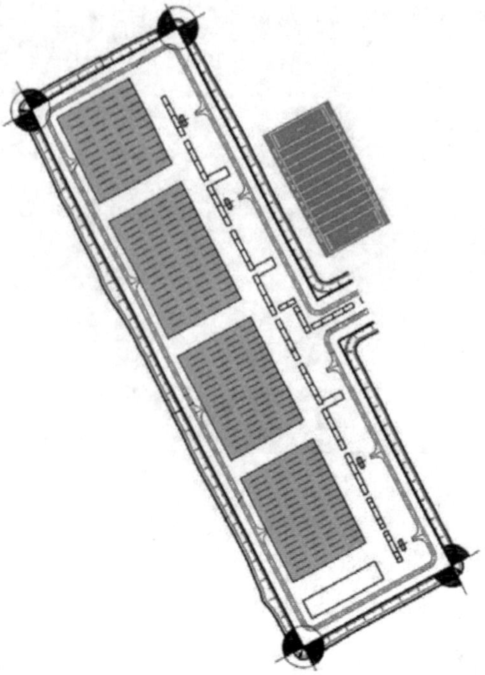

Figure 2: Size comparison between the four flare pits at the planned LNG Facility near Darwin, Australia (red), and a standard NFL football field (green).

CFD Model

In order to perform the CFD analysis, the flare pit geometry was built into a numerical model. The computational domain is centered on the flare pits and extends almost 890 meters radially from the pits. This wide domain ensures that upstream of the flare pits the wind profile is well developed, while a large distance downstream of the pits can be modeled accurately. The domain height was set to 1,250 meters, which is sufficiently high that the ceiling of the domain has no influence on the results (see Figure 3).

To simulate the boundary layer winds air enters on one side of the domain and leaves on the other side. Where air enters the domain the model specifies velocity, temperature and turbulence intensity distributions based on typical atmospheric conditions [1]. The mesh used in this analysis contain over one million cells with cell sizes varying from 3 centimeters in the boundary layers of the flare pits, to 60 meters far above the ground and away from the plume (see Figure 4).

---

[8] All CFD analyses presented in this publication were performed using the finite volume software STAR-CCM+ v.6.02.007 developed by CD-Adapco.

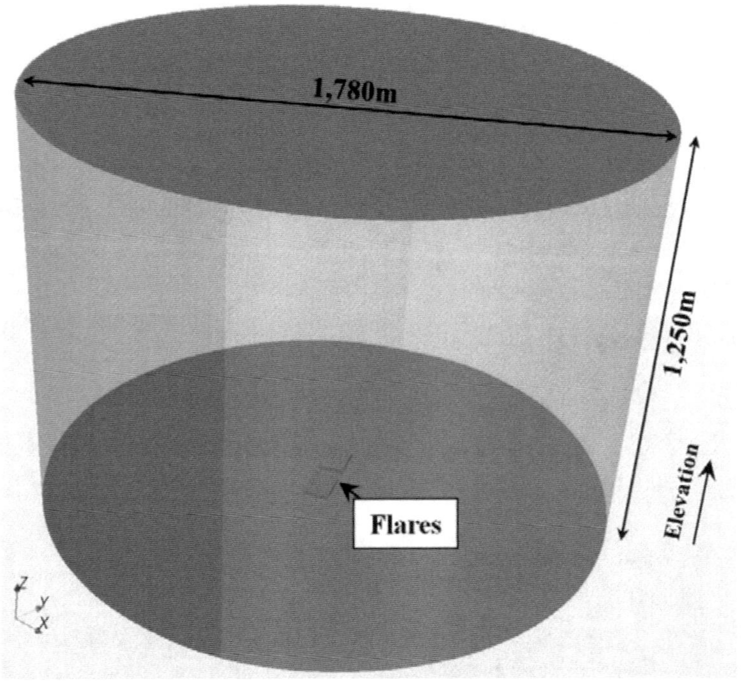

Figure 3: Typical CFD domain used in the plume modeling.

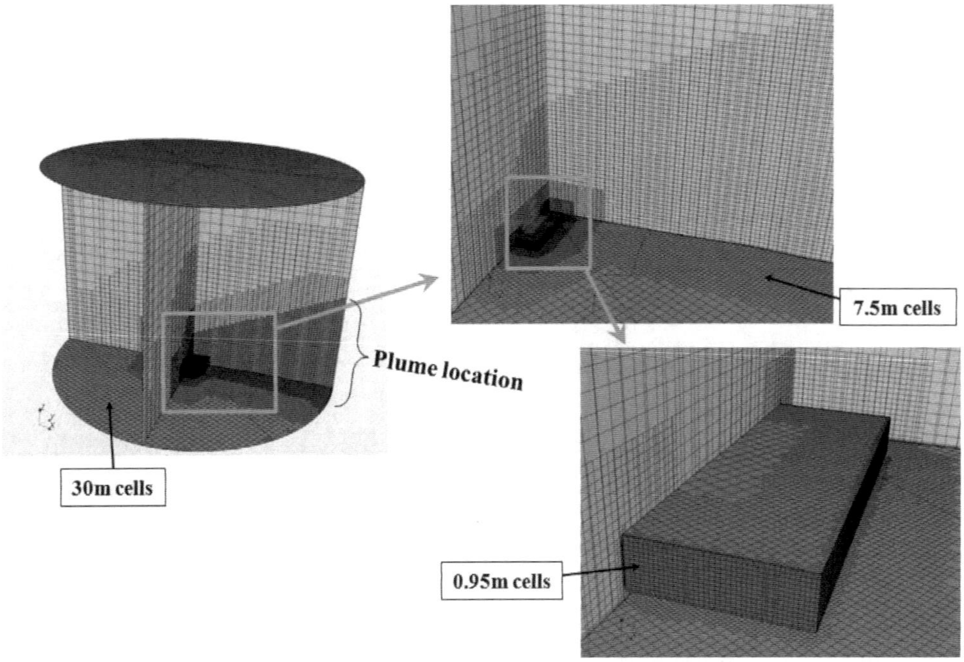

Figure 4: Typical CFD mesh.

Calculations are performed in an unsteady manner in order to capture any transient plume phenomena. The model is run for 300 seconds (5 minutes), but only the last 180 seconds are analyzed. The first two minutes (120 seconds) are therefore disregarded so that the results do not include any transient startup behaviors that are not representative of the actual plume behavior. 180 seconds is a sufficiently long duration to be able to capture all the periodic unsteady features of the plume, since these features have a characteristic period of about 10 to 20 seconds.

CFD Model Validation

In order to validate the CFD model the plume rise and temperature profile of a simple representative flare pit release was simulated using various models:

- Plume model by Chu and Goldberg [2]
- PHAST
- CALPUFF with the "Puff model" [3]
- CALPUFF with the "Slug model" [3]
- CFD (using Star-CCM+)

The model by Chu and Goldberg is a simple analytical model that simulates both the momentum driven and the buoyancy driven plume motion. The model assumes a wind speed that is invariant with elevation (in reality, the wind speed increases with elevation). To provide a relevant comparison, the wind speed at 350 meter elevation was used in the model since within the range of distances of interest; the plume motion is mostly influenced by wind speed between the 200 meter and 500 meter elevations.

PHAST is a risk analysis tool used to model the effects of hazardous releases. Since PHAST is not capable of modeling multiple rectangular flare pits, the validation case was run on a single circular flare pit. This simplification removes most of the three dimensional flow characteristics that only CFD is designed to model. The simulation parameters are summarized in Table 1.

CALPUFF is a Gaussian dispersion model commonly used in regulatory air quality modeling applications [3]. It has the capability to model an emitted plume either as a series of overlapping Gaussian "puffs" or as an elongated "slug." Both modes are tested in this validation. The flare pits have been modeled as two elevated rectangular area sources. CALPUFF contains only simple algorithms to account for the effects of obstacles. It cannot simulate the detailed flow between or around the flare pit structures.

Finally, Star-CCM+ was used to model the full three-dimensional plume. A k-ε model was used to simulate the turbulence in the wind, inside the plume and around the flare pits. In order to provide an appropriate comparison with PHAST and CALPUFF, a single circular flare pit was modeled. This simpler geometry removes some of the complex flow interactions that appear when running multiple flares pits simultaneously, and that cannot be modeled by integral models. The simulation parameters are summarized in Table 1.

Table 1: Atmospheric and flare pit running conditions used for validation purposes.

Atmospheric Stability Class	D (neutral)
10m elevation ambient temperature	27°C
10m elevation wind speed	14.9 m/s (54 km/hr)
350m elevation wind speed	24 m/s (86 km/hr)
Combusted gas flow rate	Around 200,000 tons/hrs
Combusted gas temperature	570°C

Given the fact that the plume models differ in their approach and assumptions, their results are somewhat different, especially in the vicinity of the flare pits where the details of the pit geometry are most significant. The Chu and Goldberg model does not consider any geometry details, while PHAST and CALPUFF do consider the plume source elevation. However, only CFD is able to accurately represent the details of the flow around the flare pits.

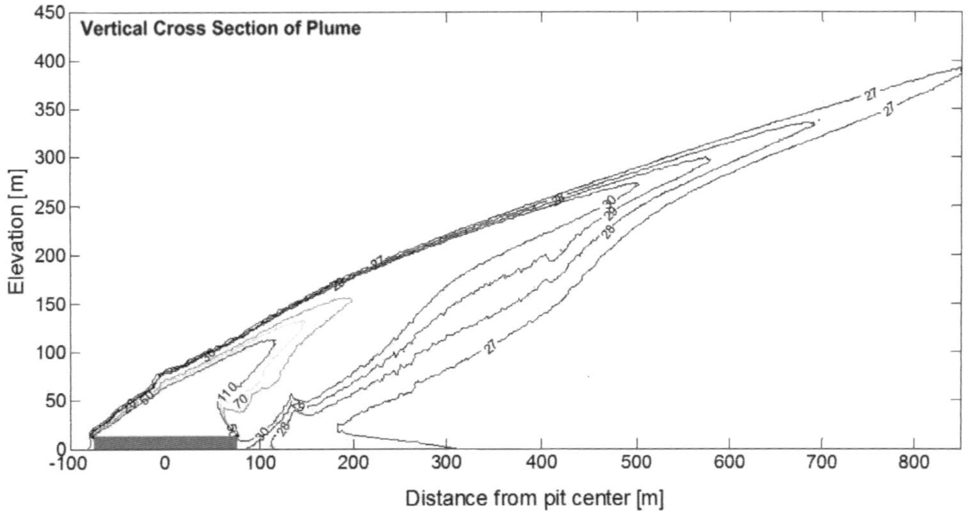

Figure 5: CFD results for the validation case conditions. Contours of equal temperatures are shown on a vertical plane aligned with the wind direction. The corresponding temperature of each contour is also shown.

An elevation view of the results of the CFD simulation is shown in Figure 5. The plume elevation and the plume temperature profiles were compared with the other models and the results of the comparison are presented in Table 2 and Figure 6. Given their differing approaches, the model results are in relatively good agreement in terms of plume rise.[9]

[9] Given all the uncertainties inherent in estimating atmospheric flows, we believe that a difference of only 20% in plume height prediction is very good agreement.

The plume temperature profile will depend on the turbulence, and on the detailed atmospheric wind speed and temperature profiles. The exact geometry of the pit also affects the temperature results since different geometries create vortices of different size and strength, which in turn result in different plume dilution rates. Given the very different assumptions included in the various models, temperature profile differences are not surprising and are expected. Note that as shown in Figure 6, the CALPUFF and the PHAST results are conservative and predict elevated ground temperatures for longer distances, when compared to the CFD results.

Table 2: Comparison of the validation test case results. In the case of CALPUFF, PHAST and CFD, at each cross-section, the plume height was determined to be the elevation of the warmest point in the plane of symmetry of the plume.

Model	Plume height, 400m downwind	Plume height, 800m downwind	2m elevation, 50°C downwind distance
CALPUFF puffs	275 m	430 m	220 m
CALPUFF slugs	275 m	430 m	170 m
PHAST	265 m		140 m
CFD	230 m	370 m	Nowhere
Chu and Goldberg	290 m	450 m	N/A

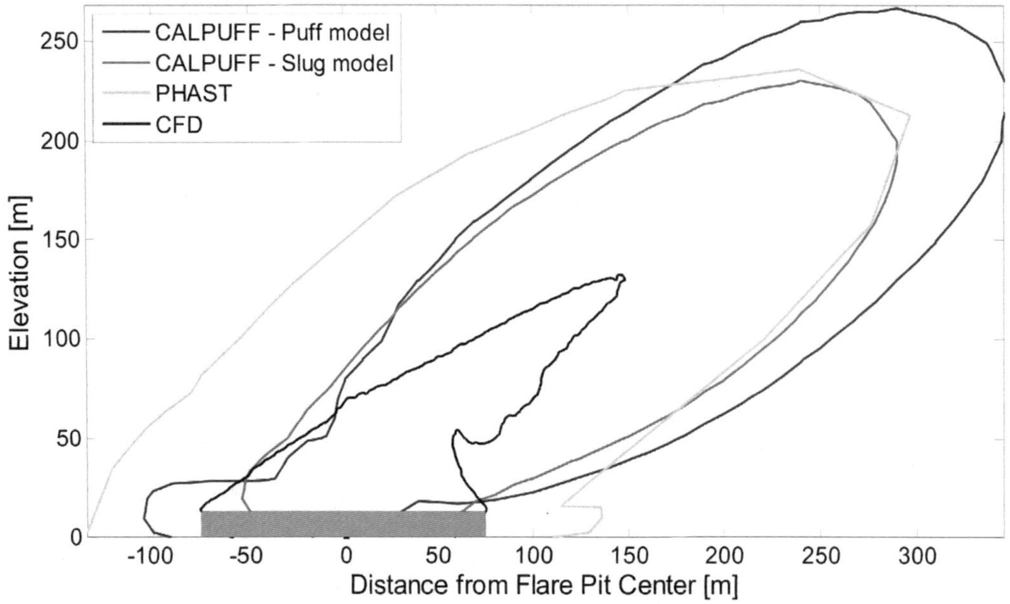

Figure 6: Vertical plume cross-section, as obtained using various models. The curves represent the location of the 70°C temperature. The Chu and Goldberg model is not included in the comparison since it is not meant to accurately represent plume temperatures.

Flow Features in a Flare Pit Plume

The CFD results obtained during the validation stage exhibit what is commonly referred to as a bifurcating plume. Because of the cross wind, the plume acts like a "jet in cross flow". The jet (*i.e.* buoyant plume of the flare pit products) enters into a cross flow (*i.e.* the wind). This phenomenon of counteracting vortices has long been recognized and is well known [4].

Bifurcating plumes have been studied in several publications [5, 6, 7]. This bifurcation and the existence of counteracting vortices has a major influence on the ground flare pit's downwind plume temperature distribution. Centrifugal forces in each of the counteracting vortices push the colder heavier gas outward, while keeping the warm lighter gas inside the vortex (see Figure 7). This results in high temperatures inside the vortices and colder gas outside of the vortices. The warmest downstream locations are therefore not on the plane of symmetry, but are symmetrically located on either side of the vertical symmetry plane.

These counteracting vortices cannot be captured by simpler integral models such as PHAST and CALPUFF and require that the full three-dimensional plume dynamics be modeled.

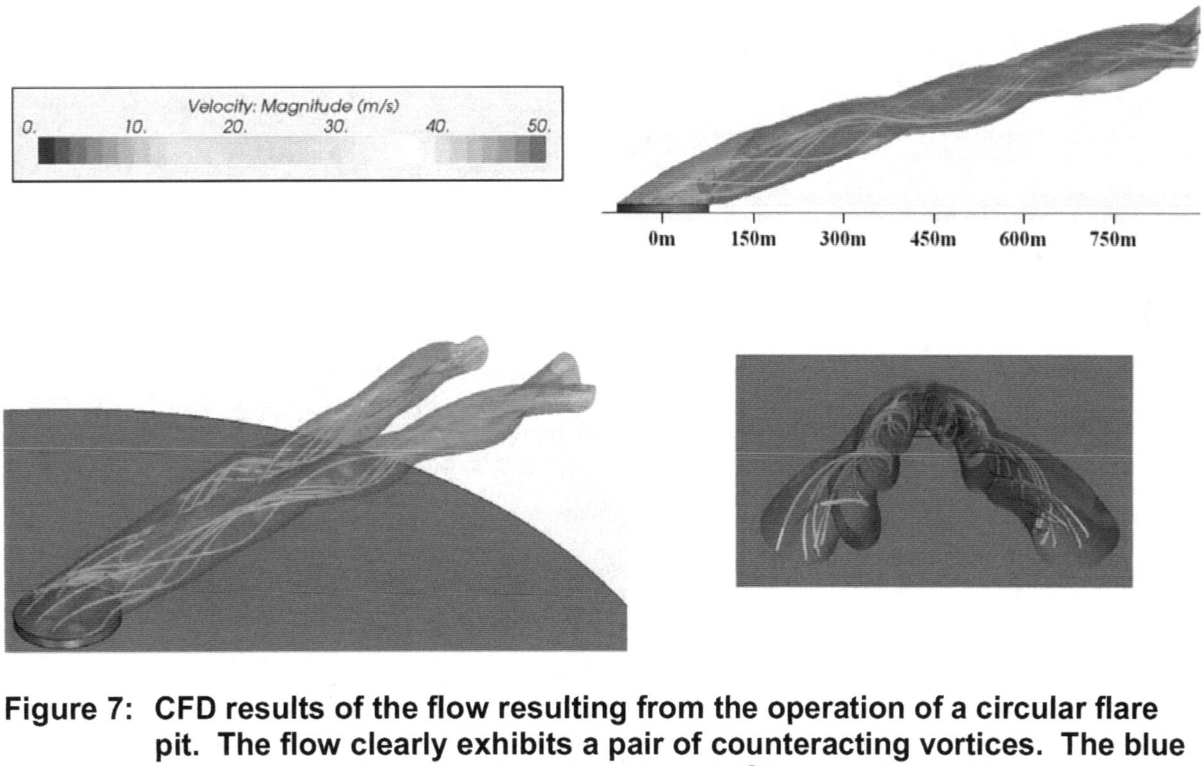

Figure 7: CFD results of the flow resulting from the operation of a circular flare pit. The flow clearly exhibits a pair of counteracting vortices. The blue surface represents the location of the 70°C isosurface. The curves represent streamlines which are color coded according to flow velocity.

Effect of Adjacent Flare Pit Operations

To investigate the effect of two flare pits, two adjacent flare pits separated by a 40 meter distance were modeled. The wind was blowing parallel to the vertical plane of symmetry separating the two pits. In this section, the CFD simulations are based on the conditions summarized in Table 3. A high wind speed and a relatively high product temperature were selected as these conditions correspond to worst-case scenarios. The results are illustrated in Figure 8 and show the existence of two pairs of counteracting vortices, one per pit.

Table 3: Atmospheric and flare pit running conditions used in the CFD model.

Atmospheric Stability Class	D (neutral)[10]
10m elevation temperature	27°C
10m elevation wind speed	14.9 m/s (54 km/hr)
Combustion gas flow rate	Around 85,000 tons/hrs
Combustion gas temperature	Around 1,200°C

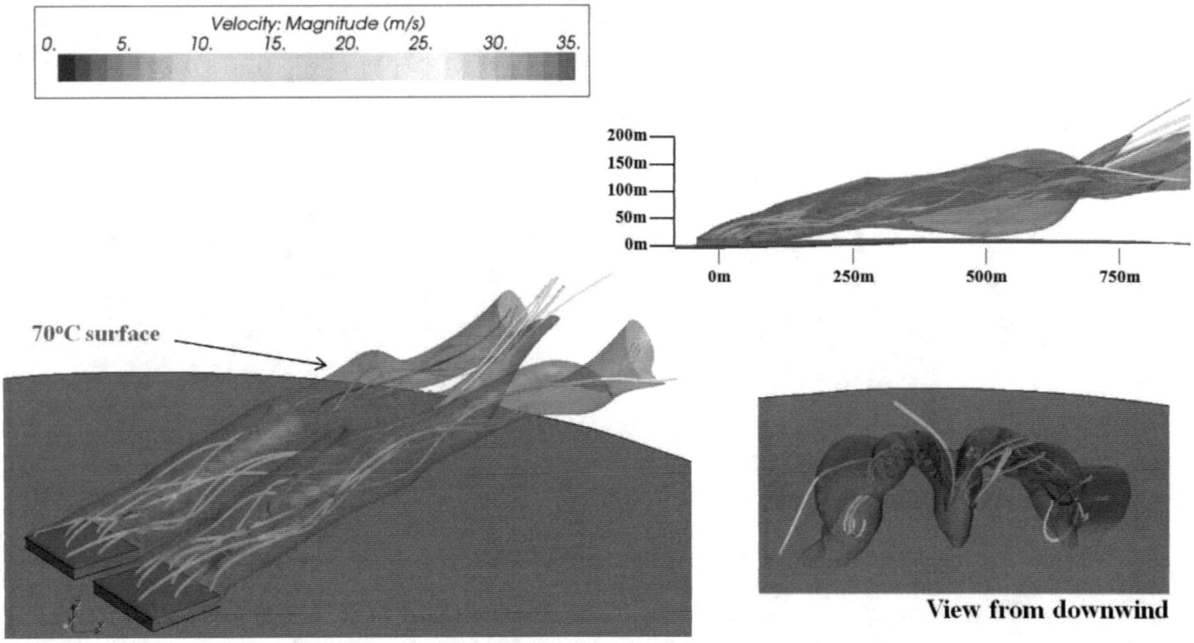

Figure 8: Instantaneous plume profile. The 70°C surface is represented in blue. Streamlines are presented with colors proportional to the local gas velocity.

[10] The selection of an atmospheric stability class D is in accordance with high wind weather conditions.

The existence of these counteracting vortices has significant implications on the ground temperatures. In the vertical plane of symmetry between the flare pits, the two middle vortices merge to create a significant downward flow that pushes the warm gases towards the ground, thereby increasing near-ground temperatures. The effect of this downward flow is most significant at about 500 meters downwind of the flare pits. A cross section of the plume, 500 meters downwind of the flare pits clearly shows the vortex structures (Figure 9). The temperatures on the vertical plane of symmetry and at 2 meter elevation are shown in Figure 10 and Figure 11. As seen in the results, due to the flow features around the walls of the flare pit, the temperature behind the pits is relatively high. In addition a peak 2 meter elevation temperature of 64°C can be observed at a 500 meter distance from the flare pits.

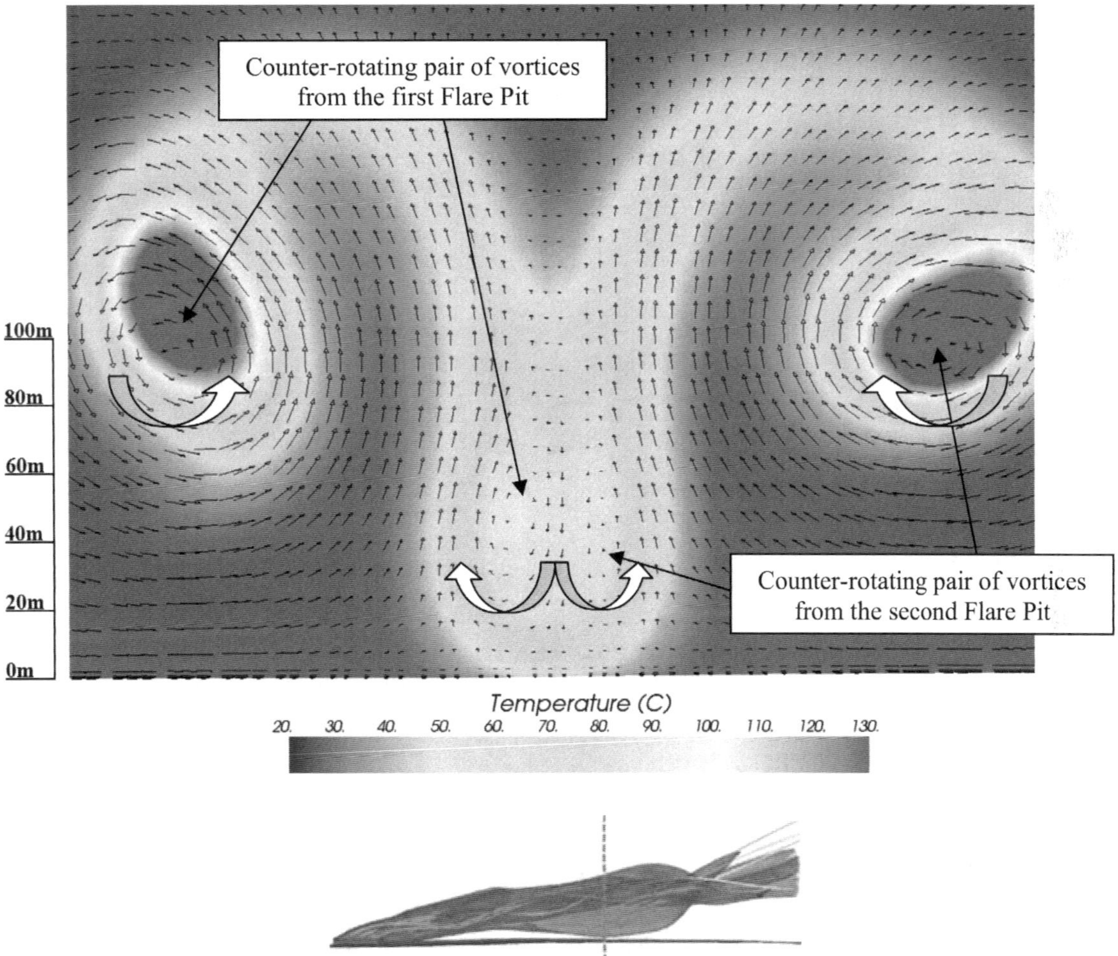

Figure 9: Plume cross-section, 500 meters downwind from the flare pits (see bottom sketch). The temperatures are represented by the colors, while the velocities are shown by the arrows. The longer the arrows, the higher the flow velocity.

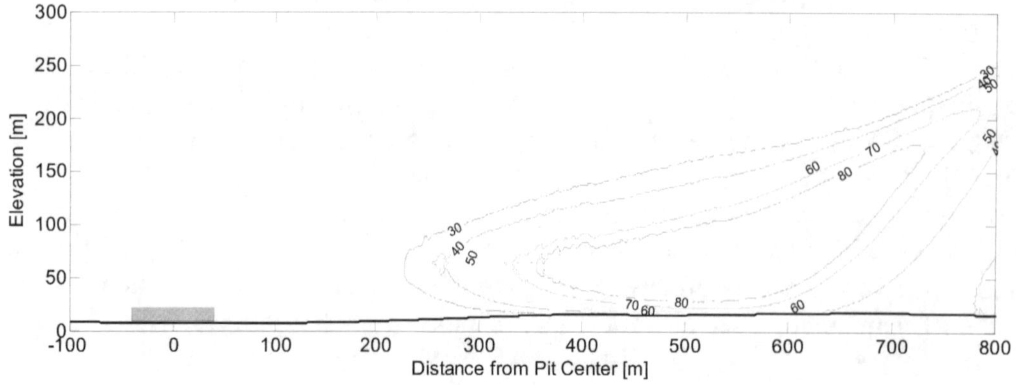

Figure 10: Average temperature distribution on the plane of symmetry (vertical plane) between the two flare pits. Temperatures are in degree Celsius.

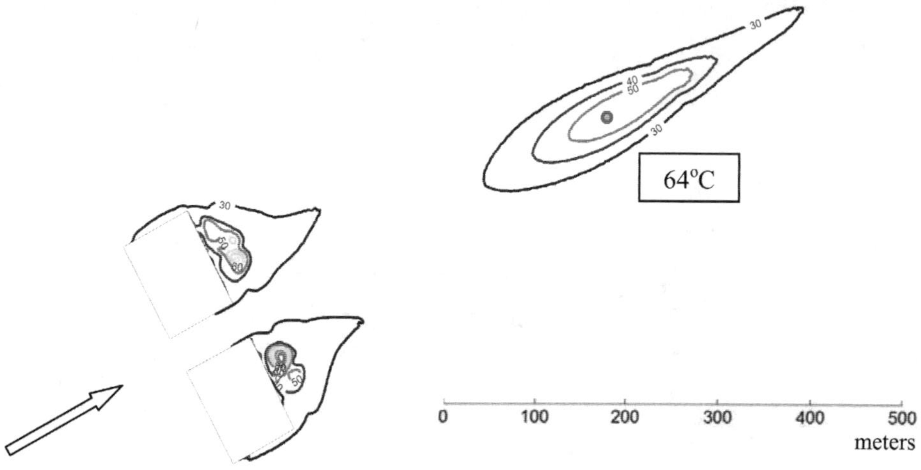

Figure 11: Average temperature distribution at 2 meter elevation. The two flare pits are represented by white rectangles. The wind direction is represented by the arrow, and the location of the maximum far-field temperature by a red circle.

Given the significant influence of the vortex interactions, the distance between flare pits is an important parameter to consider when designing or planning multi-flare pit operations. In particular, if two operating flare pits are far apart, their vortices do not interact and both flare pit plumes rise independently without merging. On the other hand, if the flare pits are touching, then the two flares act as a single larger one, and again, no vortex interaction occurs.

Between these two extremes, there is a distance between flare pits that will generate a maximum downward flow that pushes the combustion products toward the ground, thereby generating relatively high ground temperatures. In the present study, several distances were investigated and the results are presented in Table 4. These results highlight the importance of flare pit separation distance on the downwind ground temperatures.

Table 4: Maximum temperatures during steady operation of two flare pits, as obtained with CFD. These temperatures do not include the temperatures obtained in the immediate vicinity of the flare pits.

Elevation	20m pit separation	40m pit separation	150m pit separation	40m pit separation 45 deg winds
2 meters	71 °C	64 °C	<30 °C	32 °C
6 meters	84 °C	74 °C	<30 °C	32 °C
10 meters	84 °C	80 °C	<30 °C	39 °C
14 meters	90 °C	84 °C	<30 °C	40 °C
16 meters	90 °C	84 °C	<30 °C	40 °C
22 meters	100 °C	89 °C	33 °C	50 °C

The results in Table 4 represent a hypothetical worst case scenario where upset conditions would require flaring at the maximum flow rate, where the combustion gas temperature would be relatively high and where the wind conditions would be very exceptional (*i.e.* high winds blowing perpendicular to the flare pits). The likelihood that these conditions coexist is very remote, and the values in Table 4 constitute a very unlikely worst case scenario.

Note that these results use a wind that is blowing in a direction parallel to the axis of symmetry. This results in strong pairs of vortices that generate very strong downward currents when they interact. Other wind directions may generate greater temperatures in the immediate vicinity of the flare pits; however, the downwind temperatures remain significantly lower. As an example, the maximum temperature resulting from wind blowing at 45 degrees with respect to the symmetry plane is shown in the last column of Table 4.

Conclusions

CFD can play an important role in understanding the impact of large industrial ground flares and other similar releases. Simpler dispersion models will accurately calculate features such as plume height, but are not capable of resolving complex interactions between plumes, nor can they account for detailed flow around structures. CFD modeling demonstrates that there are complex interactions which can occur between adjacent flare pit plumes and which can have very significant localized effects on downwind temperatures and plume density.

The space between flare pits is shown to be a significant dimension which should be considered during plant design. With the increasing availability of CFD codes and computers capable of running them, CFD should serve as an important tool for plant design and industrial hazard planning.

References

[1] *K-ε Model for the Atmospheric Boundary Layer Under Various Thermal Stratifications*. C. Alinot, C. Masson. J. of Solar Energy Engineering. Vol 127. Nov. 2005

[2] *Fluid Mechanics for Industrial Safety and Environmental Protection*. T.K. Fannelöp. Elsevier Science. Pp 281-287. 1994

[3] A User's Guide for the CALPUFF Dispersion Model (Version 5). Earth Tech, Inc. Concord MA. 2000

[4] *Vortical Structures in the wake of a transverse jet*. T.F. Fric and A. Roshko. J. of Fluid Mechanics. Vol 279, 1-47. 1994

[5] *High-Resolution Numerical Models for Smoke Transport in Plumes from Widland Fires*. P. Cunningham and S.L. Goodrick. 2006

[6] *A Computational Model for the Rise and Dispersion of Wind-Bloan Buoyancy-Driven Plumes – I. Neutrally Stratified Atmosphere*. X. Zhang, A.F. Ghoniem. Atmospheric Environment. Vol 27A (15). 2295-2311,1993

[7] *Review of Vortices in Wildland Fire*. Journal of Combustion. Vol 2011. J.M. Forthofer, S.L. Goodrick. 2011

(122f) Anti-Surge Methods for the Base Load LNG Plant Propane Compressor

J. C. Kuo, Chevron Energy Technology Company, Houston, TX

The propane compressor represents a significant capital investment and operating expense of a LNG plant, and is an essential component in providing revenue streams of the plant. Therefore, operation and protection of the propane compressor are critical issues in the plant design and operation.

The propane compressor can operate under a wide range of conditions over the plant life cycle, especially during transient operations such as start-up, shutdown, restart and load variation. When the process condition changes, there is risk of surge where the compressor operates below its minimum stable flow, resulting in vibration and damage to the compressor and its components. Therefore, the propane compressor is required to have an anti-surge system for protection against surge.

The main difference between these methods is how the recycle stream is cooled. Different licensors/technology providers may provide different methods. This paper will summarize

- All the anti-surge methods which were used on some of the existing LNG plants. The LNG plants should be in operation or under the EPC stage, with the size varying from less than 3 to over 7 MMTPA. The driver should include GE's Frame 5, Frame 6, Frame 7, Frame 9 and the total motor driven case.
- Assessment and comparison of technical and economical benefits for different anti-surge methods,
- Indicating the advantages and disadvantages under the start-up, normal operation and trip scenarios.
- Assessment of any technical risk of using each method.

(131a) Getting the Most Out of Your Refinery Hydrogen Network

Alan Zagoria, CEOC, Des Plaines, IL; Optimization Services, UOP, Des Plaines, IL

The refinery balance is an important issue in almost all refineries. The operations of some refineries are actually constrained by the availability, the others have strong incentive to reduce the operating costs, energy consumption and CO2 emissions associated with the refinery hydrogen network. We will discuss an overall, refinery-wide approach to improving the H2 network configuration and operations. Plenty of time will be left for discussion and questions.

Heat Exchange Reforming for Increased Hydrogen Production

Niels R. Udengaard, Haldor Topsoe, Inc. Houston Texas & Jack H. Carstensen,
Haldor Topsøe A/S, Lyngby, Denmark

presented at

AIChE 2012 Spring Meeting in Houston TX on April 4, 2012

1 Introduction

Heat exchange steam reforming in hydrogen production is not a new development. Haldor Topsoe developed the first smaller scale steam reformers based on the heat exchange principle about 30 years ago for hydrogen units, where export steam was undesired. The capacity of the first hydrogen units ranged from 0.1 MM SCFD to 1 MM SCFD typically used for fuel cell applications. As the technology matured it became possible to increase the capacity and in 1997 the first industrial scale (5 MMSCFD) hydrogen units were successfully put into operation. The development of the heat exchange reformer technology has continued, and today Topsoe has licensed hydrogen plants including a heat exchange reformer with capacities of up to 185 MM SCFD. At the same time a number of different designs of heat exchange reforming technologies have been developed enabling Topsoe to tailor-make new hydrogen plants or revamp existing units to obtain maximum value for client. Today Topsoe has licensed more than 40 hydrogen units based on heat exchange reforming with more than 15 years industrial experience. Industrial feedback has confirmed that use of heat exchange reforming can save up to 20% on feed and fuel consumption (with corresponding savings in CO_2 emission) as compared to conventional steam reforming.

2 Heat exchange reforming - principles

Being an endothermic reaction steam reforming of hydrocarbons require a significant heat input to obtain the desired conversion into hydrogen. In the conventional steam reformer heat transfer takes place mainly by radiation which leads to a limited thermal efficiency as evidenced by a high flue gas temperature (typically more than 1800°F). The thermal efficiency of a conventional steam reformer is around 50% and the surplus heat is to a large extent used for steam production. Many clients have little or no use of the steam export generated in a hydrogen plant and it is therefore considered of low value.

Heat exchange reformers are characterized by being very compact and having a high thermal efficiency. In a heat exchange reformer the majority of the heat transfer takes place by convection with hot flue gas or hot process gas whereby the thermal efficiency can be increased by as much as 60-70% as compared to the radiant solution. This results in more of the waste heat being used to produce hydrogen instead of steam. Heat exchange reforming is well suited both as stand-alone units and as revamp option for capacity increase of existing plants.

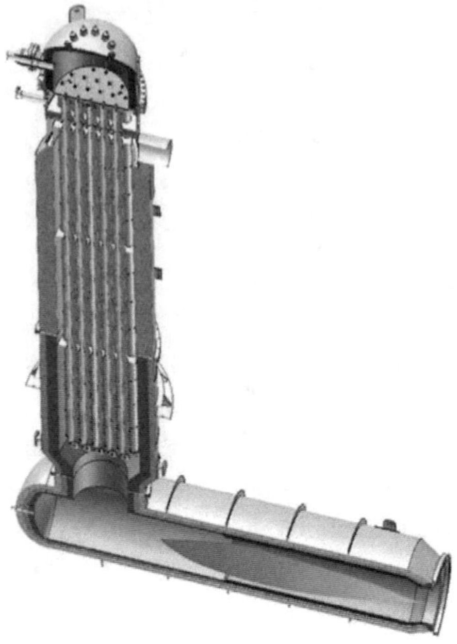

Figure 1 Heat Exchange Reformer

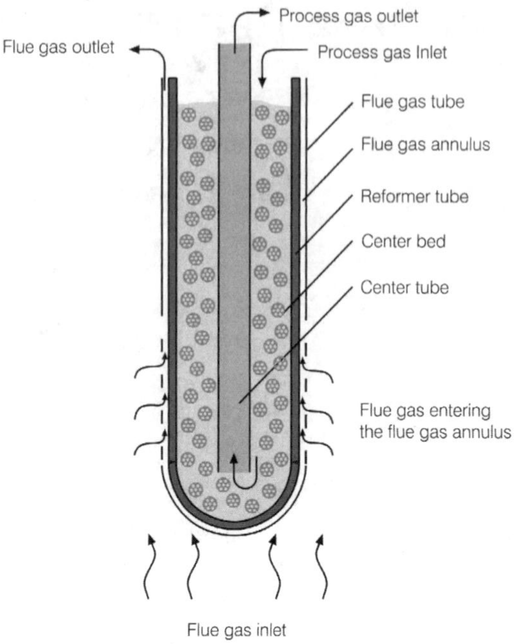

Figure 2 Heat transfer principle

3 Heat exchange reforming technologies

3.1 HTCR

The Haldor Topsoe Convection Reformer (HTCR) is a heat exchange reformer in which the process gas is heated mainly by hot flue gas. The HTCR is very compact and suited for new hydrogen units and for add-on revamps for capacity increase of existing plants.

The HTCR is shown in Figure 1 and the principle is shown in Figure 2. The reformer consists of a vertical, refractory-lined vessel, containing the tube bundle with bayonet tubes. The heat from the flue gas is transferred to the process gas inside the bayonet reformer tubes resulting in low fuel consumption and zero steam export. In the HTCR reformer the external heat input is provided by only one burner which ensures a very easy operation and a fast load response. The easy operation results in only a minimum of operator attendance is required and there are examples of unattended operation of HTCR plants.

The HTCR unit is to a high degree skid-mounted and the reformer is shop-lined minimizing installation time and cost on site..

Industrial experience with HTCR includes more than 30 plants and design capacity up to 27 MMSCFD.

Case story 1 - a grassroots 27 MM SCFD HTCR hydrogen plant

In connection with an extensive revamp of an existing refinery in the Russian Federation an analysis of the future hydrogen balance showed that additional 160 MM SCFD of hydrogen would be required. The major part of the hydrogen was needed for a new hydrocracking unit and the remaining part would be needed for hydrotreating purposes. The new hydrotreating unit was scheduled to come on-stream one year before the hydrocracking unit and the refinery decided to build two separate grassroots plants to cover the hydrogen requirement - one 27 MM SCFD unit and one 130 MM SCFD unit. For the smaller unit the HTCR technology was selected due to:

- Low feed and fuel consumption
- Fast implementation time
- No need for export steam as steam requirements would be covered by the larger hydrogen unit

Figure 3. HTCR tube bundle for a 27 MMSCFD hydrogen plant

Table 1 shows the consumption figures for a 27 MM SCFD HTCR hydrogen plant compared to a conventional SMR process typically used for this capacity range.

27 MM SCFD hydrogen plant	HTCR process	Conventional SMR process
Feed, BTU/SCF H_2	354	352
Fuel, BTU/SCF H_2	11	53
Feed + fuel, BTU/SCF H_2	365	405
Steam export, BTU/SCF H_2	0	60
Cost of feed and fuel, MM USD/year	18.5	20.5

Table 1. Consumption figures for a HTCR and a conventional 27 MMSCFD hydrogen plant

The above example clearly illustrates the advantage of the HTCR process in case of low or no value of steam export. Based on a cost of feed and fuel of 5.36 USD/MM BTU[1], the annual savings for a HTCR plant amount to USD 2 MM compared to a conventional steam reformer process. Furthermore the 11% lower consumption of feed and fuel results in a correspondingly lower emission of CO_2.

3.2 HTER

In the Haldor Topsoe Exchange Reformer (HTER) the reaction heat is provided by hot process gas. The HTER can be used in grassroots hydrogen plants in combination with a radiant wall steam reformer as well as an add-on unit for additional capacity in an existing plant.

The HTER utilizes a bayonet tube or a two-bed system (Figure 4) allowing for optimal utilization of the heat transfer area. The hot reformer effluent is used as heating medium and the high pressure results in a very compact design.

The inclusion of an HTER unit reduces fuel consumption and steam production and may increase capacity up to 30%. The HTER is well suited for both capacity revamps in existing plants as well as new units where factors such as low feed and fuel consumption and compactness are important.

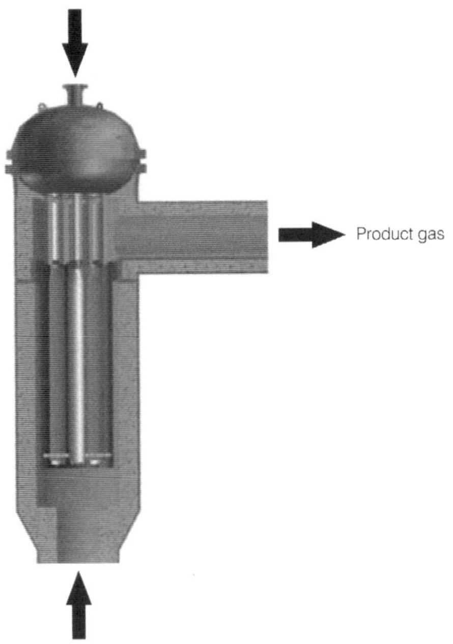

Figure 4. HTER principle

Figure 5. Plant with HTER

Since 2003 when the first HTER went into successful operation, the HTER technology has been adopted in several Topsøe licensed hydrogen plants.

Case story 2 – 3 x 185 MM SCFD grassroots hydrogen plants incorporating HTER

In connection with a large refinery project in the Middle East the refiner would need an additional 555 MM SCFD of hydrogen which was to be covered by three grassroots plants. The three plants should be designed to operate on a range of different feedstocks depending on cost and availability in the refinery.

Based on analysis of investment cost and consumption figures, the client chose a solution with three new grassroots plants of 185 MM SCFD, each plant with a state-of-the-art SMR process incorporating an HTER. Table 2 compares the key figures for the chosen HTER solution compared with an alternative without an HTER.

185 MMSCFD plant	State-of-the-art SMR process + HTER	State-of-the-art SMR process
Relative total installed cost (TIC)	100	100
Feed + fuel, BTU/SCF H_2	372	390
Steam export, BTU/SCF H_2	43	71
Feed + fuel minus steam, BTU/SCF H_2	329	319
Relative power consumption	86	100
Cost of feed and fuel, MM USD/year	129.9	136.4

Table 2. Consumption and TIC figures for a 185 MM SCFD hydrogen plant

For the same investment the solution based on an HTER results in a 5% saving in feed and fuel which, based on a price of \$5.36/MM BTU[1], corresponds to annual savings of \$6.5 MM, a correspondingly lower CO_2 emission and a lower power consumption.

3.3 TBR

The Haldor Topsoe Bayonet Reformer (TBR) is the newest addition to the Topsoe family of heat exchange reforming technologies combining the principle of convection heat transfer known from HTCR and radiant heat transfer known from the radiant wall steam reformer.

The TBR consists of a series of double tubes in a single row in a radiant furnace box similar to conventional steam reforming technology.

Feed gas flows downward through the catalyst bed located in the annulus between the two tubes as illustrated in Figure 5. At the bottom of the catalyst bed the gas turns and continues upward through the inner, empty bayonet tube. The bayonet tube arrangement is the same as that known from the HTCR technology. The gas exits at the top of the bayonet tubes and is collected in a common header.

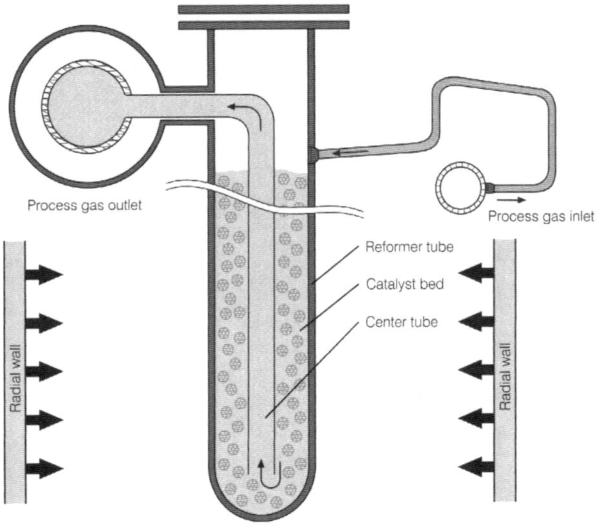

Figure 6. Simplified sketch of the TBR principle

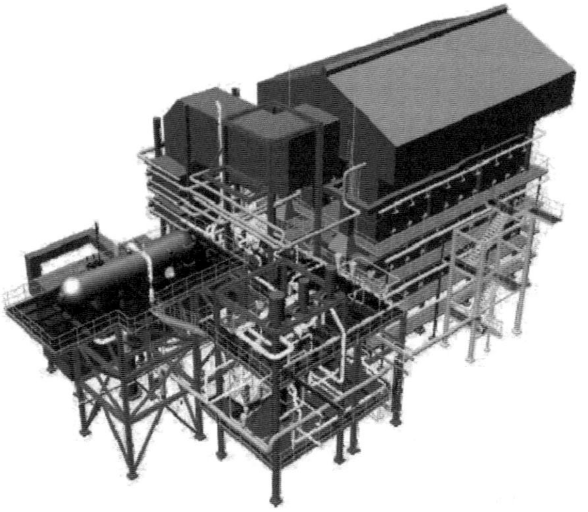

Figure 7. Hydrogen plant with TBR reformer

Improved heat utilization in combination with a high average heat flux in the TBR tubes significantly reduces the size and capital cost of the hydrogen plant and provides hydrogen production with low hydrocarbon consumption and little or no steam export.

Table 3 illustrates the main features for a 43 MM SCFD TBR hydrogen plant compared to a state-of-the-art SMR process. For the TBR process we have considered two cases: With and without steam export.

43 MM SCFD plant	State-of-the-art SMR process	TBR with steam export	TBR no steam export
Relative TIC	100	95	95
Feed + fuel, BTU/SCF H_2	390	361	350
Steam export, BTU/SCF H_2	71	30	0
Feed + fuel minus steam, BTU/SCF H_2	319	331	350
Relative power consumption	100	90	90

Table 3. Consumption and TIC figures for a 43 MM SCFD hydrogen plant based on TBR

The table illustrates that with little or no requirement for steam export, the TBR technology offers an ideal solution for the refinery with a 5% lower investment cost, and at the same time, up to 10% lower consumption of feed and fuel compared to a state-of-the art SMR process.

4 Revamping of an existing hydrogen unit

Revamping an existing hydrogen unit with an HTCR or an HTER is an economically attractive solution to increase capacity and reduce specific feed consumption.

When revamping with an add-on heat exchange reformer the capacity can typically be increased with up to 30% at a much lower investment as compared to a new plant. At the same time the compactness of a heat exchange reformer ensures that additional required plot area is kept at a minimum.

In order to determine the feasibility of a revamp Haldor Topsoe will perform a revamp feasibility study to provide the client with the optimal solution. The study will take into account the status of the existing unit, the various requirements of the client as well as prevailing conditions such as availability and price of feedstock, plot plan, available downtime for revamp etc.

4.1 Revamping with an HTCR

As a revamp option, the HTCR is a very flexible solution where the HTCR is operated in parallel with the existing reformer as illustrated in Figure 8. This lay-out will result in a revamp package consisting of few pieces of equipment, namely an HTCR with waste heat section, a process boiler and a combustion air blower that allows for independent operation of the two reformer sections.

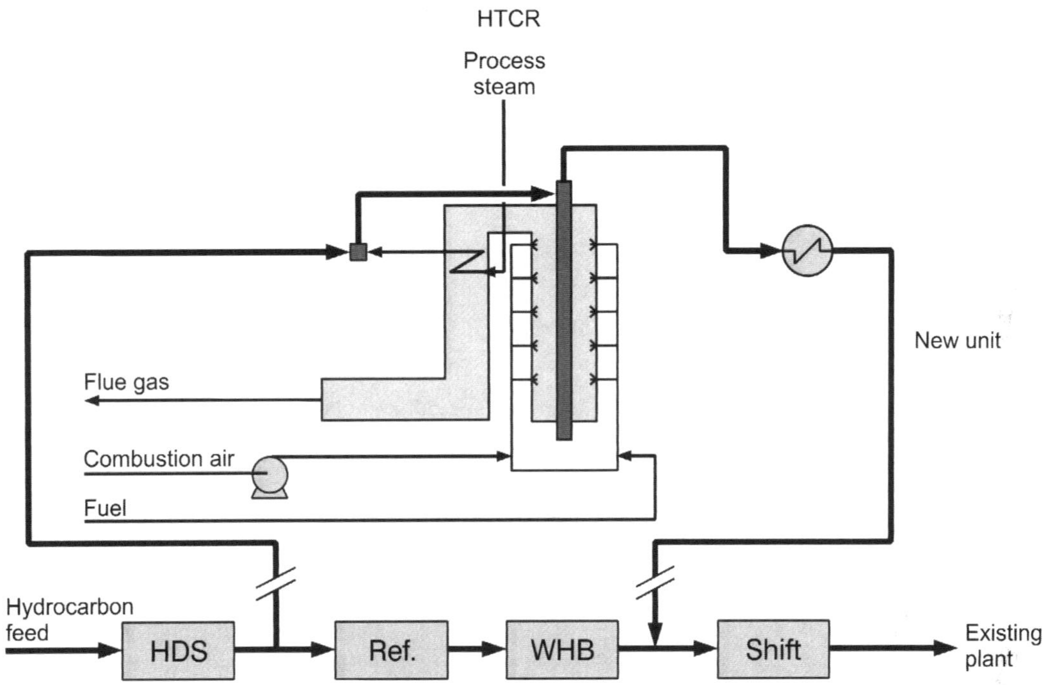

Figure 8. Schematic lay-out of an HTCR revamp solution

The on-site implementation of an HTCR revamp requires a minimum downtime as basically only three tie-ins need to be performed during a turnaround. The remaining construction can be performed during operation of the existing unit. An HTCR revamp can typically increase production capacity with up to 30%. Following a revamp the specific feed consumption will decrease as a result of the higher thermal efficiency of the HTCR.

Revamping with an HTER

Compared to the HTCR revamp an HTER revamp is completely integrated into the existing unit as illustrated in Figure 9, resulting in an economically attractive and less plot area requiring solution. The benefits of an HTER revamp are best utilized when the process gas exit temperature from the existing steam reformer is more than 1630°F. For an HTER revamp a longer downtime compared to the HTCR must be foreseen as the HTER is connected to the existing reformer.

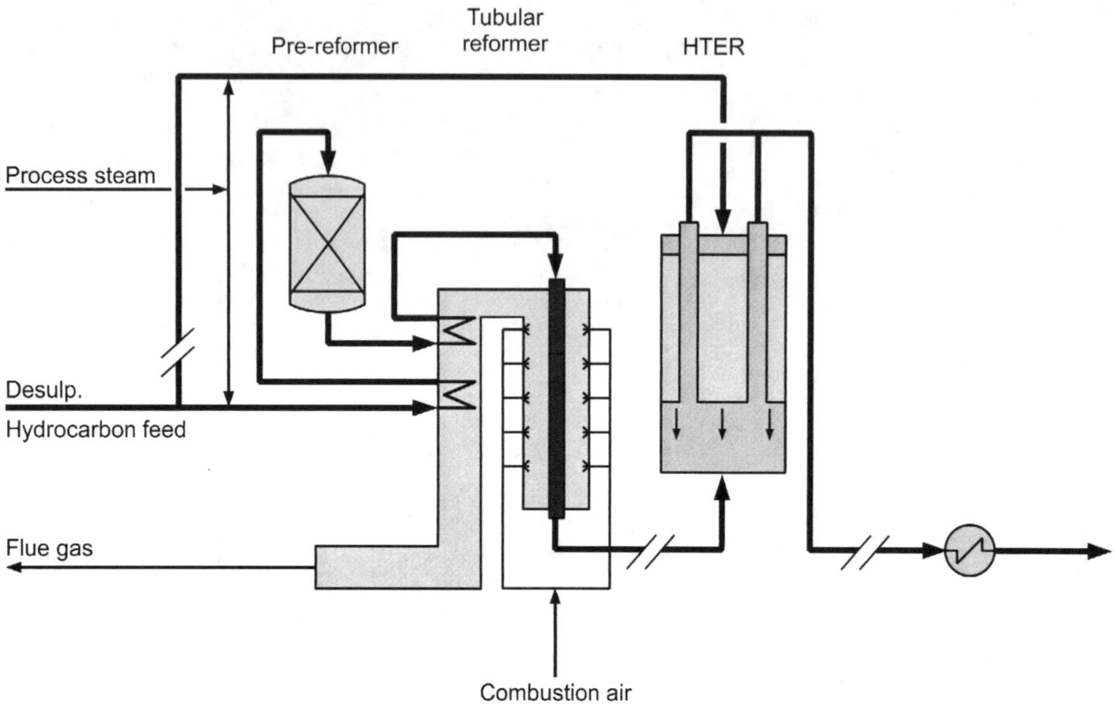

Figure 9. Schematic layout of an HTER revamp solution

An HTER revamp can typically increase production capacity by up to 30%. Integration of the highly efficient HTER results in a hydrogen plant with lower specific feed and fuel consumption and lower steam export as illustrated in the example below.

Case story 3 – increasing the capacity 25% by adding an HTER

A refinery in Asia operates a 65 MM SCFD, an 80 MM SCFD and a 110 MM SCFD hydrogen plant all designed by Haldor Topsoe. In 2009 the client wanted to increase hydrogen consumption by 27 MM SCFD. It was important that the solution could be implemented during a normal 4 weeks turnaround and that the additional plot area was minimized. The following revamp alternatives were considered:

- Adding an HTCR reformer to each of the smaller (and older) units
- Adding an HTER reformer to the 110 MM SCFD plant
- A new 27 MM SCFD grassroots plant

A feasibility study concluded that adding an HTER reformer could yield the 27 MM SCFD capacity increase at the minimum investment. The feasibility study also identified the required modifications/replacement of existing equipment necessary to achieve the desired capacity increase. Total installed cost (TIC) for the complete revamp including a new HTER and the required modifications/replacement of existing equipment amounted to approx. 60% of the TIC for a new separate hydrogen plant having the same capacity. The results in terms of energy consumption are show in Table 4 below:

110 MMSCFD plant revamp	Before revamp	After revamp
Capacity, MM SCFD	110	137 (+25%)
Feed and fuel consumption, BTU/SCF H_2	425	384 (-10%)
Steam export, BTU/SCF H_2	96	52 (-46%)
Net energy consumption, BTU/SCF H_2	329	332 (+1%)

Table 4. Consumption figures before and after an HTER revamp of a 110 MMSCFD hydrogen plant

The results show that the capacity could be increased by 25% with a 10% lower specific consumption of feed and fuel.

Figure 10. Installation of HTER during revamp

5 Conclusion

The innovative heat exchange reformer solutions developed and industrially demonstrated by Haldor Topsoe are optimal solutions for adding additional hydrogen capacity, either in the form of a stand-alone plant or as a revamp of an existing hydrogen unit. By utilizing the heat exchanger principle in hydrogen manufacture, it is possible to significantly reduce the consumption of hydrocarbon feedstock as compared to traditional radiation-based technology and reduce CO_2 emissions correspondingly. Furthermore heat exchange reforming is compact, characterized by easy operation and low investment cost. With the range of technologies presented in this paper, it is possible to provide tailor-made compact and energy efficient solutions for additional hydrogen production in all refineries.

References

1. Natural Gas Futures NYMEX price (Henry Hub) average for December 2009.
2. Dybkjaer, I., Winter Madsen, S., 'Compact Hydrogen Plants', Hydrocarbon Engineering, November 2004.
3. Gyde Thomsen, S., Han, P.A., Lock S., Werner, E., 'The First Industrial Experience with the Haldor Topsøe Exchange Reformer', AICHE, 2006.
4. Hedegaard Andersen, K., 'Hydrogen Agenda', Hydrocarbon Engineering, November 2006.
5. Winter Madsen, S., Olsson, H., 'Steam Reforming Solutions', Hydrocarbon Engineering, July 2007.

(131c) Contaminant Resistant High-Temperature Hydrogen Purification Membrane

Siu-Yue Tam[1], Andy Tsai[1,2]and Ying-Bing Jiang[2], (1)T3 Scientific LLC, Blaine, MN, (2)University of New Mexico, Albuquerque, NM

Hydrogen is a key building block of many chemical processes. Hydrogen can be produced in several ways, but economically the most important processes involve producing hydrogen from hydrocarbons. Fossil fuels including coal, oil and natural gas currently are the main sources of hydrocarbons for hydrogen production and biomass is an emerging one. Gas streams generated by hydrogen production from all of these hydrocarbon sources contain corrosive gases such as hydrogen sulfide (H_2S) and halides that attack metals on contact. Competitive hydrogen purification technologies such as pressure swing adsorption require high capital and operating cost. Membrane technology is the method-of-choice because of its low energy requirement, simplicity and a well established process. The most promising membrane materials for hydrogen purification are palladium and its alloys because of their high selectivity and permeability and high temperature tolerance. However, the critical downside is low tolerance for impurities, hydrogen sulfide, in particular. Conventional approaches to sulfur resistance improvement such as binary (e.g. Pd-Cu) or tertiary (e.g. Pd-Cu-Au) alloy often result in moderate gain at the expense of loss of permeance. The reason why these metals are active enough to transport hydrogen is also the reason why they react with reactive gases in the gas stream.

MembraGuard^{TM}technology is a different approach. MembraGuard^{TM}technology is a protective coating for hydrogen permeable metal membranes. MembraGuard^{TM}coating provides protection not only for sulfur componds, but for other reactive gases, such as halides and carbon monoxide. MembraGuard^{TM}coated membrane has high hydrogen selectivity and permeance under elevated temperature and yet is highly resistant to hydrogen sulfide. MembraGuard coated membrane was tested under elevated temperatures and pressures. Thermal resistance was demonstrated under repeated thermal cycles. Palladium membrane without any protective coating lost half of its permeability within 1 hour of contact with hydrogen sulfide. MembraGuard coated palladium membrane was demonstrated to resist hydrogen sulfide at high level with no degradation of performance for extended period. Mixed gas condition including H_2, CO, CO_2, H_2O and H_2S was also tested and MembraGuard coating was proved to be effective under the mixed gas condition.

(131d) Composite Catalytic-Permselective Membranes for H2 Purification

Daejin Kim[1], Elva Romero[1] and **Benjamin Wilhite**[2], (1)Artie McFerrin Department of Chemical Engineering, Texas A& M University, College Station, TX, (2)Artie McFerrin Dept. of Chemical Engineering, Texas A& M University, College Station, TX

Our research group is developing a novel composite catalytic-permselective membrane design approach, in which a permselective film (e.g., polymeric, metallic) is integrated with a porous catalytic film such that reaction and diffusion within the catalytic layer directly modifies the gas composition at the surface of the underlying permselective barrier [1]. In this talk, we will review general design rules for implementing this strategy for hydrogen purification from diesel reformate mixtures using water-gas-shift as the enhancing catalytic reaction. Detailed two-dimensional simulations and preliminary experimental results will be presented to demonstrate this strategy for (i) enhancing the permeance of palladium membranes, and (ii) enhancing the permselectivity of polymeric membranes.

[1] B.A. Wilhite. Composite Catalytic-Permselective Membranes: A Novel Strategy for Enhancing Selectivity and Permeation Rates via Reaction and Diffusion. *Ind. Eng. Chem. Res.*, **2011**, 50(17), 10185-10193.

(137a) Catalytic Strategies for Upgrading Pyrolysis Vapors

Shaolong Wan[1], Xiaohan Zhang[1], Daniel Resasco[2], Lance Lobban[1] and R. G. Mallinson[1], (1)Chemical, Biological, and Materials Engineering, University of Oklahoma, Norman, OK, (2)School of Chemical, Biological and Materials Engineering and Carbon Nanotube Technology Center, University of Oklahoma, Norman, OK

Pyrolysis vapors are catalytically treated directly downstream in a coupled reactor. Dual function metal/metaloxide catalysts are used to convert small oxygenates to larger carbon number molecules with less oxygen and, simultaneously, deoxygenate phenolics and transfer methoxy methyl to the ring. The ultimate goal is to maximize liquid product (and carbon) yield while minimizing hydrogen requirements.

(137c) Fischer-Tropsch Synthesis Using a Pilot Scale Fixed Bed Compact Heat Exchange Reactor

Steven J. Vallee, Zhijun Jia and Hani Gadalla, Chart Energy & Chemicals, La Crosse, WI

The Fischer-Tropsch reaction is a highly exothermic reaction (157 kJ/mol CO converted) used to convert syngas (a mixture of CO and H_2) to liquid hydrocarbons. Conventional fixed bed reactor technology suffers from heat transfer limitations. Operating conditions must be limited to prevent reactor runaway, resulting in lower productivity per unit reactor volume. Non-isothermal operating conditions make the process more difficult to start up and control and can lead to increased catalyst deactivation and unwanted methane side-product formation.

A Compact Heat Exchange Reactor (CHER) combines a reactor and compact heat exchanger into a single device, and provides a solution to the heat transfer limitations commonly observed for conventional fixed bed reactors. A pilot scale CHER was designed and constructed by Chart Energy & Chemicals in La Crosse WI, for the purpose of synthesis of diesel fuel via the Fischer-Tropsch (FT) reaction from syngas. Results are presented for a pilot scale CHER reactor for Fischer-Tropsch synthesis for >1 bpd liquid fuel production that is larger than our previously presented CHER results.

(137d) Conceptual Design and Integration of XTL Plants

Mohamed Noureldin[1], Buping Bao[1], Nimir O. Elbashir[2]and M.M. El-Halwagi[1], (1)Artie McFerrin Department of Chemical Engineering, Texas A& M University, College Station, TX, (2)Texas A& M University - Qatar

ABSTRACT

Liquid transportation fuels may be produced through gas-to-liquid (GTL), coal-to-liquid (CTL), and biomass-to-liquid "BTL" technologies which are collectively referred to as XTL. This work looks at the conceptual design and integration of XTL with focus on thermal conversion technologies. The aim of this work is to establish benchmarks for the process and innovative ways in which the various technologies can be integrated. The performance benchmarks (targets) are identified ahead of detailed design. They also allow current process performance to be gauged against the established targets and determine how far from the targets the process is operating. A base case is used to develop insights into factors which reduce the production rate of liquid fuels and overall product yield of XTL processes. Overall this work helps to identify areas for improvement and innovation including improved mass integration (e.g., feedstock utilization, reduced CO_2emissions, reduced hydrogen input) and better energy integration. A case study is used to illustrate the developed concepts.

(144a) Evaluation of Dispersion Models for LNG Siting Applications

Andrew Kohout, Office of Energy Projects, Federal Energy Regulatory Commission (FERC), Washington, DC

The federal safety standards for LNG facilities under 49 CFR Part 193 approves the use of specific dispersion models for performing vapor-gas dispersion exclusion zones. In addition, provisions allow for the use of alternative models if approved by PHMSA.

In 2010, PHMSA, in consultation with FERC, issued Advisory Bulletin ADB-10-07 to provide guidance on obtaining approval of alternative models under Part 193. The guidance is based on the Model Evaluation Protocol (MEP) described in the Fire Protection Research Foundation Reports, Evaluating Vapor Dispersion Models for Safety Analysis of LNG Facilities (2007) and Guide to the LNG Model Validation Database (2009, 2010). This method has been applied to several dispersion models to establish limitations and safety margins, which would allow it to be used in Part 193 exclusion zone calculations.

This presentation examines the results of the reviews conducted by PHMSA and FERC, including the basis for the limitations and safety margins.

AIChE 2012 Spring Conference

Topical 6: 12th Topical Conference on Gas Utilization – LNG Plant Safety and Protection I

Modeling Dynamic Heat Transfer in Piping Systems using an Enhanced ASTM C-680 Algorithm

Debby Sielegar, P.E.
Process Engineering Specialist
Bechtel Oil, Gas & Chemicals, Inc.

David Messersmith, P.E.
LNG Technology Manager
Bechtel Oil, Gas & Chemicals, Inc.

Abstract

ASTM C-680 methodology is routinely used to model heat transfer in steady state systems. But not all applications can be treated as steady state. Some applications require modeling of systems with mixed phase streams in condensing or vaporizing services that are dynamically interacting with each other.

One of the common dynamic heat transfer problems is the relief of large quantities of low temperature fluid to the dry gas flare system followed by the immediate relief of a much warmer vapor close to its saturation temperature. The warm vapor comes in contact with the cold pipe surfaces and loses heat to the piping causing condensation on the inside surfaces of the pipe. The fluid thermally contracts which may result in air ingress into the flare header that can potentially create a flammable atmosphere. It has been typical to purge the flare header and raise the temperature of the piping following the relief of low temperature fluid to prevent condensation along the walls of the header. ASTM C-680 methodology is typically used to determine the required quantity of purge gas, the duration of purging, and the metal piping temperature for each pipe segment at every time step. Other heat transfer applications include the determination of insulation requirement, sizing of condensed fluid containment, natural cooling as a cost saving approach, etc.

ASTM C-680 algorithm allows the user to iteratively calculate the temperature of all the piping layers (including pipe wall, insulation etc.), which in turn is used to calculate the inside and outside heat transfer coefficients and thermal conductivity. In this paper, a few case studies are presented using an enhanced ASTM C-680 algorithm where the piping system is divided into multiple segments for accuracy, heat transfer is time-dependent, and the fluid properties are continually updated to reflect the impact of pressure drop.

Terminology

Nu Nusselt number, dimensionless
Pr Prandtl number, dimensionless
Re Reynolds number, dimensionless
A heat transfer area, ft^2
D_i inside pipe diameter, ft
D_o outside pipe diameter (including insulation thickness, if present), ft
C_p fluid heat capacity, Btu/lb.°F
g gravitational acceleration, 32.2 ft/s^2
h_{conv} convective heat transfer coefficient, Btu/hr.ft^2.°F
h_i inside heat transfer coefficient, Btu/hr.ft^2.°F
h_o outside heat transfer coefficient, Btu/hr.ft^2.°F
h_r radiation heat transfer coefficient, Btu/hr.ft^2.°F
k fluid thermal conductivity, Btu/hr.ft.°F
k_m metal thermal conductivity, Btu-in/hr.ft^2.°F
k_{ins} insulation thermal conductivity, Btu-in/hr.ft^2.°F
L pipe length, ft
LMTD log mean temperature difference, °F
m mass flowrate, lb/hr
MW molecular weight
N number of pipe segments
P_{in} inlet fluid pressure, psig
P_{out} outlet fluid pressure, psig
ΔP pressure drop, psi
Q heat gain or heat loss, Btu/hr
r average radius, ft
r_i inside radius, ft
r_o outside radius (including insulation thickness, if present), ft
r_s outside metal pipe radius in the presence of insulation, ft
t time, s or hr
time step
t_m metal thickness, in
t_{ins} insulation thickness, in
T_{in} inlet fluid temperature, °F
T_{out} outlet fluid temperature, °F
T_b bulk fluid temperature, °F
average of inlet and outlet fluid temperatures $(T_{in} + T_{out})/2$
T_i inside surface temperature, °F
T_o outside surface temperature, °F
T_s interim temperature in the presence of insulation layer, outside metal or inside insulation surface temperature, °F
T_m metal temperature, °F
average of inside and outside surface temperatures $(T_i + T_o)/2$, or average of inside and interim temperature $(T_i + T_s)/2$ in the presence of insulation layer
T_{ins} insulation temperature, °F
average of interim and outside temperatures $(T_s + T_o)/2$ in the presence of insulation layer
T_∞ ambient temperature, °F
T_{avg} air film temperature, °F
average of outside and ambient temperatures $(T_o + T_\infty)/2$
U overall heat transfer coefficient, Btu/hr.ft^2.°F
V volume, ft^3
Wind wind velocity, mph
ρ fluid density, lb/ft^3
ρ_m metal density, lb/ft^3
μ fluid viscosity, cP or lb/hr-ft
μ_i fluid viscosity at inside surface temperature (T_i), cP or lb/hr-ft
λ latent heat of vaporization, Btu/lb
ε emissivity
σ Stefan-Boltzmann constant, 0.1713×10^{-8} Btu/hr.ft^2.°R^4

1. Introduction

1.1 Purpose

Modifying the ASTM C-680 algorithm as demonstrated in this paper is considered a simple approach to solve various heat transfer applications. As heat transfer calculation is iterative, algorithm for computer program is essential. Although some software programs provide a piping heat transfer calculation feature, options for selecting heat transfer correlations are limited. There are many recently published literatures presenting modified and proven heat transfer correlation(s) for a particular fluid and/or application.

1.2 Method of Calculation

ASTM C-680 provides an algorithm and computational methodologies to predict heat loss or heat gain in a one dimensional steady state system. The methodologies are based on standard heat transfer calculations that are commonly found in many heat transfer literatures. ASTM C-680 algorithm flowchart is shown in Figure 2. Although the algorithm may be used for both heating (heat gain) and cooling (heat loss) applications, this paper focuses on heat loss in piping systems only. Fluid flow, heat flow and temperature profile in a pipe segment are illustrated in Figure 1.

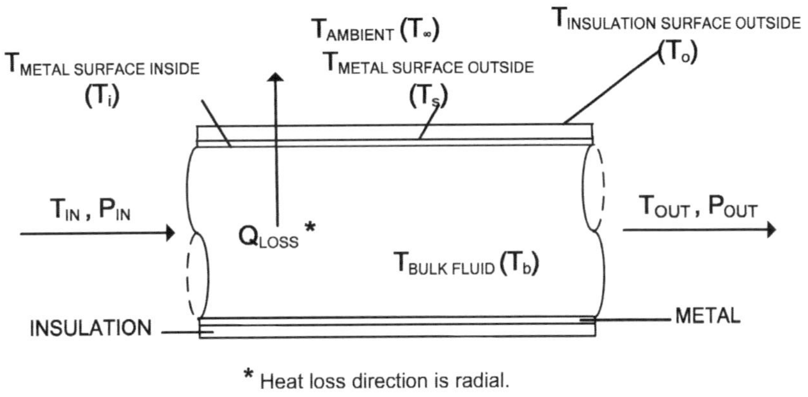

Figure 1. Heat Flow and Temperature Profile in a Piping System when $T_{AMBIENT} < T_{FLUID}$

Surface temperatures are important parameters for calculating the individual heat transfer coefficients. Individual heat transfer coefficients typically include inside heat transfer coefficient of the fluid (h_i), thermal conductivity of the metal (k_m), thermal conductivity of the insulation (k_{ins}), and outside heat transfer coefficient (h_o). Inside heat transfer coefficient (h_i) in a piping system is generally calculated based on forced convection heat transfer coefficient correlations shown in equations (1) through (4). As physical properties are directly related to temperature, the inside heat transfer coefficient is thus temperature dependent.

Seider-Tate [5]

$$Nu = 0.023\ Re^{0.8}\ Pr^{1/3}\left(\frac{\mu}{\mu_i}\right)^{0.14} \quad (1)$$

$$0.7 < Pr < 16700,\ \ Re > 10^4,\ \ \frac{L}{D_i} > 60$$

Dittus-Boelter [5]

$$Nu = 0.023\ Re^{0.8}\ Pr^{n} \quad (2)$$

where n = 0.3 for heat loss and n = 0.4 for heat gain

$$0.7 < Pr < 120,\ \ Re > 10^4,\ \ \frac{L}{D_i} > 60$$

Nusselt Film-type Condensation [7] (*)

Horizontal: $$Nu = 0.73\left(\frac{D_i^{\,3}\rho^2 g\,\lambda}{k\,\mu\,(T_b - T_i)}\right)^{1/4} \tag{3}$$

Vertical: $$Nu = 0.943\left(\frac{L^3\rho^2 g\,\lambda}{k\,\mu\,(T_b - T_i)}\right)^{1/4} \tag{4}$$

(*) Condensed liquid properties should be used. Reynolds number is based on the condensing film characteristics and is less than 2,100.

Metal and insulation thermal conductivities are also as a function of temperature.

[6] Perlite $k_{ins}(T_{ins}) = 0.4030 + 6.38E\text{-}4\ T_{ins} - 3.56E\text{-}7\ T_{ins}^{\ 2} + 3.53E\text{-}10\ T_{ins}^{\ 3}$ (5)

[6] Mineral wool $k_{ins}(T_{ins}) = 0.228 + 3.72E\text{-}4\ T_{ins} - 6.0E\text{-}7\ T_{ins}^{\ 2}$ (6)

The outside heat transfer coefficient is a function of the convective and radiation heat transfer coefficients as shown in equations (7) through (9).

Outside heat transfer coefficient [1] $$h_o = h_{conv} + h_r \tag{7}$$

Convective [1] * $$h_{conv} = 1.016\left(\frac{1}{12D_o}\right)^{0.2}\left(\frac{1}{T_{avg}}\right)^{0.181}(T_o - T_\infty)^{0.266}\sqrt{1 + 1.227\,Wind} \tag{8}$$

Radiation [1] ** $$h_r = \frac{\sigma\,\varepsilon\,(T_o^{\,4} - T_\infty^{\,4})}{T_o - T_\infty} \tag{9}$$

* For piping diameter greater than 2 ft or 24 in, use $D_o = 2$.

** T_o and T_∞ for calculating radiation heat transfer coefficients are to be in °R unit.

Once individual heat transfer coefficients are determined, overall heat transfer coefficient and heat transferred are then determined using equations (10) and (11), respectively.

Overall heat transfer coefficient $$\frac{1}{Ur} = \frac{1}{r_i h_i} + \frac{ln\left(\frac{r_s}{r_i}\right)}{k_m} + \frac{ln\left(\frac{r_o}{r_s}\right)}{k_{ins}} + \frac{1}{r_o h_o} \tag{10}$$

Heat loss $$Q = U\,(2\pi r L)\,LMTD \quad \text{where} \quad LMTD = \frac{(T_{in} - T_\infty) - (T_{out} - T_\infty)}{ln\left(\frac{T_{in} - T_\infty}{T_{out} - T_\infty}\right)} \tag{11}$$

In piping systems, the fluid outlet temperature can be calculated using equation (12).

$$Q = m\,C_p\,(T_{in} - T_{out}) + m\,\lambda \tag{12}$$

The surface temperatures can be back-calculated using the individual heat transfer coefficients and heat loss rate.

Inside metal surface temperature $$T_i = T_b - \frac{Q}{A\,h_i} \tag{13}$$

Outside metal surface temperature $$T_s = T_i - \frac{Q}{A\left(\frac{k_m}{t_m}\right)} \quad \text{or} \quad T_s = T_\infty + \frac{Q}{A\left(\frac{k_{ins}}{t_{ins}}\right)} \tag{14}$$

Outside insulation surface temperature $$T_o = T_\infty + \frac{Q}{A\,h_o} \tag{15}$$

The calculations are iteratively repeated until the calculated surface temperatures from equations (13) through (15) match closely with the inputted surface temperatures.

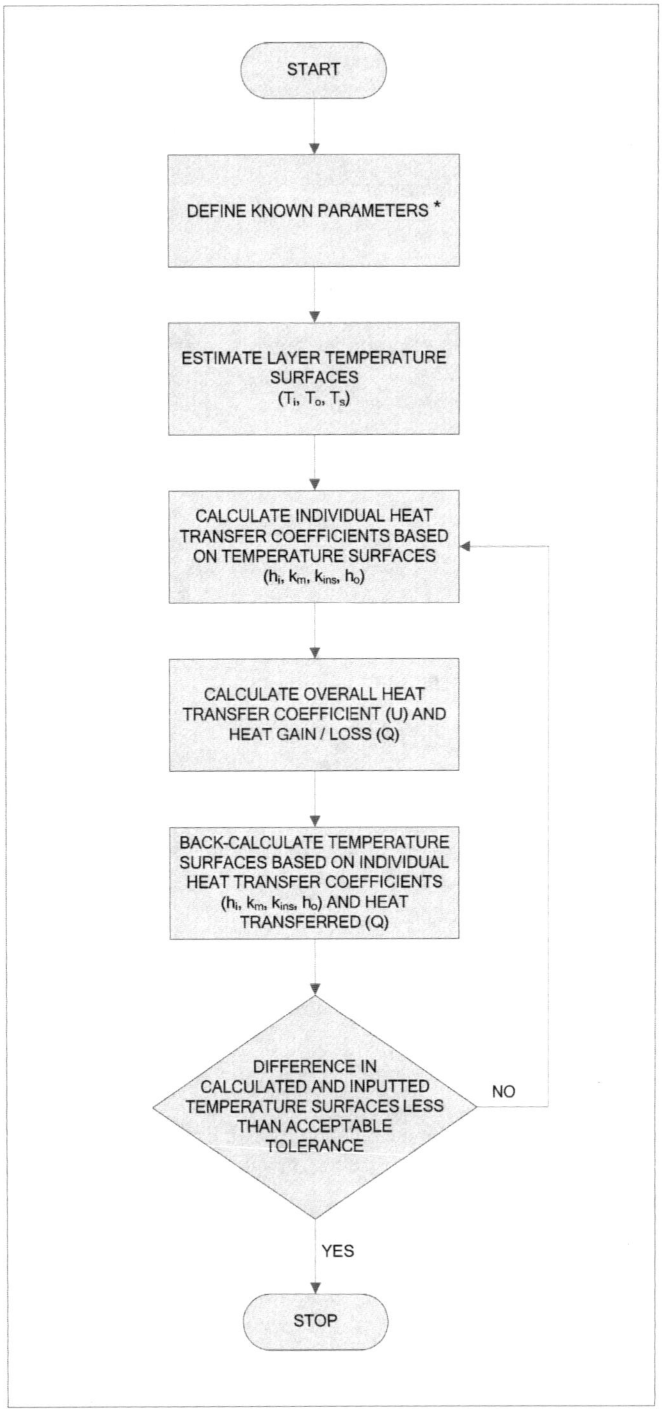

* Parameters may include initial process fluid conditions and properties, ambient temperature, system dimensions, etc.

Figure 2. ASTM C-680 Algorithm Flowchart [1]

2. Case Study Applications

In this paper, the following two case studies are presented:

1. Enhanced algorithm for a time-dependent application
2. Enhanced algorithm to quantify impact of pressure drop in a two-phase stream

2.1 Enhanced Algorithm for a Time-Dependent Application

Case study problem statement (1):

Assuming that a large quantity of low temperature methane liquid relief occurs for a considerably long period of time, the flare header metal piping temperature will drop and eventually reach the fluid temperature. If this relief event is followed by a much warmer vapor relief such as propane, the saturated warm vapor will come in contact with the cold pipe surfaces and lose heat to the piping causing condensation. This creates a safety concern since the warm vapor can thermally contract and create vacuum, resulting in air ingress into the flare header and potential for flammable atmosphere. Based on relief analysis for an LNG plant, the following parameters are adopted for this case study.

Flare header length	5,000 ft
Outside / inside pipe diameter	54 in / 53.25 in
Flare header material	Stainless Steel
Header metal density	488 lb/ft^3
Header metal heat capacity	0.11 Btu/lb.°F
Low temperature methane liquid relief or initial T_m	-259.6°F
Saturated propane vapor relief parameters:	
Relief rate	2.44×10^6 lb/hr or 503 MMSCFD
Latent heat of vaporization	183.3 Btu/lb
Inside heat transfer coefficient correlation	Nusselt film-type condensation
Defrost gas parameters:	
Inlet temperature	125°F
Molecular Weight	16.80 lb/lbmole
Inside heat transfer coefficient correlation	Dittus-Boelter

2.1.1 Displacement Gas

One of the methods to protect the flare header from vacuum conditions is to provide a displacement gas, such as fuel gas, immediately following propane vapor relief. To quantify the required displacement rate, the following assumptions are established:

- Ambient air typically contributes to heating of the flare header. For simplicity as well as conservative design purposes, the ambient factor is not considered. Heat transfer occurs only between the metal film and the bulk fluid.
- As time lag between relief of saturated propane vapor and release of displacement gas can vary, the displacement gas rate required is determined based on the maximum heat loss rate along the pipe. The maximum heat loss rate occurs when the first segmental entry of propane vapor relief either is completely condensed or reaches the end of last pipe segment. Note that heat loss and the corresponding condensation rates decrease as a function of time, therefore using the first segmental entry of propane vapor relief is a conservative approach.

Refer to Table 1 for the estimated heat loss rates and the corresponding accumulated condensed propane rates. In general, accuracy of the results can be improved with larger number of segments or in other words, shorter segment length. For complete condensation application, it is essential that the cumulative quantity of condensed fluid at the last condensing segment is equal to the total mass of fluid entering the flare header. In this case study, the optimum number of segments is determined to be 210 segments. As the quantity of condensed fluid increases, the volumetric rate and the velocity decrease, resulting in increased time for the fluid to travel to the next segment.

Table 1. Heat Loss and Accumulated Condensation Rates as Functions of Time and Segment

Heat Loss (MMBtu/hr)								Accumulated Condensation Rate x 10^6 (lb/hr)					
Time (s)	Segment						Sum (MMBtu/hr)	Segment					
	1	10	…	75	85	210		1	10	…	75	85	210
0.1	5.25						5.25	0.03					
1.0	5.21	5.25	…				57.6	0.03	0.29	…			
5.0	5.09	5.11	…				227	0.03	0.28	…			
…	…	…	…	…			…	…	…	…	…		
15	4.97	4.97	…	5.23	…		382	0.03	0.27	…	2.15	…	
20	4.95	4.95	…	5.09	…		405	0.03	0.27	…	2.13	…	
32	4.94	4.94	…	4.98	5.25	--	**422**	0.03	0.27	…	2.11	**2.44**	--

The contraction rate (dV/dt) of saturated propane vapor relief is estimated to be 15.3 million ft^3/hr, which is derived from the total maximum heat loss rate, propane latent heat of vaporization, and difference in propane liquid and vapor densities. The required quantity of displacement gas varies depending on the gas temperature and molecular weight. Based on fuel gas density of 0.055 lb/ft^3 and MW of 16.80 lb/lbmole, 455 MMSCFD of fuel gas is required for this example. Note that this requirement may be larger than the maximum operating fuel gas demand in the plant.

The rapid contraction rate of propane requires the fuel gas system to be able to supply large quantities of displacement gas, which may not lend to an economical design. In addition, the probability of occurrence of cold and warm consecutive reliefs is considerably low.

2.1.2 Purging Gas

Another method to prevent vacuum formation due to cold and warm fluid consecutive relief is to purge the flare header with high temperature gas following the relief of low temperature fluid. The purge rate and time are important parameters to be established while engineering the system.

In this case study, defrost gas at 125°F was assumed to provide purge flow until the end segment of the header approached -50°F. The target temperature of -50°F is based on providing a reasonable margin above the dew point of propane vapor. To solve this problem, the flare header was divided into multiple pipe segments, and all the header segments were assumed to be at the methane liquid relief temperature of -259.6°F. The effect of ambient heat transfer was not considered and heat transfer was assumed to occur only between the metal film and the bulk fluid.

Refer to Figure 3 for the algorithm applied to produce the temperature profiles and estimate the purging time at each defrost gas flowrate run. As an example, piping metal temperature and defrost gas outlet temperature profiles for defrost gas flow of 120,000 lb/hr are shown in Tables 2 and 3, respectively. It can be observed from Tables 2 and 3 that the defrost gas continues to lose temperature to its surroundings while the piping metal temperature is gradually raised until the last segment reaches the target temperature.

Time step is defined as the time required for the purge gas to flow from one pipe segment to another. As the gas temperature drops, the density of the gas increases, thus the velocity decreases and the time step increases. For simplification purpose, time step may be assumed to be constant based on initial, final or average of the expected purge gas temperature, considering that error introduced may only occur at the first pass of the purge gas. Once calculations for first pass are completed, determination of time step is no longer critical. In addition, the number of segments should be further optimized such that increasing the number of segments does not alter the results significantly.

Table 2. Flare Header Piping Metal Temperature Profile

Time	Pipe Segment									
Step	1	2	3	4	5	6	7	8	9	10
0	-259.6	-259.6	-259.6	-259.6	-259.6	-259.6	-259.6	-259.6	-259.6	-259.6
1	-251.6	-259.6	-259.6	-259.6	-259.6	-259.6	-259.6	-259.6	-259.6	-259.6
2	-243.8	-256.2	-259.6	-259.6	-259.6	-259.6	-259.6	-259.6	-259.6	-259.6
3	-236.1	-252.8	-256.1	-259.6	-259.6	-259.6	-259.6	-259.6	-259.6	-259.6
4	-228.5	-249.4	-252.7	-257.9	-259.6	-259.6	-259.6	-259.6	-259.6	-259.6
5	-221.2	-246.0	-249.2	-256.2	-258.7	-259.6	-259.6	-259.6	-259.6	-259.6
6	-213.9	-242.5	-245.6	-254.5	-257.8	-259.1	-259.6	-259.6	-259.6	-259.6
7	-206.8	-239.0	-242.1	-252.8	-256.9	-258.6	-259.3	-259.6	-259.6	-259.6
8	-199.8	-235.5	-238.5	-251.0	-255.9	-258.1	-259.0	-259.4	-259.6	-259.6
9	-193.0	-232.0	-235.0	-249.1	-254.9	-257.5	-258.7	-259.2	-259.5	-259.6
10	-186.3	-228.5	-231.4	-247.3	-253.9	-256.9	-258.3	-259.0	-259.4	-259.5
↓	...	...	...	...	...	...	...	...	...	...
	...	...	...	...	...	...	...	...	...	...
	...	...	...	...	...	...	...	...	...	...
261	123.9	120.1	120.1	111.8	97.6	77.0	50.1	18.4	-16.0	-50.9
262	123.9	120.2	120.2	112.0	98.1	77.6	51.0	19.5	-14.8	-49.7

Table 3. Defrost Gas Outlet Temperature Profile

Time	Pipe Segment									
Step	1	2	3	4	5	6	7	8	9	10
0										
1	-75.3									
2	-71.3	-164.2								
3	-67.3	-160.8	-162.3							
4	-63.4	-157.4	-158.9	-205.7						
5	-59.5	-154.0	-155.5	-203.2	-228.2					
6	-55.8	-150.6	-152.1	-200.8	-226.5	-240.8				
7	-52.1	-147.2	-148.7	-198.4	-224.8	-239.6	-248.1			
8	-48.4	-143.9	-145.3	-195.9	-223.1	-238.4	-247.3	-252.5		
9	-44.9	-140.5	-142.0	-193.5	-221.4	-237.3	-246.5	-251.9	-255.2	
10	-41.4	-137.2	-138.7	-191.0	-219.7	-236.1	-245.7	-251.4	-254.8	-256.8
↓	...	...	...	...	...	...	...	...	...	...
	...	...	...	...	...	...	...	...	...	...
	...	...	...	...	...	...	...	...	...	...
261	124.4	122.0	122.0	116.5	106.2	90.3	68.4	41.3	10.4	-22.4
262	124.4	122.1	122.1	116.6	106.5	90.8	69.1	42.2	11.4	-21.2

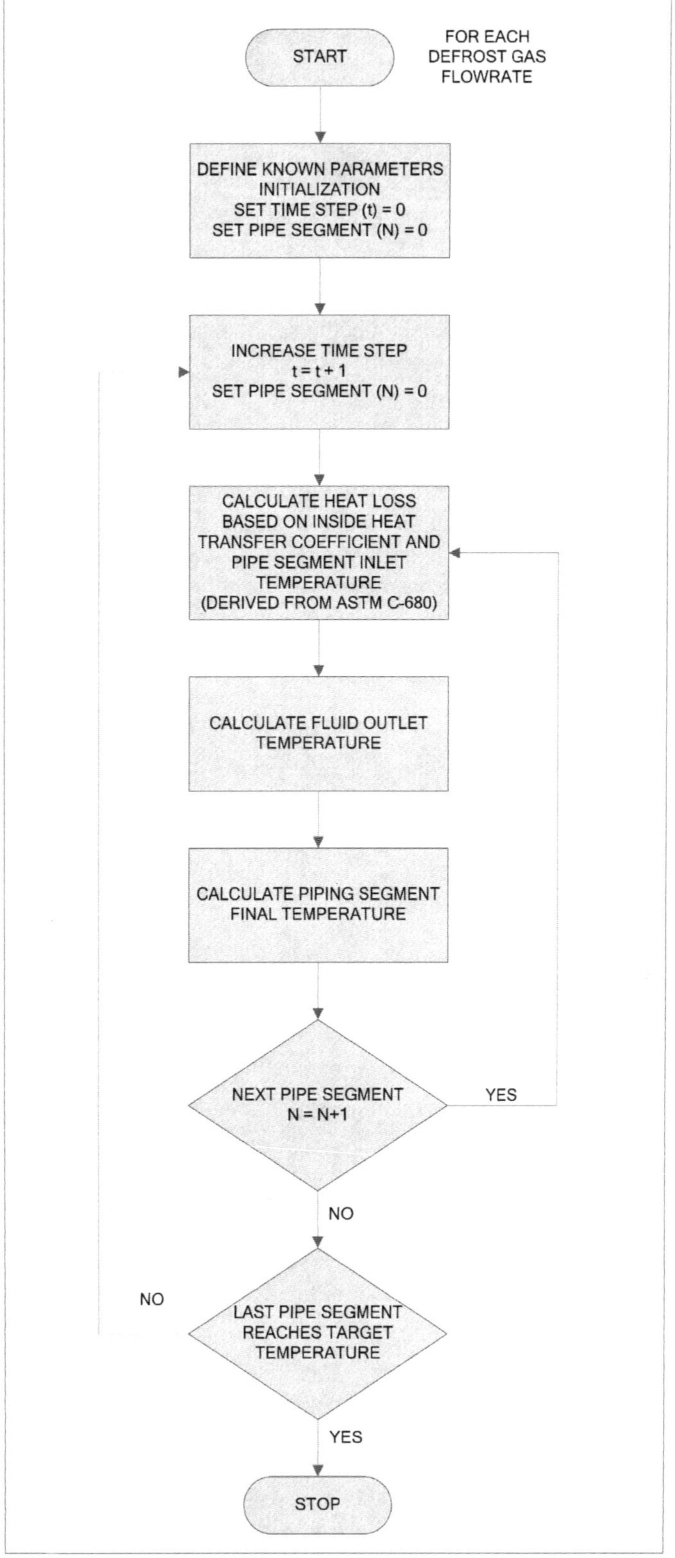

Figure 3. Flare Header Case Study with Modified Algorithm

Results from the algorithm and/or calculation methodology are summarized in Table 4.

Table 4. Purging Time and the Corresponding Purge Gas Flowrate

Gas flowrate, lb/hr	50,000	70,000	100,000	110,000	120,000	250,000	500,000
Gas flowrate, MMSCFD	27.1	37.9	54.2	59.6	65.1	135.5	271
Purging time, min	283.1	203.8	143.2	130.7	119.8	58.4	29.7

2.1.3 Discussions

Based on results of the two methods presented in the previous sections, the following conclusions can be drawn.

- Required displacement gas quantity is a function of the saturated propane vapor relief rate. At higher relief rates, propane vapor may completely condense further down the line. This increases the total heat loss rate and consequently the required displacement gas.
- Displacement gas may be introduced as early as possible to reduce the required quantity and to reduce vacuum formation potential.
- Required purging gas quantity is as a function of flare header length and is independent of the relief rates.
- Purging the flare header certainly lends to an economic design. However, a risk analysis must be performed if this method is selected. At higher purge rates, lesser time is required to warm up the flare header. The temperature profiles generated by the computer model would help with further analysis, especially for events when propane relief starts to occur even before the flare header is fully warmed to its target temperature.

2.2 Enhanced Algorithm to Quantify Impact of Pressure Drop in a Two-Phase Stream

Case study problem statement (2):

Acid gas exiting the amine absorber unit is saturated with water. Acid gas is sent to the H_2S scavenger prior to thermal oxidation. The reduction in pressure and temperature along the pipe causes water vapor condensation. The objective of this case study is to define the containment size required to capture the condensed fluids. The following parameters are adopted for this case study.

Flowrate	100,000 lb/hr
Composition	74.7 mol% CO2, 25.0 mol% H2O, 0.3 mol% hydrocarbon light ends with traces of H_2S and benzene
Inlet pressure and temperature	15 psig and 179.44^{0}F
Length of pipe	1,000 ft
Outside / inside pipe diameter	18 in / 17.25 in (uninsulated)
Pipe roughness	0.0002 ft
Wind speed	10 mph
Ambient temperature	40^{0}F
Inside heat transfer coefficient correlation	Nusselt film-type condensation

The modified ASTM C-680 algorithm for this case study is presented on Figure 4. For comparison purposes, results from a steady state model are compared with pseudo steady state results in Table 5.

Table 5. Comparison Methods for Cumulative Condensed Fluid at End of Pipe Segment

		End of Pipe Segment							
	Unit	1	2	3	. . .	8	9	10	% difference
Steady State	lb/hr	--	--	--	--	--	--	**2,486**	52.14
No ΔP correction	lb/hr	176.7	352.2	526.6	. . .	1,382	1,550	**1,717**	5.05 (Note 1)
ΔP correction	lb/hr	170.2	338.9	506.0	. . .	1,319	1,477	**1,634**	--
HTFS Hysys (Note 2)	lb/hr	171.4	341.3	509.9	. . .	1,332	1,493	**1,652**	1.08

Notes:

1. The percentage difference between pseudo steady state pressure drop and non-pressure drop corrected results varies with each case. Therefore, the percentage difference shown in Table 5 should not be used as an underlying basis for the accuracy of the two methods.
2. HTFS, the inside heat transfer coefficient correlation available in HYSYS for two-phase applications, was run for this case study.

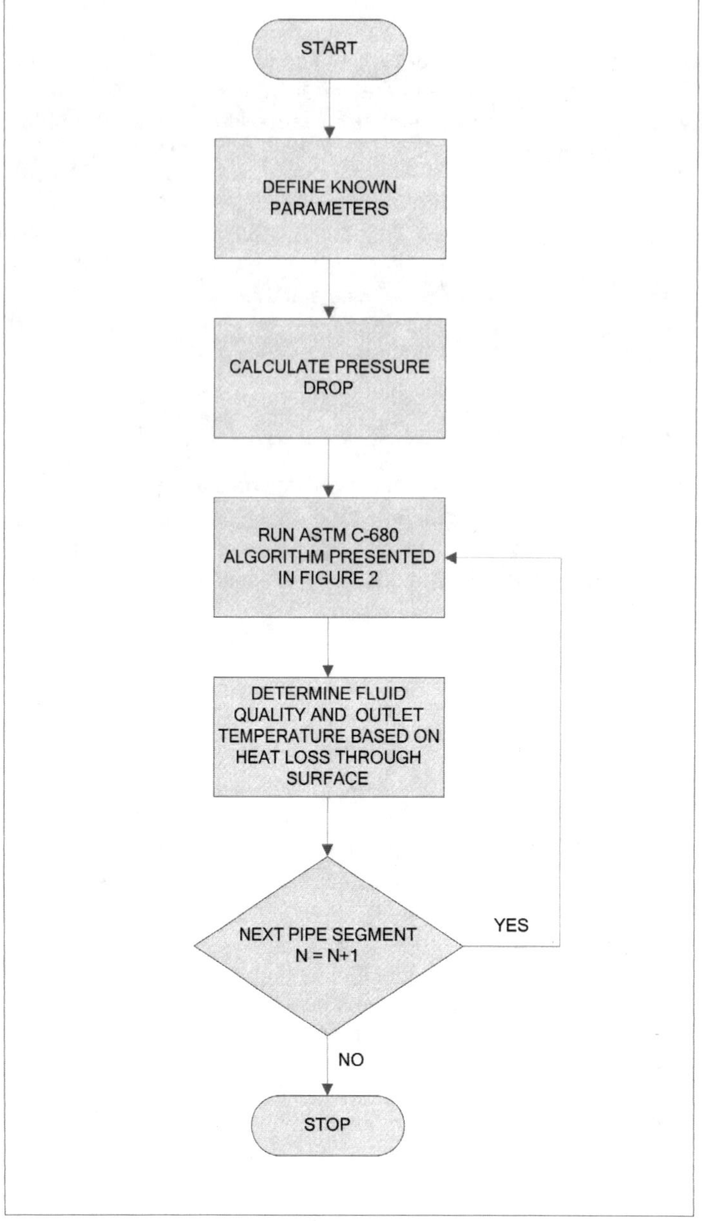

Figure 4. Modified Algorithm with Pressure Drop Factor

3. Conclusions

Based on the two case studies outlined above, it is demonstrated that ASTM C-680 algorithm can be modified to solve iterative heat transfer calculations in a piping system for various applications, such as for a time-dependent application and for a system requiring continuous update to its equilibrium state.

4. Acknowledgement

The authors would like to thank Vinod Rajkumar for his continuous support and constructive inputs. The authors would also like to thank Terry Timmerman for reviewing the paper.

5. References

1. ASTM C680, Standard Practice for Estimate of the Heat Gain or Loss and the Surface Temperatures of Insulated Flat, Cylindrical, and Spherical Systems by Use of Computer Programs, 1971 - 2010
2. Bornt, B., Spreadsheets for Heat Loss Rates and Temperatures, 1995
3. Goudar, C.T., Comparison of the Iterative Approximations of the Colebrook-White Equation, 2008
4. Konak, A.R., Calculate the Heat Loss from Pipes, 1993
5. Lindeburg, M. R., Chemical Engineering Reference Manual, 2004
6. Nayyar, M.L., Piping Handboook, 1999
7. Perry, R. H., Perry's Chemical Engineering Handbook, 2007

6. Biography

Debby Sielegar, P.E. is a licensed engineer in the state of Texas. She completed her bachelor's degree in Chemical Engineering from Purdue University in 2007 prior to joining Bechtel OG&C, Inc. While working for Bechtel as a process engineer, she pursued her Master's degree in Chemical Engineering from University of Houston and completed the degree in 2011. In Bechtel, she has satisfactorily completed assignments on various projects such as lube oil refinery, LNG liquefaction and offshore crude production. In addition, she is also a certified six sigma yellow belt.

David B Messersmith, P.E. is Manager of the LNG Technology Group – for Bechtel Corporation in Houston, Texas. He has served in various lead roles in the LNG Projects for the 15 of the last 21 years. This work included the Atlantic LNG Project conceptual design through start-up as well as many other LNG studies, FEEDs and projects. Experience includes 11 years with Kellogg and 21 years with Bechtel in various LNG and Ethylene assignments. His LNG experience includes everything from conceptual design through commissioning, start-up and testing. He holds a Bachelor's of Science degree in Chemical Engineering from Carnegie-Mellon University and is a registered Professional Engineer in the state of Texas.

Theoretical Study of Expansion Foam Application on LNG Pool Using Computational Fluid Dynamics

Bin Zhang, Xinrui Li, Ammar Alkhawaldeh and M. Sam Mannan, Mary Kay O'Connor Process Safety Center, Artie McFerrin Department of Chemical Engineering, Texas A&M Univerisity, College Station, TX

LNG facilities could harm the personnel on site and jeopardize local communities nearby, due to the flammability and dense gas behaviors of LNG. Nowadays, the growing concern with chemical process safety leads to the timely establishment of regulations and standards regarding to consequence mitigation measures, for instance, NFPA 59 A requires that LNG storage tanks should have a dike or impoundment around to constrain the LNG spill and form a LNG pool, and the expansion foam has been one of the proposed safety provisions to minimize the consequence of LNG spill on the ground. However, because of the high cost and potential risk to people of performing field tests to evaluate effectiveness of foam application on LNG pool, just a few of experiments have been conducted, whose results vary as a result of uncertainties in experimental conditions.

In this research, computational fluid dynamics modeling was employed to model LNG spill scenarios on the ground with a dike, and mainly focused on the effect of expansion foam on the LNG vapor dispersion. The effectiveness was evaluated by studying several key parameters in the LNG spill and dispersion scenario. The simulation results were compared with medium-scale LNG spill field test results conducted by Mary Kay O'Connor Process Safety Center (MKOPSC) for validation. Moreover, a sensitivity analysis was conducted to a number of essential parameters of foam application. The outcomes of the analysis were used to guide the design of foam mitigation system and application method, which hopefully could be integrated into relevant regulations and standards for the sake of safer practice in the industry.

Guidelines for Jetting and Flashing LNG Vapor Exclusion Zone Analysis

Delmar "Trey" Morrison, Ph.D., P.E.
Exponent, Inc.
4580 Weaver Parkway, Suite 100, Warrenville, Illinois 60555
tmorrison@exponent.com

Ryan J. Hart, Ph.D.
Exponent, Inc.
4580 Weaver Parkway, Suite 100, Warrenville, Illinois 60555
rhart@exponent.com

Harri Kytömaa, Ph.D., P.E.
Exponent, Inc.
9 Strathmore Road, Natick, Massachusetts 01760
hkytomaa@exponent.com

Prepared for Presentation at
2012 AIChE Spring Meeting
Houston, TX
April 1-4, 2012

UNPUBLISHED

Guidelines for Jetting and Flashing LNG Vapor Exclusion Zone Analysis

Delmar "Trey" Morrison, Ryan J. Hart, and Harri Kytomaa
Exponent, Inc.
E-mail: tmorrison@exponent.com

Keywords: LNG Jetting and Flashing; LNG Release

Abstract

Liquefied Natural Gas (LNG) has been a growing energy resource in the United States, spurring significant import and storage infrastructure development over the past decade. The infrastructure for liquefaction and export has also been developing as a result of recent shale gas finds in North America. In the U.S., the construction and operation of LNG facilities are regulated by the Pipeline and Hazardous Materials Safety Administration (PHMSA) of the U.S. Department of Transportation (DOT), and the regulations are administered by the Federal Energy Regulatory Commission (FERC). As part of the regulatory process (Title 49, Code of Federal Regulations, Part 193.2059), regasification and liquefaction facility owners are required to perform prescriptive calculations to estimate the potential consequences of major releases of LNG prior to permitting. One of the criteria is that the ½-LEL boundary of resulting flammable vapor clouds cannot cross the property boundaries (exclusion zone). The consequence analysis requirements have recently been broadened to include new spill scenarios in addition to the previously required LNG spills into impoundment or "sumps" and to allow the use of alternative computational methods to calculate the ½-LEL exclusion zones for these scenarios. Specifically, in 2010 FERC issued a written interpretation regarding the flammable vapor exclusion zone requirements requiring analysis of "jetting and flashing" phenomena accompanying a pressurized release of cryogenic LNG, as well as spills into containment trenches and troughs. This paper focuses on the new requirement for the analysis of jetting and flashing events.

Jetting and flashing events can be modeled with relative ease by using integral models; however, the selection of appropriate scenarios can add complexity to the analysis. This paper presents typical scenarios that are required for this analysis, including high pressure flange and valve packing failure scenarios. We then parametrically present the consequence studies of a broad range of LNG jetting and flashing release scenarios to provide facility owners with a basis for selecting appropriate scenarios and estimates of the consequences of such releases. Finally, we review the benefits and limitations of the simplifying assumptions associated with integral models. The data presented in the paper can be used by facility owners as a preliminary aid in the decision making process for facility upgrades and modifications.

1 Introduction

Liquefied Natural Gas (LNG) is a primary energy resource in the U.S. and internationally. LNG has a long history of international import/export via ship tanker operations. As of January, 2012, the Federal Energy Regulatory Commission (FERC) listed thirteen existing U.S. import terminals, see Figure 1 [1]. Three of these terminals are currently authorized to re-export delivered LNG (Freeport LNG, Sabine Pass LNG, and Cameron LNG). From reviewing the Federal Dockets and industry news, several terminals are considering or actively pursuing approval from FERC to construct facilities to liquefy and export natural gas. The relatively recent non-conventional gas supply growth in North America has dropped the price of natural gas in the U.S., thus making export to other regions of interest to many terminals.

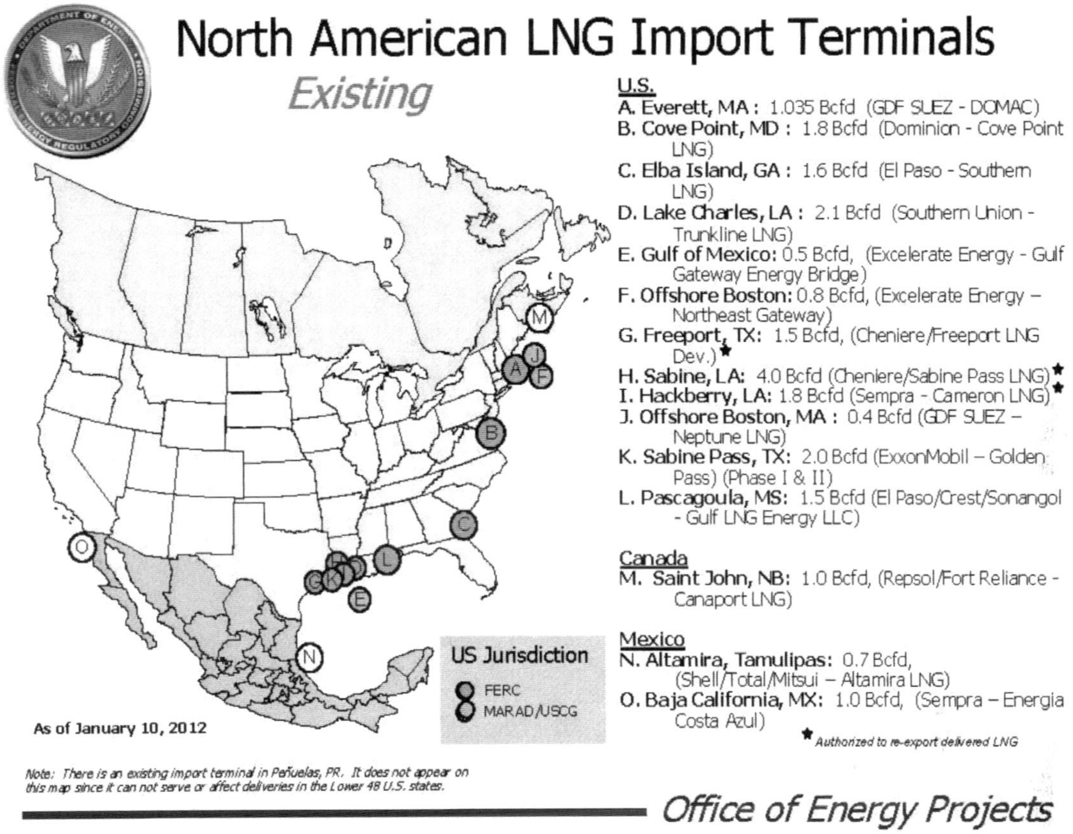

Figure 1. Existing North American LNG import terminals as of Jan. 2012.

In the U.S., the construction and operation of LNG facilities are regulated by the Pipeline and Hazardous Materials Safety Administration (PHMSA) of the U.S. Department of Transportation (DOT), and the regulations are administered by FERC. As part of the regulatory process (Title 49, Code of Federal Regulations, Part 193.2059), regasification and liquefaction facility owners are required to evaluate the potential consequences of major releases of LNG. This analysis may dictate equipment siting changes or design modifications within the proposed project. The regulations require a flammable vapor exclusion zone for flammable vapor clouds that falls within the property boundaries. The flammable range for methane is 5% by volume at the Lower Flammable Limit (LFL) and 16.5% by volume at the Upper Flammable Limit (UFL). The concentration of flammable vapors must remain below 50% of the LFL (i.e., 2.5% methane by volume) at the property boundary according to the regulations.

Modeling of the flammable vapor dispersion from a spill or release is used to demonstrate that the flammable vapor exclusion zone is maintained within the property boundaries that can be built upon (e.g., neglecting clouds over bodies of water). Only a few numerical models have been approved by PHMSA for this type of analysis. This paper applies one of those approved models [2], Process Hazard Analysis Software Tool (PHAST) Version 6.7, to a subset of release scenarios: high pressure jetting and flashing. Jetting and flashing events can be modeled with relative ease by using integral models; however, the selection of appropriate scenarios can add complexity to the analysis. In this paper, we present typical scenarios that are required for this analysis, including high pressure flange and valve packing failure scenarios. We then parametrically present the consequence studies of a broad range of LNG jetting and flashing release scenarios to provide facility owners with a basis for selecting appropriate scenarios and estimates of the consequences of such releases. Finally, we review the benefits and limitations of the simplifying assumptions associated with integral models. The data presented in the paper can be used by facility owners as a preliminary aid in the decision making process for facility upgrades and modifications.

2 Background

The regulatory requirements for consequence analysis have recently been broadened by FERC to include new spill scenarios in addition to the previously required LNG spills into impoundments. Specifically, in 2010 PHMSA issued a written interpretation to Downeast LNG regarding the flammable vapor-gas exclusion zone requirements calling for analysis of "jetting and flashing" phenomena accompanying a pressurized release of LNG [3]. Subsequently, in late 2011, FERC provided additional guidance on requirements for evaluating the potential for jetting and flashing of LNG releases in an Engineering Information Request to Sabine Pass LNG [4].

In an LNG terminal, LNG may be present in piping systems at high pressure in the vaporizer area where the pressure has been increased to meet pipeline pressure (e.g., on the order of 1000 psig) after vaporization and in the liquefaction area where natural gas has been compressed as part of the refrigeration and liquefaction process.

Saraf, et al, analyzed the jetting and flashing problem for LNG releases [5]. They analyzed unconfined jet releases from a 2-inch diameter orifice at 100 psig and 1000 psig source pressure with two-phase flashing flow of LNG. They found that immediate ignition of these releases could create jet fires of approximately 50-ft and 150-ft, respectively. Gavelli, et al, identified potential limitations of applying integral models to non-circular jet releases and in congested plant areas where obstacles may influence vapor cloud dispersion [6]. Computational fluid dynamics (CFD) codes are better-suited to model complex scenarios such as these, but with more refined models the level of conservativeness drops with modeling efforts.

2.1 Jetting and Flashing Basis by PHMSA/FERC

In the PHMSA July 16, 2010 letter of interpretation, the regulators identified jetting and flashing of LNG as a specific concern for modeling - in addition to liquid spills. Because LNG is a cryogenic liquid, low pressure/low momentum releases have historically been treated as liquid spills. PHMSA identified the concern that with a high pressure high momentum release, a jet of LNG may be propelled beyond the impoundment system (i.e., sprayed over trenches, dike walls, or other containment means). A jetting release of LNG can also result in mechanical fragmentation of the fluid which can then lead to aerosolization and conveyance of the fluid beyond the impoundment system. Concurrently, the flashing (rapid or instantaneous vaporization) of discharged LNG will contribute to fragmentation and aerosolization of the LNG; again, leading to potential conveyance of the fluid beyond the impoundment. The concern of certain release scenarios potentially leading to discharged material superseding impoundments has resulted in FERC and PHMSA to require exclusion zone analyses that consider possible jetting and flashing scenarios.

To guide in selecting appropriate potential jetting and flashing scenarios, FERC and PHMSA have identified low probability/high consequence releases as a conservative target for evaluation. The FERC staff has examined typical failure rates for piping and determined the following single accidental leakage sources are necessary to show compliance with 49 CFR Part 193:

a. For piping less than or equal to 6 inches in diameter, a full pipe rupture at any point along the line; and
b. For piping greater than 6 inches in diameter, the greater of:
    i. a full rupture of a connection less than or equal to 6 inches in diameter; or
    ii. a hole of 2 inches in diameter.

PHMSA has concurred with these criteria, and has requested that applicants provide a revised vapor dispersion analysis, including both flashing and jetting, based on single accidental leakage sources which meet the above criteria and yields the most significant vapor dispersion zone with respect to the plant property lines. Furthermore, the regulators have suggested that the analysis also consider release location, elevation, orientation, pressure, temperature, and flow rate.

2.2 Jet & Flash PHAST model

Physics of a flashing release – A pressurized flashing release of LNG from an orifice is a complex process that may involve at least the following sub-processes [7]:

- Droplet atomization
- Atmospheric expansion
- Two-phase dispersion
- Liquid rainout
- Pool formation
- Pool evaporation

PHAST Version 6.7 is commercial software developed and distributed by Det Norske Veritas (DNV). PHAST is widely used for the calculation of hazard distances from the release of several hazardous substances, including LNG. The PHAST code uses the Unified Dispersion Model (UDM), an integral model, to calculate the dispersion following a pressurized release. PHAST contains the following sub-models that are linked in the software to calculate complex release phenomena:

- Near-field jet dispersion;
- Non-equilibrium droplet evaporation and rainout, touchdown;
- Pool spread and vaporization;
- Heavy gas dispersion; and
- Far field passive dispersion.

The UDM allows for continuous, instantaneous, constant finite-duration and general time-varying releases. Additionally, the UDM allows for possible plume lift-off when a grounded plume becomes buoyant. The UDM has been validated extensively with experimental data and is the subject of several peer-reviewed scientific papers [8]. PHAST has received approval for use in evaluating jetting and flashing release scenarios necessary to comply with the FERC/PHMSA jetting and flashing requirements. However, the approval did specify limitations of using the PHAST-UDM for evaluating jetting and flashing dispersion [2]:

- the PHAST UDM is limited to circular LNG pools and impoundments with low-aspect ratios;
- the PHAST UDM is appropriate only for releases along the downwind direction;
- the PHAST UDM may not be used for scenarios with multiple coincident releases that may influence one another;
- and, in some cases, PHAST UDM may not be appropriate for the analysis of dispersion over varying or sloped terrain and instances where large obstructions are present.

Furthermore, it was decided that a recommended safety factor of 2 be applied to the LFL in exclusion zone analysis (i.e. the ½ LFL concentration will be used) in order to account for uncertainties that may arise from potential turbulence, source term specifications, dispersion over water, and the uncertainties associated with the experimental studies submitted for validation.

The regulators did determine that pure methane may be suitably used as a conservative substitute for LNG composition. Additionally, they determined that the guidance provided in 49 CFR 193.2059, concerning the ambient conditions to be used in the model, were appropriately conservative. The atmospheric modeling guidelines provided by 49 CFR 193 (and NFPA 59A) include [9, 10]:

- Pasquill-Gifford atmospheric stability: Category F
- Surface roughness factor of 0.03 m
- 50% humidity
- Average ambient air temperature for the region
- Ground surface temperature, assumed to equal ambient temperature
- Wind speeds that are exceeded less than 10% of the time, or 4.5 mph (2 m/s), whichever results in the longest predictable downwind dispersion distance

3 Parametric Analysis

In order to provide guidelines for rapidly screening possible LNG Jetting and Flashing scenarios, we performed parametric analyses on input parameters using PHAST Version 6.7. The parameters adjusted in this study include: source pressure, source temperature, leak (orifice) diameter, release elevation, release angle, and wind speed. To be conservative, the properties of pure methane were used in the model since it will have a slightly higher vaporization rate than LNG, which is a mixture of hydrocarbons. A summary of the range of parameters and values examined in this study is presented in Table 1.

Table 1. The parameters varied during the LNG Jetting and Flashing sensitivity analyses and the range of values examined in this study.

Source Pressure		Source Temperature		Leak Diameter		Release Elevation		Wind Speed	
(psig)	(bar)	(°F)	(°C)	(in)	(cm)	(ft)	(m)	(ft/s)	(m/s)
50	3.45	-265	-165	1	2.54	1.64	0.5	3.28	1
100	6.89	-256	-160	**2**	**5.08**	6.56	2	**6.56**	**2**
250	17.2	**-247**	**-155**	3	7.63	**16.4**	**5**	9.84	3
500	34.5	-229	-145	4	10.2	32.8	10	13.1	4
750	51.7	-211	-135	5	12.7	65.6	20	16.4	5
1000	68.9			6	15.2			23.0	7
1250	**86.2**							32.8	10
1500	103								

The values emphasized in Table 1 represent the "base case" model used in this study, for which all sensitivity parameters are deviated. The base case was chosen as being reasonably representative of a typical LNG Jetting and Flashing scenario. The remaining model boundary conditions are listed in Table 2.

Table 2. Baseline model boundary conditions.

Averaging Time	18.75
Orifice Discharge Coefficient	0.6
Pasquill Stability	F
Surface Roughness Length (mm)	30
Surface Roughness Parameter	0.068857
Atmospheric Temperature (°F)	68
Surface Temperature (°F)	68
Relative Humidity Fraction	0.7

The leak diameters were varied from 1 to 6 inches. The PHAST program was allowed to calculate the mass flow rate through each orifice size as a function of pressure. Figure 2 graphs the mass flow rate versus pressure for each orifice size. For higher pressures and larger orifices, the flows depicted may be unrealizable or unsustainable at a terminal if the pumping capacity is insufficient to maintain the source pressure. In these cases, the loss of backpressure on the pump system may lead to rapid depressurization of the system followed by low pressure high mass flow of LNG. This situation may require more detailed analysis than that presented in this paper. The results presented in this paper do not include these effects on jetting and flashing.

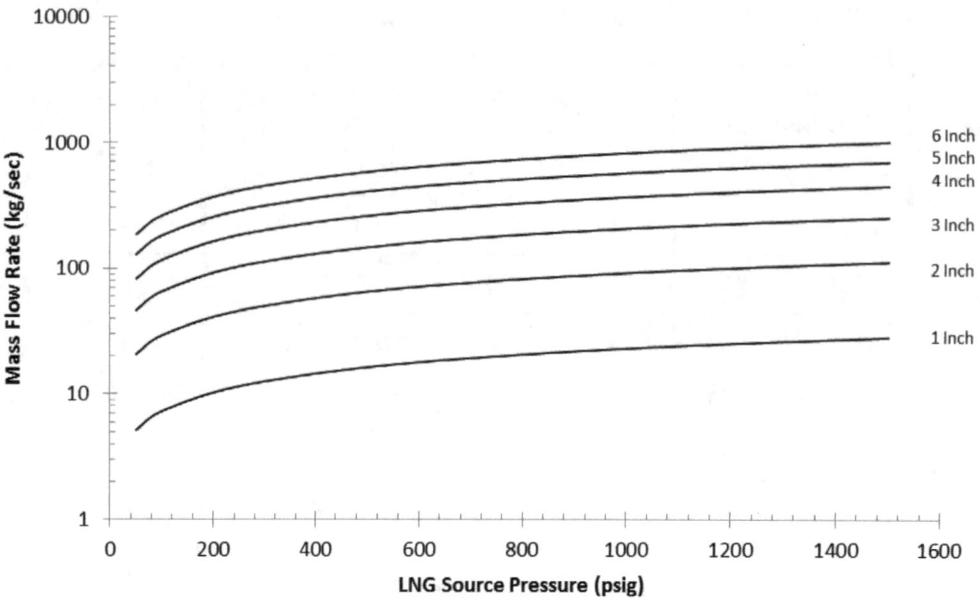

Figure 2. LNG mass flow rate through example orifice sizes at a source temperature of -247°F.

3.1 Droplet Formation

During a release of LNG from a pressurized source, the liquid jet will be fractured into a spray of droplets. PHAST uses built in correlations for calculating droplet diameter due to aerodynamic (mechanical) breakup and flashing breakup. For the leak model employed here, the program calculated the drop size according to both mechanisms and used the smaller of the two. From inspection of the data (see Figure 3), obvious trends in the droplet diameter were not apparent, which is a consequence of the program switching between mechanical and flashing breakup correlations. Droplet size was not affected by orifice size; it proved to only be a function of pressure and temperature in the models. Liquid rainout is affected by the rate of evaporation of droplets, which is related to release elevation, droplet size, and ambient temperature. The liquid rainout fraction is listed in the figure as x_L for the three cases where it occurred. For high values of rainout, collection and conveyance of liquid may be necessary; thus, simple jetting and flashing models such as that employed in PHAST may not be appropriate.

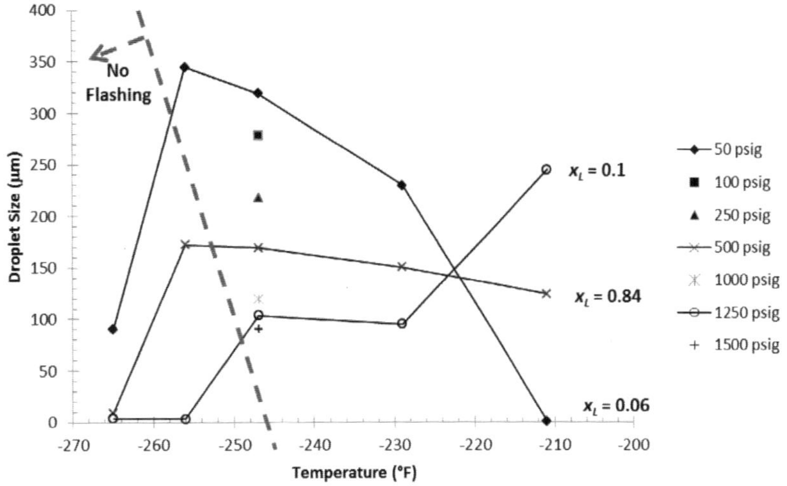

Figure 3. Methane droplet diameter versus temperature.

3.2 Variation of Orifice Size with Pressure

A parametric analysis was performed to vary leak orifice size and pressure within the ranges listed in Table 1. These releases were modeled at a height of 5 m and source temperature of -247°F. Figure 4 provides ½ LFL plume contours at 0.5 m elevation for a subset of the cases shown as the maximum steady dispersion plume. For the lower pressure, lower mass flow rate releases, the plumes demonstrate a wider aspect, thus potentially a larger flammable mass. In terms of the hazard posed by the vapor cloud, the measurable target is the boundary of the facility; thus, the flammable mass and overall size of the cloud is less important than the length. The representative centerline side view contours for the clouds shown in Figure 5 indicate that higher pressure releases create taller clouds due to the expanding jet. Figure 6 presents a summary of downwind dispersion distance for the ½ LFL boundaries for a range of orifice sizes at increasing pressure, fixed source temperature, and the baseline model/weather conditions listed in Table 2. These maximum distances are reported at an elevation of 0 m, indicating that the clouds remain close to the ground. An interesting feature of the result is the initial hump in dispersion distance at 50 psig that tails-off and appears to be independent of the source pressure. The longer distances for the 50 psig release pressure with leak diameters of 4, 5, and 6 inches is related to liquid rainout that occurs for these scenarios. The pool formation is followed by evaporation that results in cold dense vapors close to the ground.

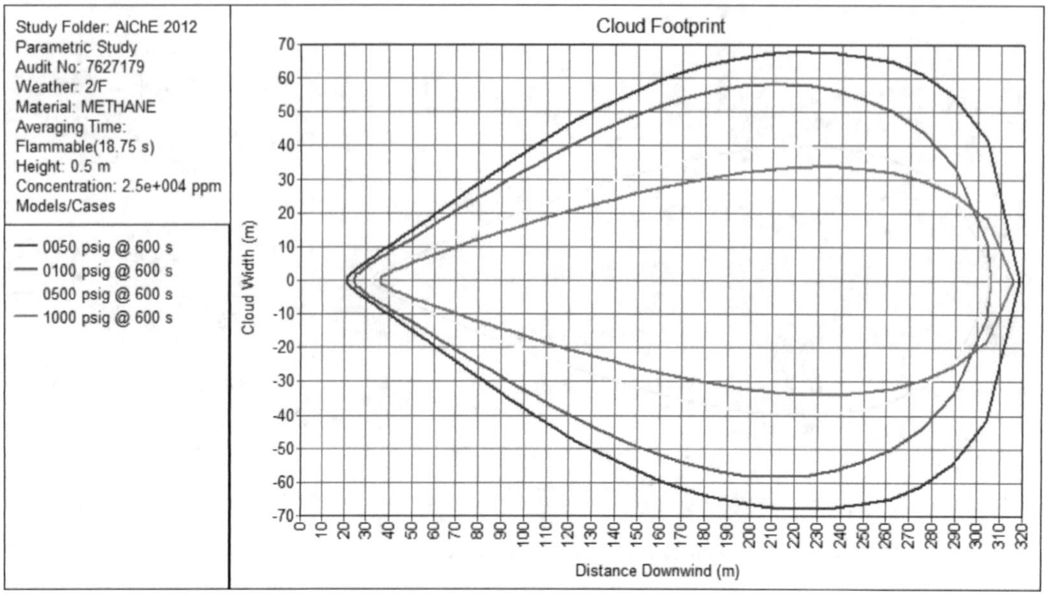

Figure 4. Representative plume contours for 2-Inch orifice at 5 m elevation and -247°F.

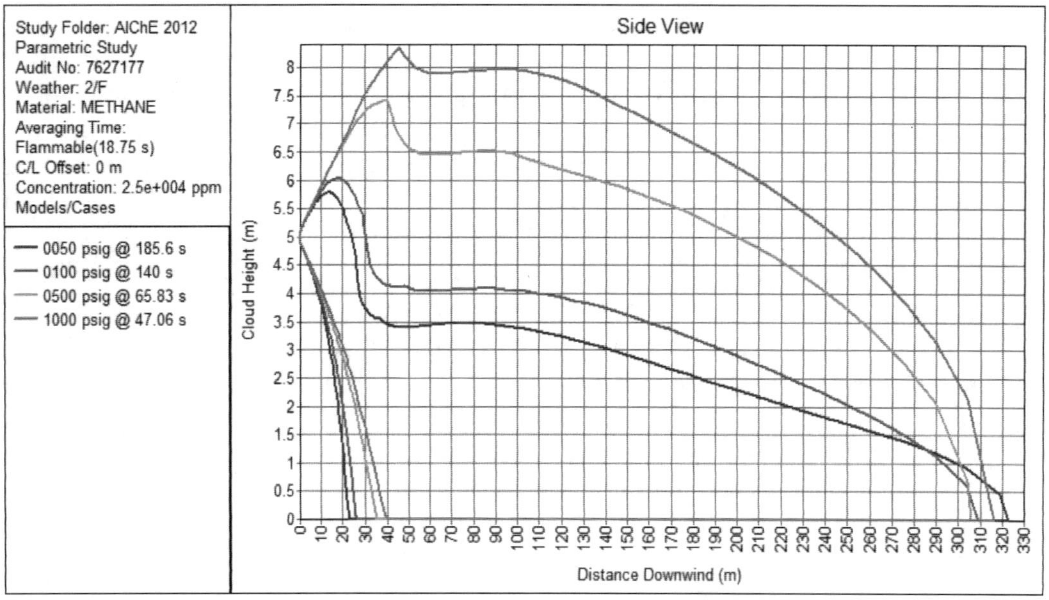

Figure 5. Representative centerline side view contours for 2-Inch orifice at 5 m elevation and -247°F.

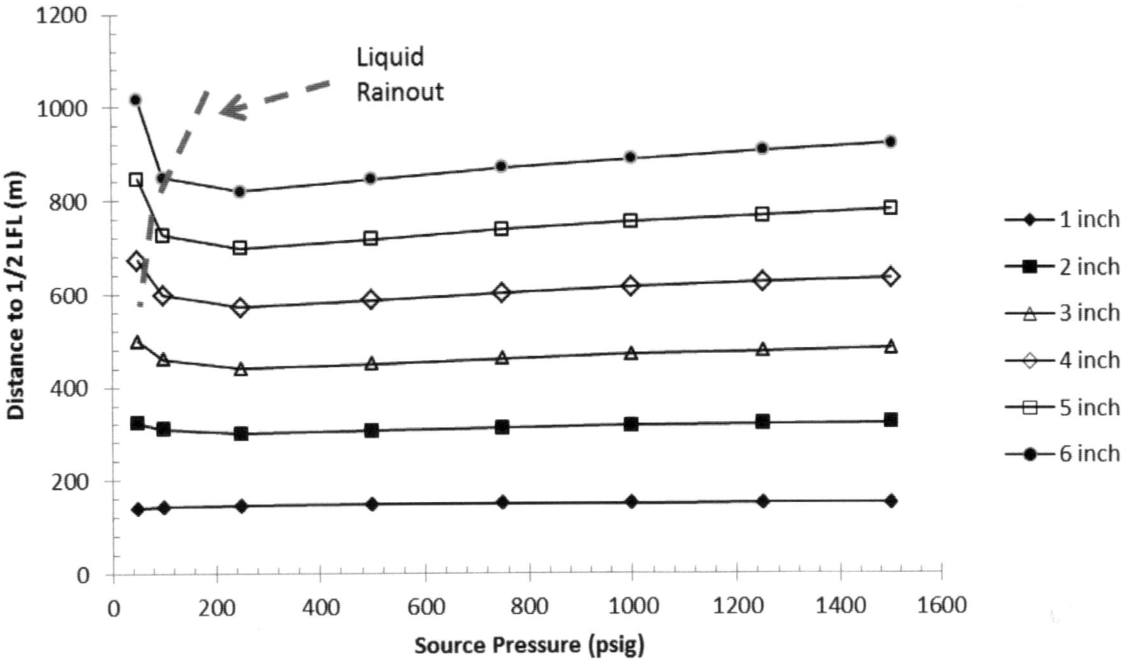

Figure 6. Dispersion distance to ½ LFL at -247°F and baseline weather conditions.

3.3 Effect of Release Elevation

The elevation above ground level for a release, in addition to the other release parameters, will greatly affect the size and length of the vapor cloud. Since LNG vapors are colder and denser than ambient air, low elevation releases may be inferred to create longer cloud lengths than high elevation releases. To test this inference, a series of elevations were evaluated for a range of source pressures at -247°F. The elevations spanned the range from 0.5 m (e.g., ground level piping), 5 m (e.g., low pipe racks), to 20 m (e.g., storage tank roof piping). The results of this parametric analysis were unusual. From Figure 7, for all pressures the downwind dispersion distance decreases with increasing elevation. However, there is slightly different behavior observed for the 100 psig case at the lower elevation release; the behavior then matches the trends observed with the other release pressures for elevations greater than 5 m.

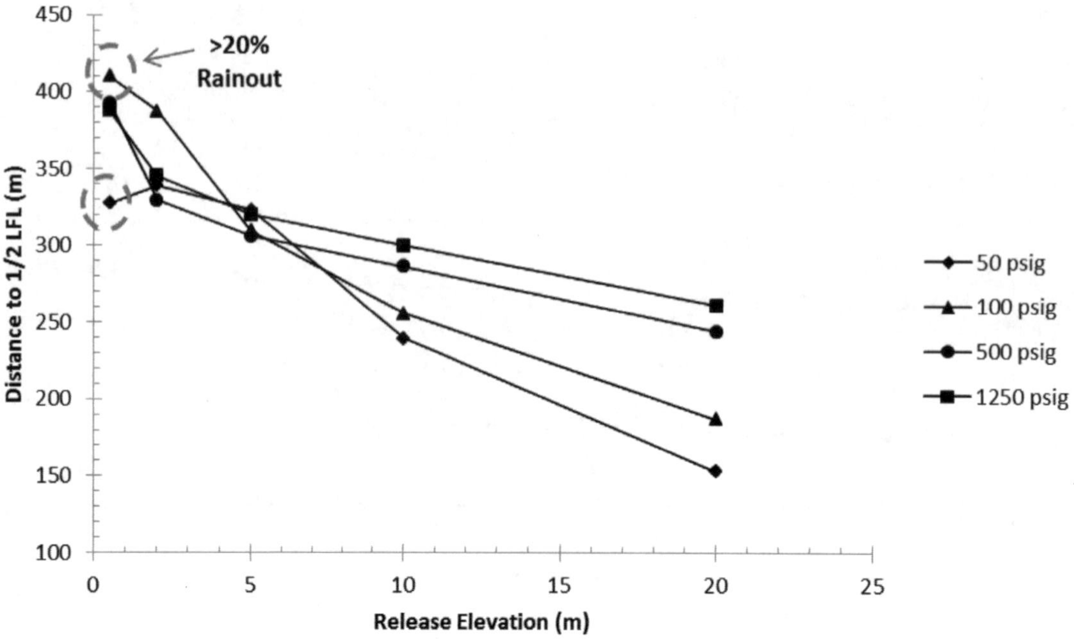

Figure 7. Trends in ½ LFL cloud downwind dispersion distance for a range of source pressures versus source elevation.

3.4 Effect of Release Angle

Three constant source pressure cases (50 psig, 500 psig, and 1000 psig) were considered to evaluate the effects of release angle on downwind dispersion. Four release angles were evaluated: down, horizontal, 45 degrees from horizontal, and vertical upwards. These are sketched in Figure 8 below. The example discussed here is for the following base model conditions: a source temperature of -247°F, at 5 m elevation, with 2-inch orifice, 2 m/sec wind speed, and F stability class. The relation between the maximum dispersion distance for the ½ LFL vapor cloud are consistent with experience and intuition for cold, dense vapor clouds (see Figure 9). For upward directed releases, complete evaporation occurs for all pressure cases with no liquid rainout predicted. Air entrainment and dilution subsequently prevents touchdown of the ½ LFL vapor cloud and diminishes the downwind dispersion distance (see Figure 10). The horizontal releases also don't predict any rainout at this temperature; however, the cold, dense cloud remains close to the ground and travels quite a distance. The downward oriented releases all produce significant rainout and liquid pooling. No bund wall was implemented in the model; consequently, the pool sizes grew to a size constrained by the vaporization rate. Due to the significant rainout, the downward facing jets do not meet the jetting and flashing criteria advanced by PHMSA/FERC. Instead, the effect of trenches, troughs, and impoundments must be considered on these liquid spills, vapor generation, and vapor dispersion. Depending upon the conveyance and impoundment design at the release location, PHAST may not be the best model for this analysis.

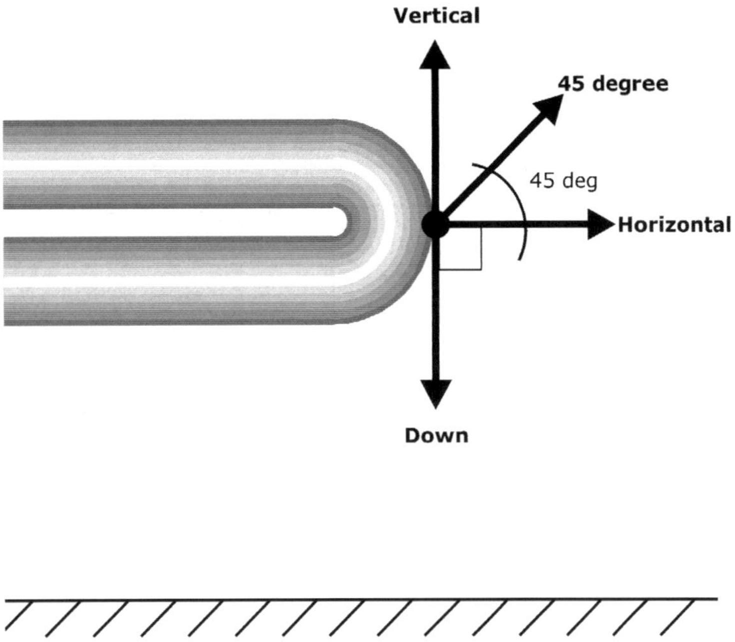

Figure 8. Release directions evaluated in the model.

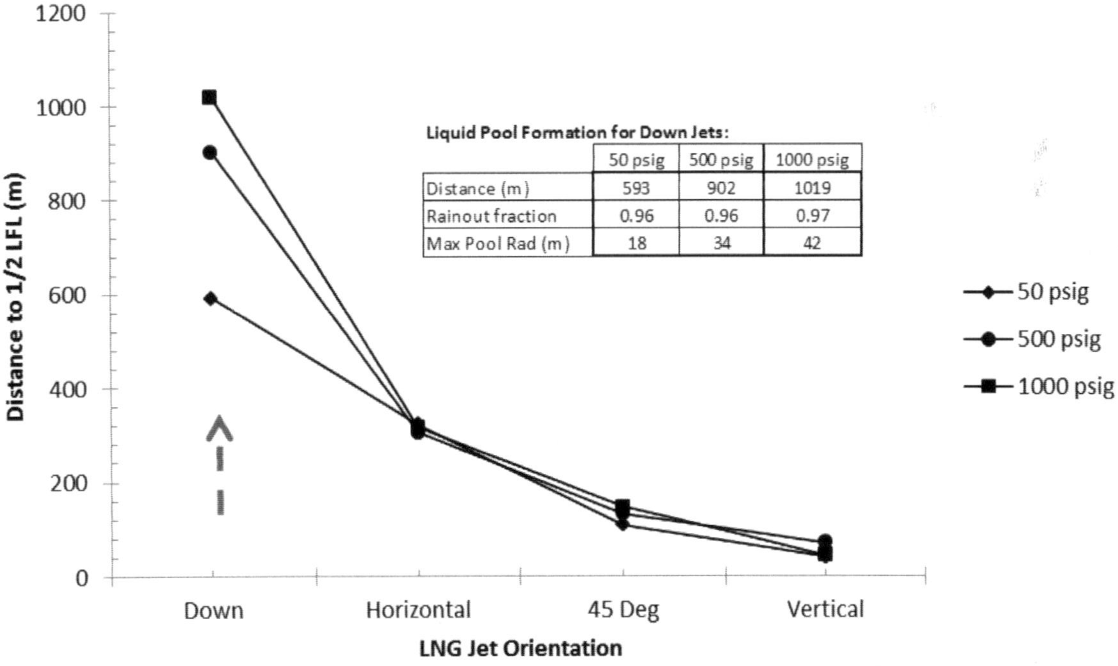

Figure 9. Maximum downwind dispersion distance to ½ LFL for varying release angle and source pressure.

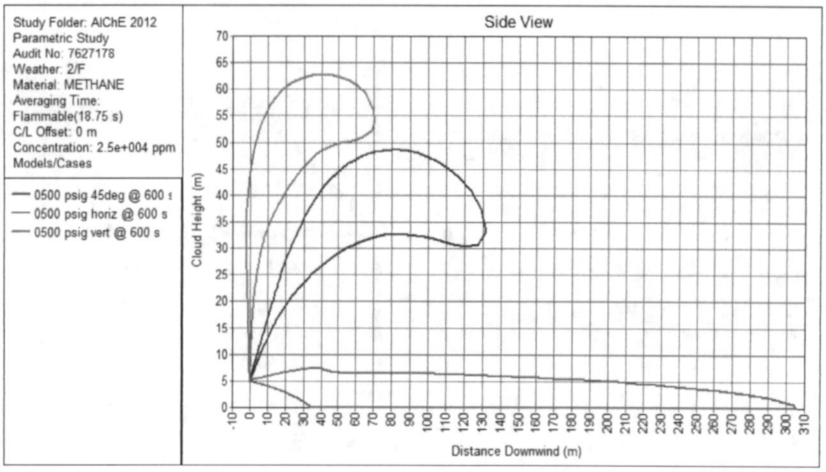

Figure 10. Comparison of side views for horizontal, 45 degree, and vertical releases at 500 psig.

3.5 Effect of Wind Speed

During a release, the vapor cloud dispersion will be influenced by the wind speed and stability class. The wind speed was evaluated at 2 m/sec and F stability class for most of the parametric studies. The regulations provide a choice between these conditions and those prevalent at the site; however, FERC has recently required a wind sensitivity analysis to demonstrate that dispersion results are reported with conservative wind conditions. Higher wind speeds can conceivably disperse a cloud faster than low wind speeds thus decrease the ½ LFL boundary distance. The wind conditions were varied from 1 m/sec to 4 m/sec for releases at -274°F to demonstrate the downwind dispersion distance versus source pressure. The results are depicted in Figure 11 for a 2-inch diameter orifice and in Figure 12 for a 4-inch diameter orifice. These results demonstrate an unusual trend. At a wind speed of 2 m/sec, the dispersion distance drops sharply to reach a minimum around 250 psig source pressure. At the 3 and 4 m/s wind speeds examined, the opposite behavior occurs with a maximum dispersion distance at this source pressure. A smoothly increasing function versus pressure at the low wind speed of 1 m/sec was also observed. All of the results appear to be converging upon a maximum dispersion distance versus pressure.

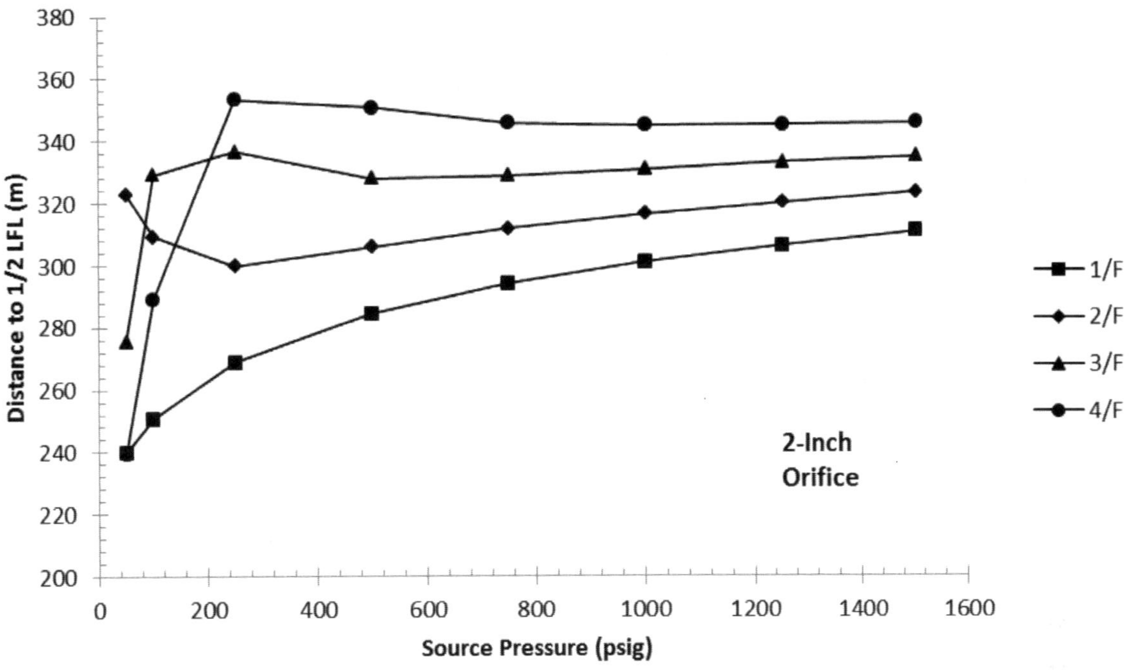

Figure 11. Variation of dispersion distance for a 2-inch orifice with baseline conditions as a function of wind speed.

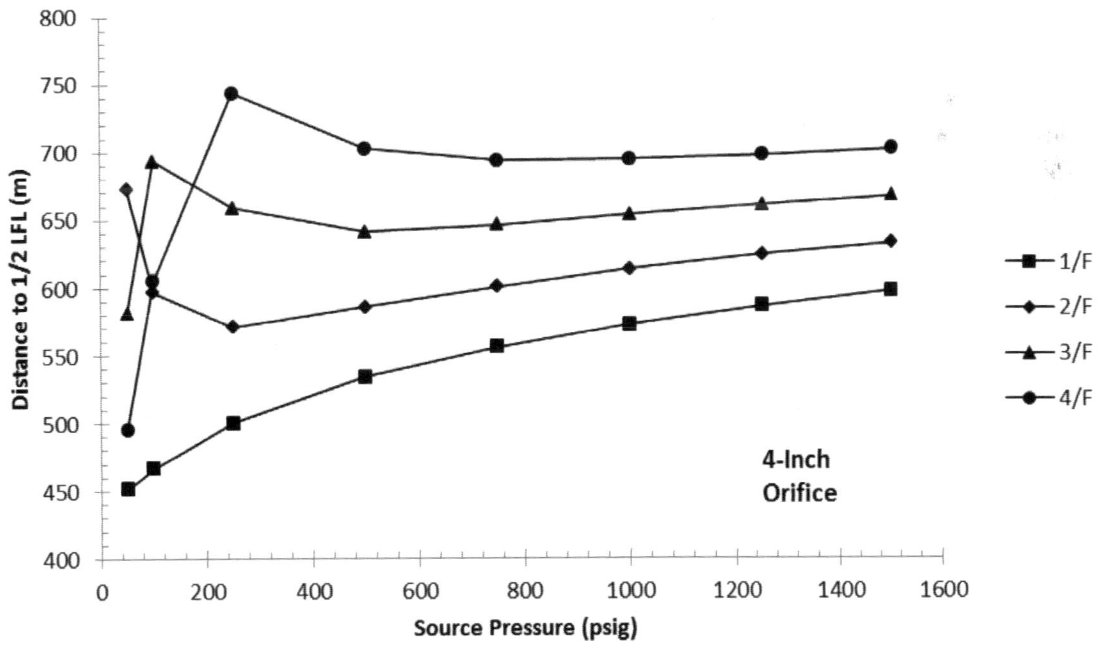

Figure 12. Variation of dispersion distance for a 4-inch orifice with baseline conditions as a function of wind speed.

3.6 Effect of Source LNG Temperature

To demonstrate the effect of LNG temperature on the downwind dispersion, a parametric study of source temperature versus source pressure was performed for a 2-inch diameter orifice at 5 m elevation above ground level. As the release temperature was raised for the higher pressure releases, the downwind dispersion distance dropped as would be anticipated due to lower cloud density (see Figure 13). Figure 14 provides the centerline side view contours for three temperature values in the 50 psig case. At high temperature, the jet completely disperses at a fraction of the distance for the colder cases.

These same source pressure and temperature cases were modeled at 1 m elevation to investigate what effect that elevation would have on the dispersion distance. The results are presented in Figure 15. For the low pressure case at 50 psig, the dispersion distance dropped with the lower elevation release likely due to warming of the cloud by the ground. Three cases were found to have significant liquid rainout of greater than 10% of the release mass; however, this did not appear to significantly affect the maximum dispersion distance. The dispersion distances for the higher pressure cases increased with a decrease in elevation, likely due to the reduction in air entrainment due to cloud location at ground level.

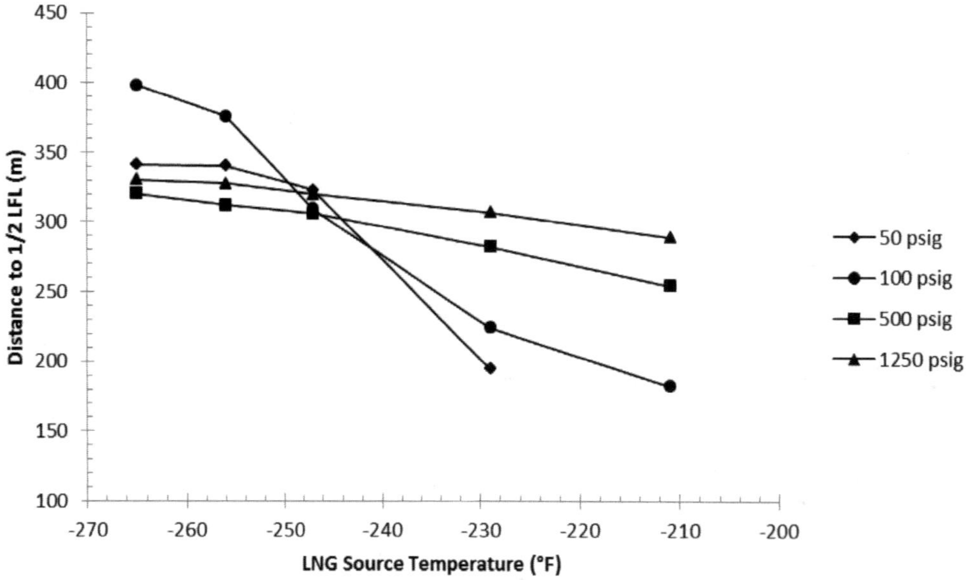

Figure 13. Comparison of distance to ½ LFL for a range of source temperature and source pressure values for a source at 5 m elevation. Note that cloud maximum distance for 50 psig/-211°F release does not reach ground level (see Figure 14).

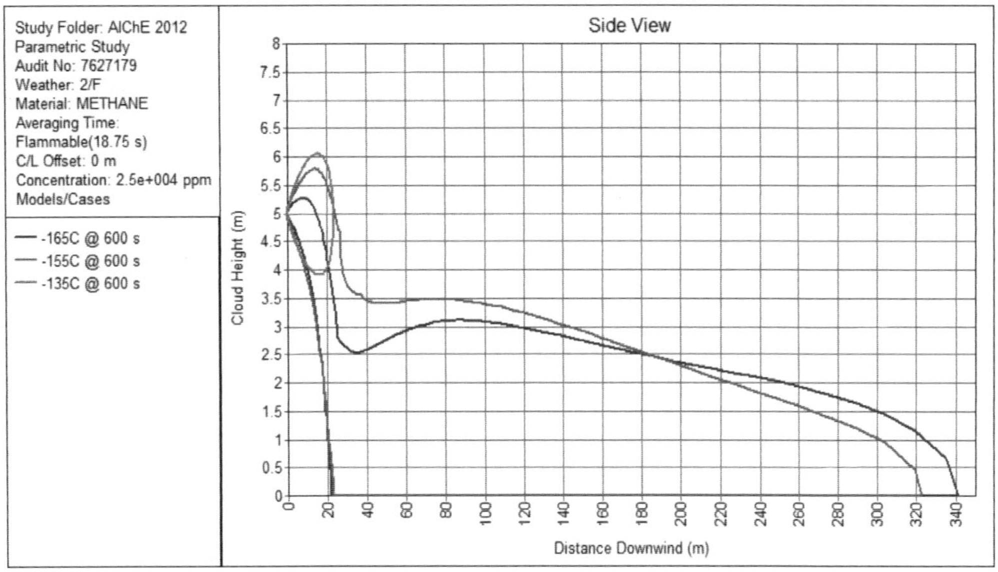

Figure 14. Representative side view contours for 2-Inch orifice at 5 m elevation and a series of source temperatures at 50 psig.

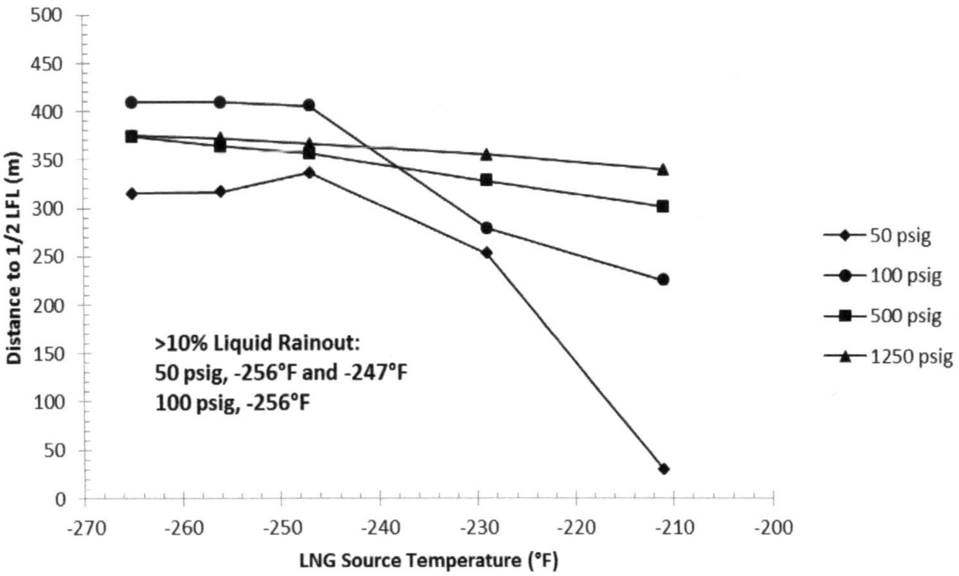

Figure 15. Comparison of distance to ½ LFL for a range of source temperature and source pressure values for a source at 1 m elevation.

4 Conclusions & Recommendations

This paper has presented the results of a series of parametric studies on pressurized LNG releases to aid facility owners in evaluating the maximum downwind dispersion distance of the vapor cloud. The PHAST program was used to perform the analyses. We found that the downwind dispersion distance was not dominated by any individual parameter, such as release temperature. In addition, we found that intuitive trends in release behavior may not be correct; thus, modeling for the specific conditions of the release is recommended. The results presented in this paper will give the analyst an indication of the release and dispersion behavior for conditions similar to those presented; however, the authors suggest that siting decisions be based upon modeling of the facility using appropriate parameters for the exiting/planned process operations.

5 References

[1] North American LNG Import Terminals, downloaded from http://www.ferc.gov/industries/gas/indus-act/lng.asp, January 10, 2012.

[2] PHMSA Docket No. 2011-0075, Final Decision on petition for approval for PHAST UDM Version 6.6 and 6.7, October 7, 2011.

[3] PHMSA Interpretation #PI-10-0005, Jul 16, 2010.

[4] FERC Engineering Information Request, Docket No. CP11-72-000, November 8, 2011.

[5] Saraf S,Medhekar, S.,Yang, E-H, “Have You Considered High-Pressure LNG Releases?” AIChE Spring Meeting, April 2009, 9th Topical Conference on Natural Gas Utilization, Tampa, Florida, April 26-30, 2009.

[6] Gavelli, F., Davis, S.G., and Hansen, O.R., “CFD Simulation of Vapor Dispersion from LNG Jetting and Flashing Releases,” Paper presented at AIChE Spring meeting, Natural Gas Utilization Topical Session, San Antonio, USA, March 22-24, 2010.

[7] FLASHING LIQUID JETS AND TWO-PHASE DROPLET DISPERSION, Henk Witlox and Mike Harper, DNV Software, London, UK , www.dnvsoftware.com, Phil Bowen and Vincent Cleary, Cardiff University, UK, American Society of Safety Engineers Middle East Chapter, 7th Professional Development Conference & Exhibition, Kingdom of Bahrain, March 18-22, 2005.

[8] Witlox, H.W.M. and Holt, A., 1999, A unified model for jet, heavy and passive dispersion including droplet rainout and re-evaporation, International Conference and Workshop on Modeling the Consequences of accidental Releases of Hazardous Materials, CCPS, San Francisco, California, September 28 – October 1, pp. 315–344.

[9] 49 CFR 193.2059 “Flammable vapor-gas dispersion protection”

[10] NFPA 59A "Standard for the Production, Storage, and Handling of Liquefied Natural Gas (LNG)" (2009) Section 5.3.3.6

(144e) Effect of LNG Chemical Composition on Consequence Assessment

Xiaodan Gao[1], Yi Liu[1], Luc Vechot[2], Tomasz Olewski[2]and M. Sam Mannan[1], (1)Mary Kay O'Connor Process Safety Center, Department of Chemical Engineering,Texas A& M University, College Station, TX, (2)Chemical Engineering, Texas A& M University at Qatar, Doha, Qatar

Over the last ten years, LNG has become a popular fuel and its demand has been keeping the increasing trend. This leads to the construction of new LNG facilities around the world for import, storage and re-gasification. The formation of a vapor cloud following the LNG accidental spill presents potentially significant flammable hazards to both the facility and the surrounding communities.

Correctly predicted exclusion zones during the risk assessment process are critical to effectively mitigate the consequence of a LNG spill and protect the public from the flammable vapors. The accuracy of every step of consequence modeling, including LNG spilling, spreading, vaporization and vapor dispersion, influence the precision of exclusion zone estimation. Numerous publications have advanced mechanisms of the above mentioned steps to obtain a certain level of confidence, including comprehensive Computational Fluid Dynamics models to describe complex geometry and turbulence. Nevertheless, most of the available studies assume pure methane as the analog of LNG. This assumption may not be correct, since the vaporization process of mixture as LNG may be significantly different from the pure methane liquid. As a matter of fact, the methane fraction of LNG can vary from 83 to 99% mol, and neglecting this factor may lead to the inaccuracy of all the following steps. In 2007, the US Government Accountability Office had reported: the Public safety consequences of a liquefied natural gas spill need clarification (Report 07-633T, March 21st2007). Under this report, the expert panel evaluated several research areas, one of which is evaluation of the effects of LNG composition on vaporization rate, thermal radiation from fire and explosive behavior.

This paper has investigated the influence of multi-component factor on the vaporization rate as well as vapor dispersion modeling. The studied mixtures involve five components: nitrogen, methane, ethane, propane and butane. Suitable equations of state and vapor-liquid equilibrium of LNG have been adopted to predict the vaporization rate, followed by the vapor dispersion modeling. The behavior of pure methane liquid and several different LNG compositions are compared. The results confirmed the significant impact of composition, which has emphasized the necessity of this factor to be involved in LNG consequence assessment process.

(144f) Source Term Modeling of Vapor Cloud Formation During a Cryogenic Liquid Spill Based on a Boiling Model by CFD Method

Yi Liu[1], Tomasz Olewski[2], Luc Vechot[2], Xiaodan Gao[1] and Sam Mannan[1], (1)Mary Kay O°Connor Process Safety Center, Texas A& M University, College Station, TX, (2)Chemical Engineering, Texas A& M University at Qatar, Doha, Qatar

The formation of a flammable vapor cloud following a spill of cryogenic liquids, such as ammonia and LNG (liquefied natural gas), on a ground or water includes significant toxic, fire or explosion hazards. The ability to predict the consequences of such an event is crucial to perform proper risk assessment. Numerous researchers have been focused on the modeling of the vapor cloud dispersion. However there is a lack of accurate source term model describing the vapor cloud source, which has obviously an effect on the predicted exclusion zones.

Webber et al. [Webber, DM; Gant, SE; Ivings, MJ; Jagger, SF. *LNG source term models for hazard analysis: A review of the state-of-the-art and an approach to model assessment.* 2010] prepared a wide literature review of LNG source term models as well as the experimental data available to validate them. Eighteen models, most of which are integral and two of them are computational fluid dynamics (CFD) codes (FLACS by GexCon and Liquid Spill Modeling System by Cambridge Environmental Research Consultants) are briefly reviewed. All models differ primarily regarding assumptions, simplifications or/and correlations.

A boiling model, as one part of the source term modeling of cryogens, has been developed by using CFD code. Different boiling phenomena, including film boiling, transition boiling and nucleate boiling, are identified by the presented model. The vaporization rate can be calculated depending on the ground temperature. The boiling model by CFD method can give a more accurate result than integral models or empirical correlations, especially when a complex geometry of the spill and surrounding is considered.

Based on the boiling model, the pool formation and vaporization during cryogenic liquid spill on ground and/or water are studied in this paper. Liquid nitrogen is chosen as a safe analogue. The effects of the surface type (ground or water), hot surface conditions (temperature, slope, etc), weather conditions and two-phase boiling flow rate are discussed and compared against experiments or/and literature data.

AUTHOR INDEX

AUTHOR INDEX

AUTHOR INDEX